Biomedical Signals Based Computer-Aided Diagnosis for Neurological Disorders

M. Murugappan
Rajamanickam Yuvaraj
Editors

Biomedical Signals Based Computer-Aided Diagnosis for Neurological Disorders

Springer

Editors
M. Murugappan
Electronics & Communication Engineering
Kuwait College of Science and Technology
Doha, Kuwait

Rajamanickam Yuvaraj
Nanyang Technological University (NTU)
Singapore, Singapore

ISBN 978-3-030-97847-1 ISBN 978-3-030-97845-7 (eBook)
https://doi.org/10.1007/978-3-030-97845-7

This Springer imprint is published by the registered company Springer Nature Switzerland AG
The registered company address is: Gewerbestrasse 11, 6330 Cham, Switzerland

Preface

Computer-aided diagnosis (CAD) tools have witnessed tremendous growth in recent years. These tools can be used to diagnose a wide variety of diseases, such as diabetes, epilepsy, seizure, mental stress, anxiety, attention deficit disorder, and others. CAD is most commonly used for people suffering from neurological disorders (NDs) such as epilepsy, Alzheimer's disease, dementia, stroke, schizophrenia, Parkinson's disease, migraine, and multiple sclerosis. NDs commonly affect people over the age of 60, but in recent years they have also begun to affect younger age groups. Researchers are focusing on developing intelligent diagnostic aids for NDs patients to improve the accuracy of diagnosis in order to improve the quality of life of those living with the disease. The number of casualties from neurodegenerative disorders is one of the biggest factors contributing to the total number of casualties in the world. Most existing CAD tools are based on either biosignals or biomedical images. Compared to biomedical images, biosignals can be used to develop intelligent CAD tools, which are usually preferred by researchers. In the past several years, biosignals such as electrocardiogram (ECGs), electromyogram (EMG), electroencephalogram (EEG), and electrooculograms (EOG) have emerged as popular biosignals in ND detection, diagnosis, prognosis, and rehabilitation. Research has proposed numerous methods and algorithms to extract meaningful information from these biosignals for diagnosis and to combine these with artificial intelligence (AI) to develop intelligent-assistive devices for establishing a CAD system.

Several researchers from different countries have been involved in the book *Biomedical Signals based Computer-Aided Diagnosis System for Neurological Disorders*. Twelve chapters are ultimately accepted, after a double-blind review by 25 reviewers. There are several topics covered in the book, including the development of new methods for analysing biosignal data, processing biosignal data, diagnosing human disabilities, modelling biosystem characteristics, and several applications based on the above. In specific, the book covers a detailed review of hardware tools used for a neurodiagnostic, a case study of different types of non-linear measures in EEG signals for brain–computer interface design for NDs, EEG signals-based seizure and epilepsy detection using machine and deep learning algorithms, biosignals-based emotion recognition in autism spectrum disorder, EMG signals used for intramuscular activities detection, and a real-time wearable system for seizure detection. It is a great honour for us, the editors of this book, to thank the authors for correcting and modifying their papers in accordance with the feedback from our reviewers, as well as the reviewers who were able to allot a time slot within their schedules to review the papers and comment critically on the contents with questions for improvement. We are very grateful to each author and reviewer who have taken the time to contribute to this book.

Doha, Kuwait M. Murugappan
Singapore Rajamanickam Yuvaraj

Acknowledgements

S. No	Reviewer Names and Affiliation
1	Dr. John Thomas, Montreal Neurological Institute, McGill University, Montreal, Canada.
2	Dr. Surabi Bhatia, Department of Information System, College of Computer Science and Information Technology, King Faisal University, Saudi Arabia.
3	Dr. P. A. Karthick, Department of Instrumentation Engineering, National Institute of Technology, Trichy, Tamil Nadu, India.
4	Dr. Usman Akram, College of Electrical and Mechanical Engineering, National University of Sciences and Technology, Pakistan.
5	Dr. Jerritta Selvaraj, Vels University, Chennai, India.
6	Dr. Muhammad Enamul Hoque Chowdhury, College of Engineering, Qatar University, Qatar.
7	Dr. Ali K. Bourisly, Biomedical Engineering Unit, Department of Physiology, Faculty of Medicine, Kuwait University, Kuwait.
8	Dr. N. Punitha, Department of Biomedical Engineering, SSN College of Engineering, Chennai, India.
9	Dr. N. P Guhan Seshadri, National Institute of Technology, Raipur, India.
10	Dr. Bikesh Kumar Singh, Department of Biomedical Engineering, National Institute of Technology, Raipur, India.
11	Dr. Wan Khairunizam, Faculty of Electrical Engineering Technology, Universiti Malaysia Perlis (UniMAP), Malaysia.
12	Dr. Saidatul Adreenawati Awang, School of Mechatronic Engineering, Universiti Malaysia Perlis (UniMAP), Malaysia.
13	Dr. Anandhi, Department of Biomedical Engineering, Agni College of Engineering, Chennai, India.
14	Dr. Yogesh Sharma, National Institute of Technology, Raipur, India.

15 Dr. Sumithra Palanisamy, Dr.N.G.P. College of Technology, India.
16 Dr. Ramanathan Subramanian, University of Canberra, Australia.
17 Dr. Mufti Mahmud, Nottingham Trent University, United Kingdom.
18 Dr. Ramaswamy Palaniappan, University of Kent, United Kingdom.
19 Dr. Sivaramakrishnan Rajaraman, National Institute of Health, USA.
20 Ms. Muthumeenakshi Subramanian, Case Western Reserve University, USA.
21 Dr. Hanif Heidari, Damghan University, Iran.
22 Dr. M. Murugappan, Kuwait College of Science and Technology, Kuwait.
23 Dr. Yuvaraj Rajamanikam, Nanyang Technological University, Singapore.
24. Dr. S. Arun, Vels University, India.
25. Dr. P. Karthikeyan, Madras Institute of Technology, Anna University, India.

Contents

Abnormal EEG Detection Using Time-Frequency Images and Convolutional Neural Network

Rishabh Bajpai, Rajamanickam Yuvaraj, A. Amalin Prince, and M. Murugappan

Abstract In the process of diagnosing neurological disorders, neurologists often study the brain activity of the patient recorded in the form of an electroencephalogram (EEG). Identifying an abnormal EEG serves as a preliminary indicator before specialized testing to determine the neurological disorder. Traditional identification methods involve manual perusal of the EEG signals. This method is relatively slow and tedious, requires trained neurologists, and delays the treatment plan. Therefore, the development of an automated abnormal EEG detection system is essential. In this study, we propose a method based on short-time Fourier transform (STFT), which is a time-frequency (TF) representation, and deep convolutional neural network (CNN) to detect abnormal EEGs. First, the filtered time-series EEG signals are converted into TF images by applying STFT. Then, the images are fed to three popular configurable CNN structures, namely, DenseNet, SeizureNet, and Inception-ResNet-V2, to extract deep learned features. Finally, an extreme learning machine (ELM)-based classifier detects the input TF images. The proposed STFT-based CNN method is evaluated using the Temple University Hospital (TUH) abnormal EEG corpus, which is available under the public domain. The experiment showed that the combination of the SeizureNet-ELM model achieved an average (fivefold cross-validation) accuracy, specificity, sensitivity, and F1-score of 85.87%, 88.43%,

R. Bajpai · A. A. Prince
Birla Institute of Technology & Science Pilani, K K Birla Goa Campus, Zuarinagar, Goa, India
e-mail: amalinprince@goa.bits-pilani.ac.in

R. Yuvaraj (✉)
Nanyang Technological University (NTU), Singapore, Singapore
Office of Education Research (OER), National Institute of Education (NIE), Singapore, Singapore
e-mail: yuvaraj.rajamanickam@nie.edu.sg

M. Murugappan
Department of Electronics and Communication Engineering, Kuwait College of Science and Technology, Al-Jahra, Kuwait
e-mail: m.murugappan@kcst.edu.kw

M. Murugappan, R. Yuvaraj (eds.), *Biomedical Signals Based Computer-Aided Diagnosis for Neurological Disorders*, https://doi.org/10.1007/978-3-030-97845-7_1

83.23%, and 0.858, respectively. The results demonstrate that the proposed framework may aid clinicians in abnormal EEG detection for the early treatment plan.

Keywords Pathology diagnosis · EEG · Convolutional neural network · Extreme learning machine · Abnormal EEG corpus

1 Introduction

The human brain is a highly complex organ that uses millions of neurons communicating with each other to control the human body. Each neuron communicates by sending ions through its axons and dendrites. This ion movement can be recorded and plotted against time in an electroencephalogram (EEG). Hans Berger recorded the first-ever EEG in the year 1924 [1]. His setup consisted of two silver wires as electrodes placed inside the patient's scalp, with a capillary electrometer to record the signals. As a result, the patient was only given alpha-wave readings (now also known as Berger waves). These days, EEG signals are primarily acquired in two ways. The first method is to place electrodes directly inside the brain to catch some of the ions emitted by the neurons. The second, more popular noninvasive method is to place electrodes at the scalp. These electrodes detect charged waves that reach the scalp due to the movement of ions in the brain. The current scalp EEG, unlike Hans', uses more than two electrodes. Typically, a mesh of 19 electrodes with tips made of conductive gel is placed on the scalp to capture the signals. It also captures a wider range of frequencies in general. The majority of EEG devices maintain a sampling frequency of 256–512 Hz. EEG readings provide a very high temporal resolution as compared to other brain imaging methods. This, combined with its low cost of equipment, makes it a popular alternative to newer yet more expensive technologies like positron emission tomography (PET) and functional magnetic resonance imaging (fMRI) scans [2]. EEG is used to diagnose a condition such as sleep disorders, epilepsy, brain death, Parkinson's disease, etc. [3–6]. EEG is used in many brain-computer interface projects, where the brain signals are used to control computer actions such as flying a drone, controlling a prosthetic limb, etc. [7, 8]. EEG signals are also used by neurologists to investigate emotional and behavioral patterns. EEG interpretation typically involves recording signals for a period of time followed by a manual evaluation by a neurologist. This evaluation can only be done by a trained neurologist and requires an understanding of patterns of brain wave behaviors associated with different pathologies. This process of pattern matching is tedious, time-consuming, and prone to error. Furthermore, numerous studies have found low inter-rater and intra-rater reliability (I&IR) in EEG interpretation [9, 10]. As people try to extract more information from EEG and increase the scope for its diagnosis, this manual pattern-matching method will get more difficult. Consequently, if a machine handles some extent of this process, it could ease the neurologist's burden and make the process more reliable and quicker.

In recent times, machine learning (ML) is the most appropriate solution to a problem that requires model building and pattern matching. Machine learning algorithms effectively use a computer's ability to process vast amounts of data of information quickly. In a traditional manual EEG evaluation, the neurologists first determine if a particular EEG is abnormal or normal [11]. The process entails looking for general anomalies such as aberrant wave sharpness or height, differences in activity in different lobes of the brain, and so on. A neurologist's average time to read an EEG is roughly 12 min [12]. If an automated system was present to aid in this first step, the neurologist would rule out normal EEGs much quicker and read abnormal EEGs more efficiently. Similar approaches have been used in radiology to reduce the time it takes to diagnose strokes with machine learning for computed tomography (CT) scans [13–15]. Hence, the automation of the first step of EEG interpretation is the motivation of this study. The present study utilizes deep learning algorithms instead of the traditional machine learning approaches. Traditional ML requires feature extraction from the EEG signal manually, resulting in an information loss [16]. Furthermore, handmade features may not capture the high amount of variance shown between EEGs of different subjects. In the past decade, deep learning has gained a huge following, as it has obtained astonishing results in a wide range of applications [17]. The low dependence on human subject expertise due to automatic feature extraction is prevalent among researchers. Besides, features learned automatically are often more robust than handmade ones. Many EEG-related applications, such as neurodegenerative disease diagnosis, epileptic seizure detection, brain-computer interface tasks, and sleep stage detection [3–6, 18–21], have been applied deep learning. In this work, we propose a deep learning-based system to detect abnormal EEG from normal EEGs. In particular, we use a convolutional neural network (CNN) feature extractor followed by an extreme learning machine (ELM) decision layer to classify EEGs. We investigate three popular CNN structures, namely, Inception-ResNet-V2, DenseNet, and SeizureNet. The CNN extracts the information dense feature maps from the time-frequency (TF) image. These feature maps are then fed into an ELM-based classifier. The Temple University Hospital (TUH) abnormal EEG corpus is used to perform the experiments [22]. To the best of our knowledge, this is the first attempt to investigate the use of the ELM after the CNN in the abnormal EEG detection system. The primary contributions of this study include (i) introducing the application of ELM-based classifiers in the detection of abnormal EEG and (ii) the prediction time improvement compared to previous methods of detection on the same corpus.

The chapter is structured as follows. Section 2 describes the existing literature in the scope of this study. Section 3 presents a dataset detail, pre-processing steps, and the proposed methodology, including a description of the deep learning components employed. Section 4 presents and interprets the study's findings. This section also compares the results with the existing literature. Finally, Sect. 5 elucidates the conclusions of this study along with the drawback and future scope.

2 Related Studies

Only a few papers have successfully validated their deep learning approaches on the TUH abnormal EEG corpus for automated abnormal EEG detection. The baseline approach was published in [11]. This method employed a random forest (RF) classifier trained on mel-frequency cepstral coefficients (MFCCs) feature array to obtain an error rate of 31.7% for abnormal EEG classification. One of the first successful deep learning applications was authored in [23]. Here, the authors evaluated a CNN model with a multilayer perceptron decision layer (CNN-MLP) and achieved an accuracy of 78.8%. In 2017, Schirrmeister et al. achieved 85.4% accuracy with another CNN-based abnormal EEG detection system [24]. One-dimensional CNNs (1D-CNNs) were applied in [25, 26], which achieved accuracies of 79.34% and 82.24%, respectively. Later, a novel architecture that combined the features of popular CNN networks like Inception Net, and DenseNet along with gated recurrent units (GRU) called ChronoNet, achieved a classification accuracy of 86.57% for abnormal EEG detection [27]. Another study [28] employed pretrained CNN architectures such as AlexNet and VGG16 for abnormal EEG detection. After pretraining, the fully connected decision layers were replaced with a support vector machine (SVM), resulting in a classification accuracy of 86.59% and 87.32%, respectively. The authors of [29] followed a peculiar approach of dividing the signal into three parts and feeding each part to a separate pretrained AlexNet CNN network, before concatenating their outputs and passing them on to an MLP decision layer. This model, termed as the fusion-based model, achieved 89.13% as its classification accuracy. Among the recent literature, Gemein et al. [30] proposed a network called the temporal convolutional network (TCN) for abnormal EEG detection, which was 86% accurate during its evaluation. R. Singh et al. [31] applied short-time Fourier transform (STFT) transformations of the signals to the VGG19 model for extracting feature maps and subsequently used an RF classifier to reach 88.04% accuracy for classification. Finally, Bajpai et al. [32] proposed the use of an existing network, SeizureNet (first used in seizure detection in [33]), along with an SVM decision layer to reach an overall accuracy of 96.6%. The approach described in this study follows the lead of [32] and explores other impactful types of decision layers for the same problem.

Tables 1 and 2 show a summary of the performance measures from the aforementioned studies. There is a general lack of variability in decision layers that may be seen in these studies. Majority of the work focuses on either traditional machine learning models such as SVM, RF, or MLP. Some of them directly feed their outputs to a SoftMax layer. There is plenty of room to experiment with different types of decision layers. Furthermore, the input modality used in most papers is also very similar, a 2D matrix of signals where signal data points make up the columns and channels make up the rows. The other two commonly used modalities are the raw signal itself and STFT transform. Additionally, only a few papers have published a cross-validated result, the best of which is from ChronoNet in [27]. With these research gaps in mind, we investigated deep learning solutions with an ELM

Table 1 A review of previous work with cross-fold validation

Reference	Method	Input modality	Accuracy (%)	Sensitivity (%)	Specificity (%)
Roy et al. (2018) [27]	ChronoNet	Raw signal	86.60	–	–
Gemein et al. (2020) [30]	TCN	2D signal matrix	86.16	–	–
Schirrmeister et al. (2017) [24]	Shallow/ deep CNNs	2D signal matrix	85.40	75.1	94.1

Table 2 A review of previous work with single-fold validation

Reference	Method	Input modality	Accuracy (%)	Sensitivity (%)	Specificity (%)
Bajpai et al. (2021) [32]	SeizureNet + SVM	STFT plot	96.6	100	90.48
Alhussein et al. (2019) [29]	AlexNet + MLP	2D signal matrix	89.13	–	–
Singh et al. (2020) [31]	VGG19 + random forest	F7 + STFT	88.04	84.61	93.45
Amin et al. (2019) [28]	AlexNet + SVM	2D signal matrix	87.32	78.57	94.67
Shukla et al. (2021) [26]	1D CNN	Raw signal	82.24	–	–
Yildirim et al. (2018) [25]	1D CNN + MLP	Raw signal	79.34	–	–
Lopez (2017) [23]	2D CNN + MLP	2D feature matrix	78.80	75.4	81.9

decision layer. We also utilized all the training channels by creating a 22-channel image instead of the typical 3-channel RGB image. With this approach, we achieved a cross-validation accuracy of 85.87%, which is comparable to the state-of-the-art cross-validation accuracy.

3 Materials and Methods

3.1 Data Preparation

The EEG signals employed for training and evaluation purposes in this study were obtained from Temple University Hospital (TUH) abnormal EEG corpus database (v2.0.0) [22]. It is a subset of the larger TUH EEG corpus containing more than 30,000 recordings since 2002. The dataset is maintained by the Neural Engineering Data Consortium (NEDC) at TUH and is being updated regularly to include more information. Each recording contains a manual label specifying whether it is

Table 3 Label-wise breakdown of TUH abnormal EEG corpus

Dataset	Type	Patients	Recordings
Training set	Abnormal	893	1346
	Normal	1237	1371
	Total	2130	2717
Evaluation set	Abnormal	105	126
	Normal	148	150
	Total	253	276

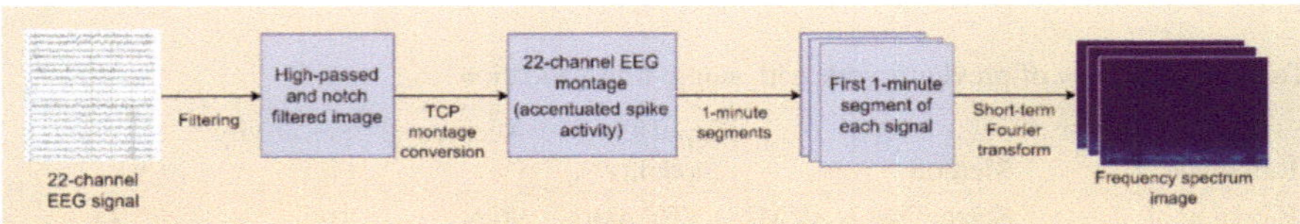

Fig. 1 Data pre-processing pipeline for the proposed system

clinically abnormal or not. The dataset was already partitioned into two sets, namely, train and evaluation. This is done so that all evaluations are based on the same set of data and hence comparable. Care is taken to ensure no data is common between these two sets. Table 3 shows the number of patients placed in each subdivision of the dataset along with the number of recordings for the same. The recordings in the dataset had the varying number of electrodes, with a minimum of 21, and a maximum of 31 EEG electrodes. All the recordings acquired from the standard international 10–20 positions F8, F7, F4, F3, Fz, FP2, FP1, A2, A1, C4, C3, Cz, P4, P3, Pz, O2, O1, T6, T5, T4, and T3 were considered.

The preparation of data for abnormality detection is illustrated in Fig. 1. The pre-processing step involved passing each signal through a fourth-order Butterworth filter of 1 Hz to remove baseline fluctuations in the signals. It is then passed through a fourth-order notch filter of 60 Hz to remove electrical interference in the data. The filtered 21-channel EEG signal is converted to a transverse central parietal (TCP) montage to accentuate the spike activity in the signal [11, 34], resulting in the 22 montage signals: F8-T4, F7-T3, F4-C4, F3-C3, FP1-F3, FP1-F7, FP2-F8, FP2-F4, A1-T3, C4-T4, C3-CZ, C3-P3, C4-P4, CZ-C4, T3-T5, T3-C3, P3-O1, P4-O2, T6-O2, T5-O1, T4-A2, and T4-T6. While most recordings were originally sampled at a frequency of 250 Hz, some had to be from a higher frequency to match the rest of the dataset. Neurologists can diagnose the abnormality of an EEG session by inspecting just the early part of the signal [11]. Since the goal is to automate the process without adding additional dependencies, we segmented all the signals down to the first 1-min segments. This leaves us with 15,000 time points per recording. Then, each 1-min segment from each channel was transformed into a TF image.

3.2 Time-Frequency Representation-Based Abnormal EEG Detection

In this paper, a time-frequency representation-based approach is presented to distinguish normal and abnormal EEG signals. The proposed method involves obtaining TF representation of normal and abnormal EEG segments marked by the neurologists. In our proposed approach, short-time Fourier transform (STFT) is used to generate TF images for CNN inputs.

STFT is a traditional TF representation method which relies on the processing of nonstationary signals in short windows of time to capture the time-varying frequency content. STFT uses a sliding window operation to partition the signal into short and generally overlapping segments. Fourier transforms of the segments are calculated, and process is repeated until the sliding window reaches the end of the signal. The square of the magnitude of an STFT is referred to as the "spectrogram" and is widely used to estimate how the energy of a signal is distributed the TF plane:

$$X(w) = \int_{-\infty}^{\infty} x(t)e^{-jwt}dt \tag{1}$$

$$X(\tau, w) = \int_{-\infty}^{\infty} x(t)\omega(t-\tau)e^{-jwt}dt \tag{2}$$

where $X(\tau, w)$ is the STFT, $x(\tau)$ is the analyzed signal, and $\omega(\tau)$ is the window function. STFT of a given signal $x(\tau)$ and windowing function $\omega(\tau)$ can also be written in terms of their Fourier transforms $X(\omega)$ and $W(\omega)$ as

$$X(t, \omega) = \frac{1}{2\pi}\int_{-\infty}^{\infty} X(\varepsilon)W(\omega-\varepsilon)e^{jet}d\varepsilon \tag{3}$$

Therefore, the spectrogram of the signal is defined as

$$S_{SP}(t, \omega) = |X(t, \omega)|^2 \tag{4}$$

Over traditional Fourier transform, STFT offers a better balance between time-domain resolution and frequency-domain resolution, given the suitable parameters. Essentially, STFT works by breaking the signal into smaller windows and applying Fourier transform on each of them to extract their frequency information. Then these windows of frequency information are plotted against time, creating the spectrum plot that we feed into the CNN. In this study, we have created an STFT plot for each channel of the first 1-min signal, creating a 22-channel image. The parameters used in this study are a segment length of 500 points, an overlap length of 250 points, and a zero-padding set to 100 (i.e., nfft = 600). This results in a spectrogram of size 127 × 60, where the first dimension refers to the frequency resolution and the second

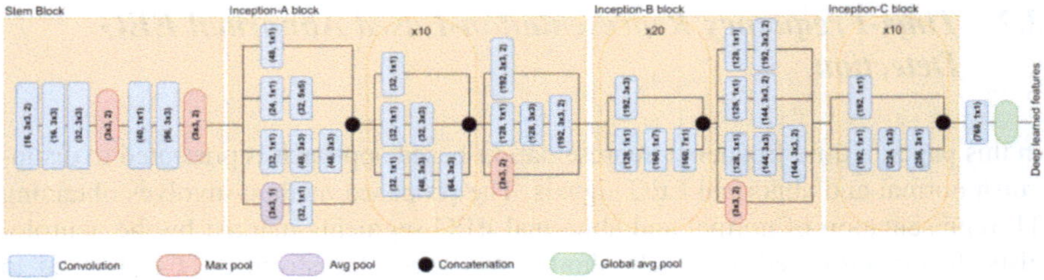

Fig. 2 Architectural overview of Inception-ResNet-V2

dimension refers to the time resolution of the image (Fig. 2). The 22 spectrograms generated for each recording this way are converted into gray scale shape (288, 432) and concatenated along the z-axis to form a 22-channel 3D representation of each recording.

3.3 *Convolutional Neural Networks*

Recently, convolutional neural networks (CNN or ConvNets) are one of the most widely used deep learning algorithms for object recognition, image classification, disease detection, and so on. CNNs replace the tedious and slow conventional feature extraction and classification algorithms. A CNN automatically learns how to extract meaningful features from the input and classify them. CNN is comprised of an input layer, multiple hidden layers, and an output layer. The hidden layer of a CNN is typically composed of a convolutional layer and a pooling layer. The output of the hidden layers is then flattened and passed to a fully connected layer, which forms the output layer. The function of each layer is explained as follows.

Convolutional layer: The convolution operation is applied between the input and a kernel (weight matrix) in this layer. The kernel is usually much smaller than the input and acts like a sliding window to cover the entire input. The output of this layer is the activation or feature map. The stride controls how much the kernel convolves with the input sample in each step, and how much overlap lies between each step. This operation acts as a feature extractor by learning from the diverse input signals.

Pooling layer: This is a down-sampling layer where the pooling operation is employed to reduce the feature map's dimensions, while retaining the significant information. This layer also helps to regulate overfitting. The pooling operation can be performed in different ways, and this study employed the average pooling operation (Eq. (4)).

Fully connected layer: The output of the final pooling layer is flattened into a one-dimensional array and passed to a fully connected layer. This layer applies an activation function to a weighted sum of the inputs. The final sigmoid layer outputs a single class (normal or abnormal) for the given input. The sigmoid layer is removed after training so that the model can work as a feature extractor.

Let X be the input image. For a given layer, the kth feature map fm_k can be obtained through Eq. (2). The feature map's dimensions can be further reduced in the pooling layer, which is average pooling in this study. Once convolutions and pooling have created a sufficiently small and dense representation of the input, it can be flattened and passed into the FC layers as mentioned previously.

Typically, an activation function like rectifier linear unit (ReLU) is used to introduce nonlinearity in the features. Dropout layers introduce randomness during the training, which in turn lowers the amount of overfitting of the model. For an activation function $g(x)$, the convolution and pooling operations can be summarized as

$$fm_k = g(w_k * X + b_k) \tag{5}$$

$$pm_k = \text{Average pooling}(fm_k, l, s) \tag{6}$$

where w_k denotes the filter matrix applied to the image, b_k denotes the bias added to kernel, * denotes the convolution operation, l denotes the size of the window that is pooled, and s refers to the length of the stride, or movement of the window across the image (decides the overlap).

Using the aforementioned layers, one can generate their own CNN. It becomes a question of tuning the number and order of convolutional, pooling, and fully connected layers, until the network achieves the desired performance. With recent advances in CNN, many pretrained deep CNNs have been used for machine learning problems. In this study, the performance of three popular configurable CNN architectures, namely, Inception-ResNet-V2, DenseNet, and SeizureNet, is investigated. These models have been applied to various real-world applications such as image classification, emotion recognition, schizophrenia detection, and seizure detection.

3.3.1 Inception-ResNet-V2

This model is the latest revision of a combination of the best features from the Inception family of CNNs and the ResNet family of CNNs. The concept of inception blocks used in this model allows the model to use different filter sizes in its building blocks instead of restricting itself to a single size [35]. Figure 2 shows the overview of the Inception-ResNet-V2 model. We have presented a less detailed version of the model diagram to highlight only the significant parts of the model architecture.

As depicted above, the model's building blocks run multiple sizes of filters on the same input and then concatenate the result before passing it ahead. This allows for more robust pattern matching. Along with that, this model uses residual or "skip" connections. Traditional feed-forward networks connect the output of the ith layer as input to the $(i + 1)$th layer [36], which gives rise to the transition:

$$X_i = H(X_{i-1}) \tag{7}$$

Residual connections bypass the nonlinear transformations with an identity function:

$$X_i = H(X_{i-1}) + X_{i-1} \tag{8}$$

These connections allow back-propagated gradients to flow without getting diminished by the large number of layers in this model. In other words, this solves the vanishing gradient problem.

3.3.2 DenseNet

Introduced by Huang et al. in 2017 [37], DenseNet is a convolutional neural network which applies the notion of skip connections of residual networks in a novel way. DenseNet is essentially made up of Dense blocks and transition blocks. In DenseNet blocks, every layer has a direct connection to every subsequent layer inside a single block. These are called Dense connections. This converts the Eq. (4) to

$$X_i = H_i([X_0, X_1, .., X_{i-1}]) \tag{9}$$

Not only does this ease the problem of vanishing gradients, but it also makes training times much faster for these networks (Fig. 3).

3.3.3 SeizureNet

SeizureNet is a CNN network introduced by Asif et al. [33] that was created with EEG spectrograms in mind. It essentially utilizes the same building blocks as that of DenseNet, with slight modifications for this domain such as different filter sizes and dropout parameters. The original model is an ensemble of three separate branches

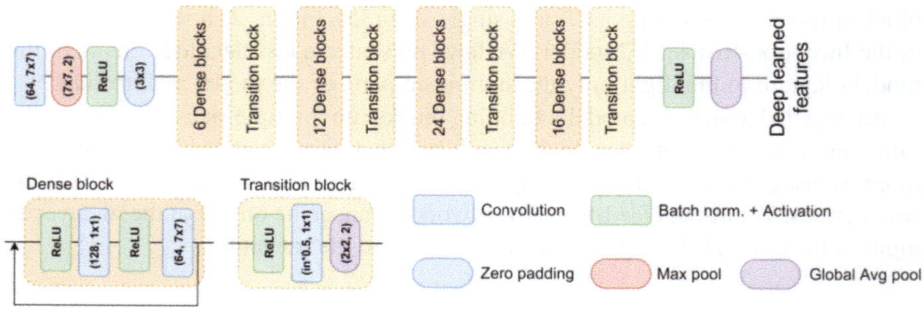

Fig. 3 Architectural overview of DenseNet

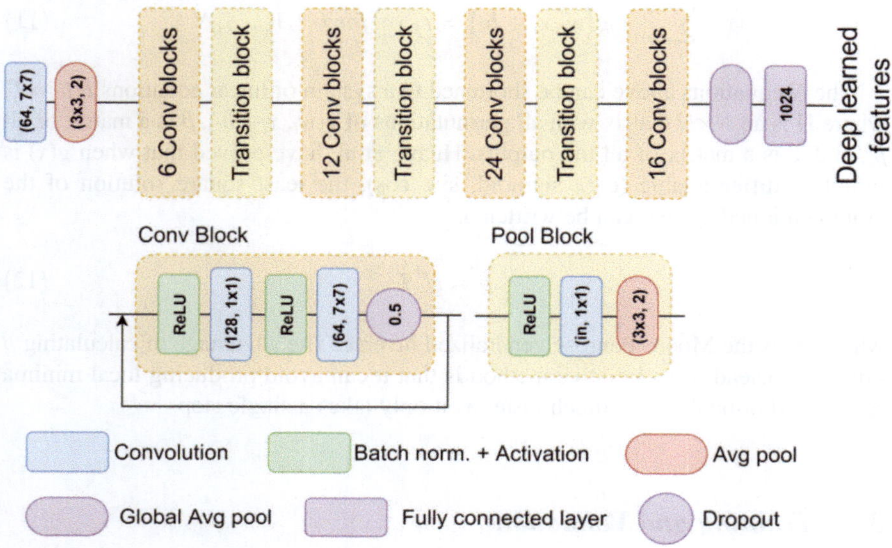

Fig. 4 Architectural overview of SeizureNet

with different numbers of dense blocks and pooling layers. The model used in this study is a slightly modified version of SeizureNet, using only one branch of the original network. This decision was inspired by the size of the dataset in use (Fig. 4).

3.4 Extreme Learning Machine

The decision layer for every model used in this study is a single-hidden layer feed-forward neural network (SLFN) which is trained using the extreme learning machine (ELM) method. Introduced by Huang et al. [38], this method of learning weights involving random weights and biases lowers the risk of overfitting of the decision layer. Additionally, the training speed of ELMs is very short. For N distinct samples (x_i, t_i) where x_i is the input vector and t_i is the output vector, a standard feed-forward network can be modeled as.

$$\sum_{i=1}^{N} \beta_i g_i(x_j) = \sum_{i=1}^{N} \beta_i g(w_i . x_j + b_i) = o_j \quad \text{for } j = 1, \ldots, N \tag{10}$$

where w_i and β_i are vectors of the weights from the input nodes to the ith hidden node and from the ith hidden node to the output nodes, respectively. b_i denotes the threshold of the hidden layer, $g(x)$ is the activation function, and w_i. x_j denotes the dot product of w_i and x_j. For an SLFN that approximates the output with a zero error, this equation becomes

$$\sum_{i=1}^{N} \beta_i g\left(w_i . x_j + b_i\right) = t_j \qquad for\, j = 1, \ldots, N \tag{11}$$

The N equations above can be shortened to a system of linear equations $\boldsymbol{H\beta} = \boldsymbol{T}$, where **H** is an $N \times N$ matrix with all permutations of $g(w_i.\ x_j + b_i)$, β is a matrix of all β_i, and **T** is a matrix of all the outputs. Huang et al. have proved that when $g(x)$ is infinitely differentiable (e.g., sigmoid, sin, cos), the least square solution of the aforementioned system can be written as

$$\beta = H^{\dagger} T \tag{12}$$

where $H^{\dagger}$ is the Moore-Penrose generalized inverse. The advantage of calculating β this way instead of an iterative method is that it can avoid producing local minima results. Additionally, it is much faster as it only takes a single step.

3.5 Training and Validation

The training-validation methodology as well as the various hyper-parameter details is elaborated in this section. The CNN-based deep learning models were implemented in Python (v3.6) using the Keras module of the TensorFlow library. All operations were performed on a workstation powered by an Intel (R) Core (TM) i7-6700HQ CPU @2.59 GHz with an NVIDIA GeForce 1080 Ti GPU running on the Ubuntu operating system. The hyper-parameter optimization for all models in this study was done manually. Table 4 list the hyper-parameters employed in this study.

The performance measures for this study were obtained using two approaches that allow us to compare with literature and a wider audience. The first approach follows a single-fold validation scheme, similar to the papers mentioned in Table 1. Here, the training subset of the corpus is further divided into two subsets. The first subset contains 80% of the training set and is used for model training. The second subset containing the remaining 20% of the data is used for validation. The model

Table 4 Hyper-parameter values

Hyper-parameters	Values
Optimizer	Adam, $\beta_1 = 0.9$ and $\beta_2 = 0.999$
Learning rate	0.001
Batch size	16
Dropout	25% [SeizureNet] None [DenseNet] None [inception]
Epoch	50
Decay	None
Metrics	Accuracy
Loss function	Binary cross-entropy

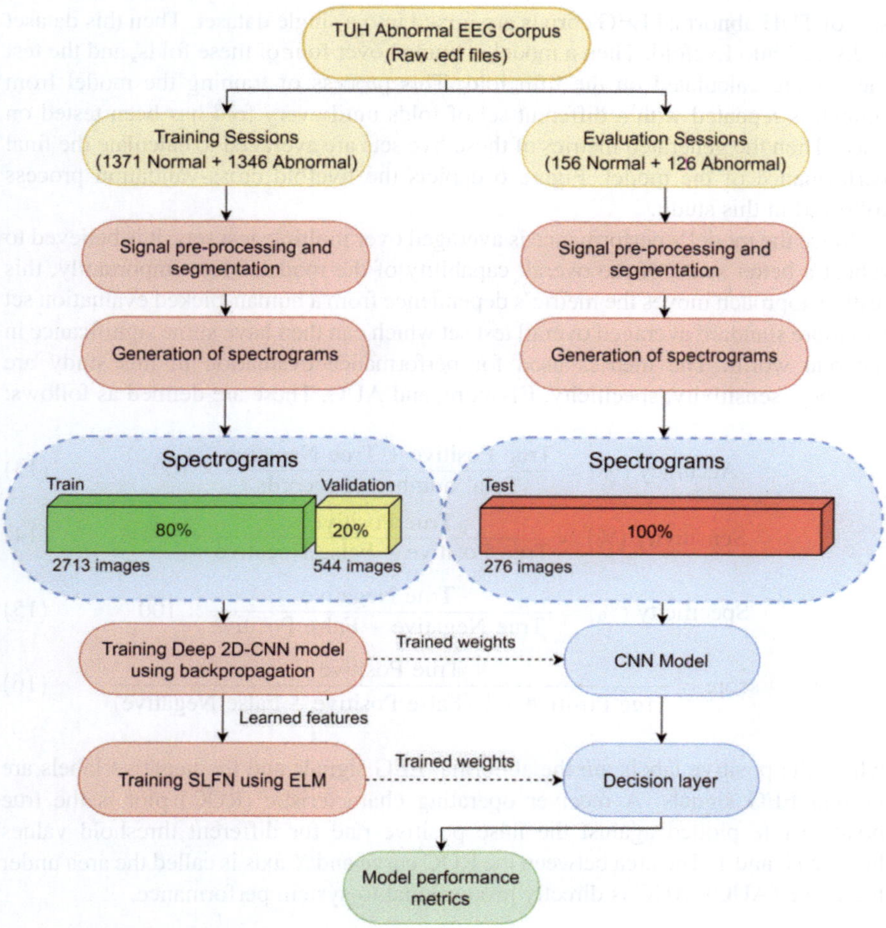

Fig. 5 Pre-split testing method on TUH abnormal EEG corpus

that has been trained as a result of this is then tested on the TUH corpus evaluation set. By testing on the standard evaluation set, one can easily compare the performance of multiple models in a reliable standardized way. Figure 5 shows a high-level flow of this method of training and evaluation. It is evident through the diagram that the models in this study were trained over 2713 STFT images, validated over 544 STFT images, and finally tested over 276 STFT images.

While this method creates easily reproducible results through a seed value, the results of the single-fold method are highly dependent on the breakup of the training set. Even with a balanced and curated evaluation set like this, a single test set cannot emulate the real-world scenario. Hence, metrics calculated on these do not represent the average performance of the model in a clinical setting. For this reason, we have the second approach of testing. In the second approach, the training and evaluation sets of TUH abnormal EEG corpus are mixed into a single dataset. Then this dataset is divided into fivefold. Then a model is trained over four of these folds, and the test metrics are calculated on the fifth fold. This process of training the model from scratch is repeated with a different set of folds until every fold has been tested on once. Then the generated metrics of these five sets are averaged to calculate the final performance of the model. Figure 6 depicts the fivefold cross-validation process followed in this study.

Since the model's performance is averaged over multiple test sets, it is believed to reflect a better sense of the overall capability of the model. More importantly, this testing approach moves the metric's dependence from a human-picked evaluation set to a more standard averaged overall test set which can then have some significance in the real world. The metrics used for performance evaluation in this study are accuracy, sensitivity, specificity, F1-score, and AUC. These are defined as follows:

$$\text{Accuracy }(\%) = \frac{\text{True Positive} + \text{True Negative}}{\text{Total number of records}} \times 100 \tag{13}$$

$$\text{Sensitivity }(\%) = \frac{\text{True Positive}}{\text{True Positive} + \text{False Negative}} \times 100 \tag{14}$$

$$\text{Specificity }(\%) = \frac{\text{True Negative}}{\text{True Negative} + \text{False Positive}} \times 100 \tag{15}$$

$$F1 - \text{score} = \frac{\text{True Positive}}{\text{True Positive} + \frac{1}{2}\,(\text{False Positive} + \text{False Negative})} \tag{16}$$

where the positive labels are the abnormal EEG signals and the negative labels are normal EEG signals. A receiver operating characteristic (ROC) plot is the true positive rate plotted against the false positive rate for different threshold values between 0 and 1. The area between the ROC curve and *X*-axis is called the area under the curve (AUC). AUC is directly proportional to system performance.

4 Results

This section lists and compares the performance metrics of each of the EEG models for abnormal EEG classification. From Table 4, we can see that each model was trained for a maximum of 50 epochs. The best models were usually saved before the completion of 50 epochs through the checkpointing method of TensorFlow. From Table 5, it can be observed that the SeizureNet-ELM system performed the best, achieving an accuracy, sensitivity, specificity, and F1-score of 84.06%, 80.16%, 87.33%, and 0.8402, respectively. The other models achieved performed very similar to each other with slight differences in sensitivity and specificity, which are small enough to be attributed to the randomness involved in the training set creation.

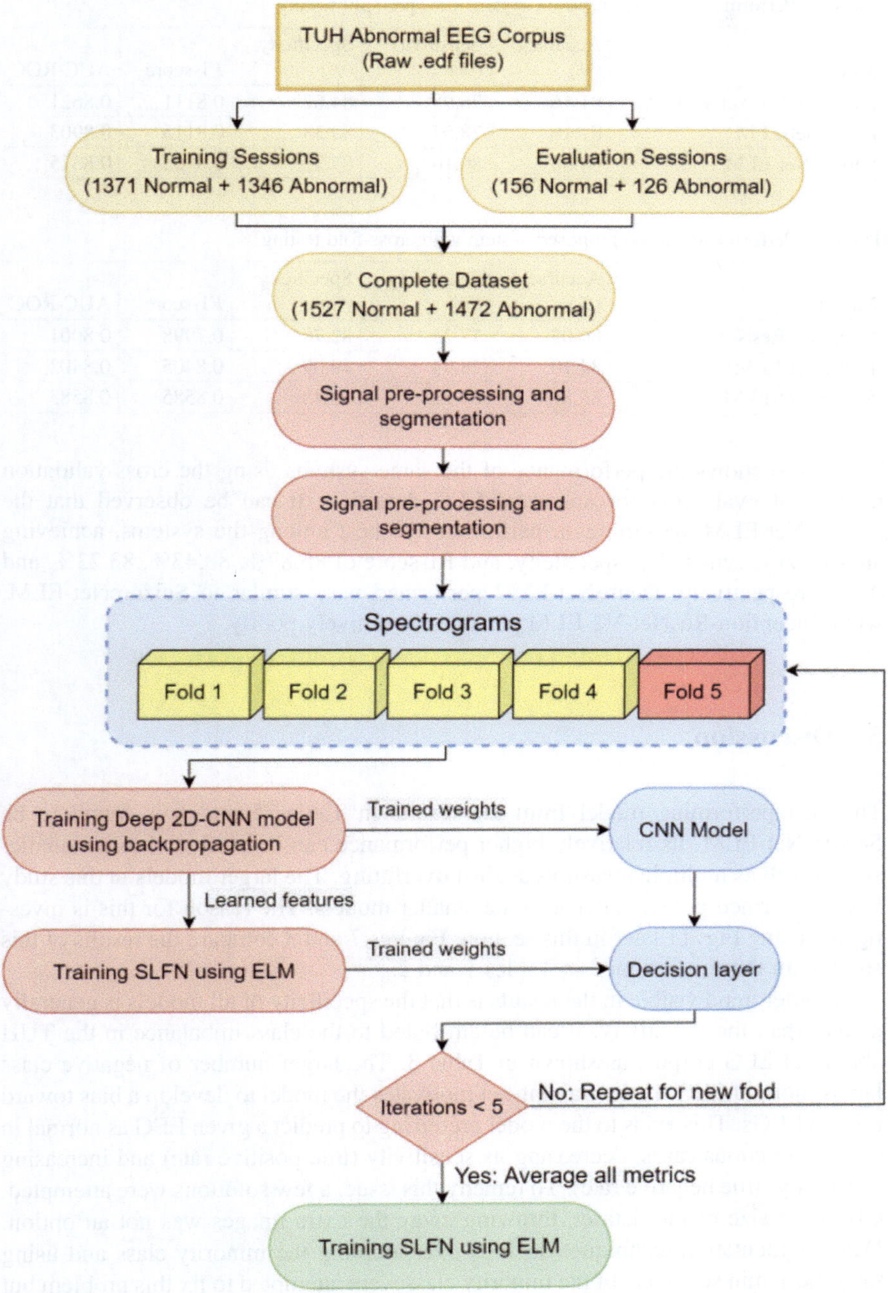

Fig. 6 Fivefold cross-validation method on TUH abnormal EEG corpus

Table 5 Performance of the proposed system with pre-split testing

Model	Accuracy (%)	Sensitivity (%)	Specificity (%)	*F*1-score	AUC-ROC
Inception-ResNet-v2-ELM	81.16	76.98	84.67	0.8111	0.8621
DenseNet-ELM	81.16	78.57	83.33	0.8115	0.8903
SeizureNet-ELM	84.06	80.16	87.33	0.8402	0.8375

Table 6 Performance of the proposed system with cross-fold testing

Model	Accuracy (%)	Sensitivity (%)	Specificity (%)	*F*1-score	AUC-ROC
Inception-ResNet-v2-ELM	80.05	77.25	82.78	0.7998	0.8001
DenseNet-ELM	84.10	78.94	89.09	0.8405	0.8402
SeizureNet-ELM	85.87	83.23	88.43	0.8585	0.8583

Table 6 shows the performance of the same systems using the cross-validation method of evaluation for abnormal EEG detection. It can be observed that the SeizureNet-ELM system again performed the best among the systems, achieving an accuracy, sensitivity, specificity, and F1-score of 85.87%, 88.43%, 83.23%, and 0.858, respectively. DenseNet-ELM performed very similar to SeizureNet-ELM, while Inception-ResNet-V2-ELM performed relatively poorly.

5 Discussion

The best performing model from the results in Tables 5 and 6 is found to be SeizureNet-ELM. Its relatively higher performance can be attributed to its smaller size as well as multiple measures against overfitting. The larger models in this study have performed poorly relative to the smaller models. The reason for this is investigated using Fig. 11 later in this section. Figures 7 and 8 compare the results of this study with those mentioned in Tables 1 and 2.

Another trend visible in the results is that the specificity of all models is generally greater than the sensitivity. It can be attributed to the class imbalance in the TUH abnormal EEG corpus, as shown in Table 3. The larger number of negative class labels (normal EEG) in the training set motivates the model to develop a bias toward normal EEGs. This leads to the model preferring to predict a given EEG as normal in some ambiguous cases, decreasing its sensitivity (true positive rate) and increasing specificity (true negative rate). To remedy this issue, a few solutions were attempted. Given the size of the dataset, throwing away the extra images was not an option. Data augmentation techniques such as oversampling the minority class and using multiple 1-min segments of the minority class were attempted to fix this problem but to no avail. More extensive data augmentation methods such as generative adversarial network (GAN)-based augmentation, noise addition, and Fourier transform-based augmentation were not applied in this study. From the above graphs, we can

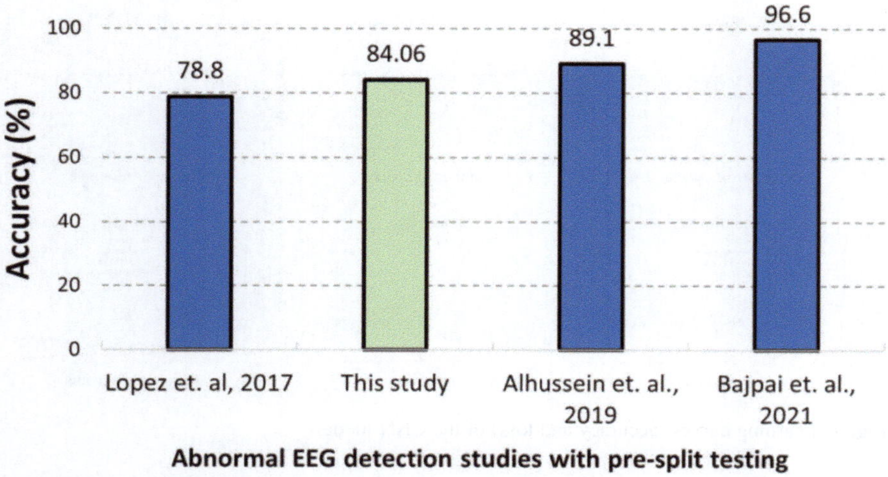

Fig. 7 Performance of proposed model compared to pre-split results in literature

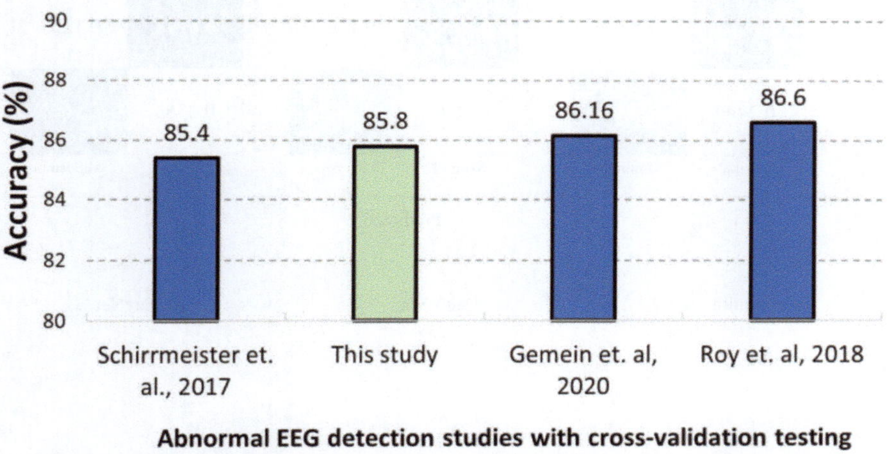

Fig. 8 Performance of proposed model compared to cross-validation results in literature

see that the results achieved in this study are comparable to the state-of-the-art in terms of cross-validation metrics (Fig. 9). Figure 10 shows the learning curves for the models. The dotted lines represent accuracy and losses for individual folds in the cross-validation method, while the solid line depicts the average performance.

Looking at the gap between the average training accuracy/loss and the average validation training accuracy/loss, it is evident that the problem of overfitting is a major player in the reduced accuracy of this approach. Even with the early stopping, the point with the least distance between training and validation accuracy/loss has enough gap to show that the models have not generalized effectively. A few reasons

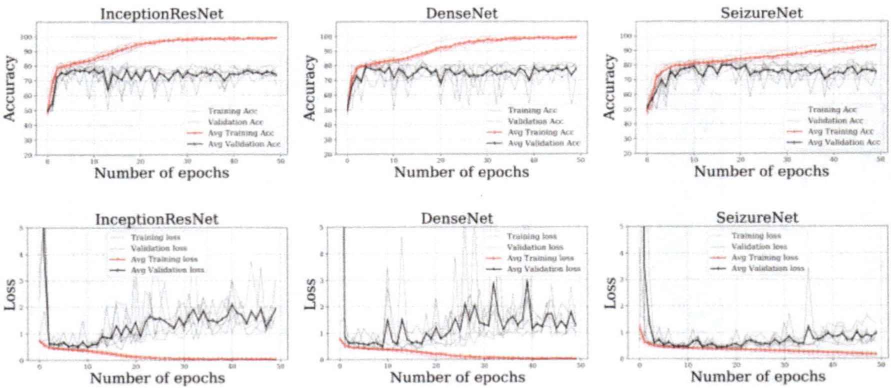

Fig. 9 Learning curves (accuracy and loss) of the CNN models

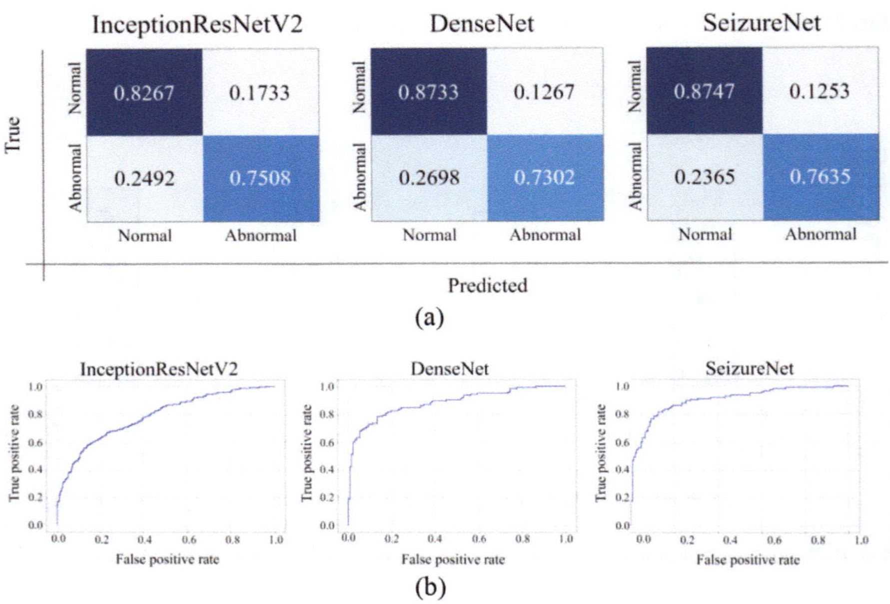

Fig. 10 Confusion matrices (**a**) and ROC (**b**) of the CNN models

for this could be the limited amount of variety in data. Generally, deep learning models (especially CNNs) are trained on millions of data points before validation and testing. In this case, we only have a limited number of patients. In our attempts to improve the accuracy, using more segments of the same patient's recording led to the model learning patient-specific features and overfitting the same/more instead of learning general features. Another reason this approach did not work as well as that of SVM in Bajpai et al. [32], despite using the same models, is because unlike ELMs,

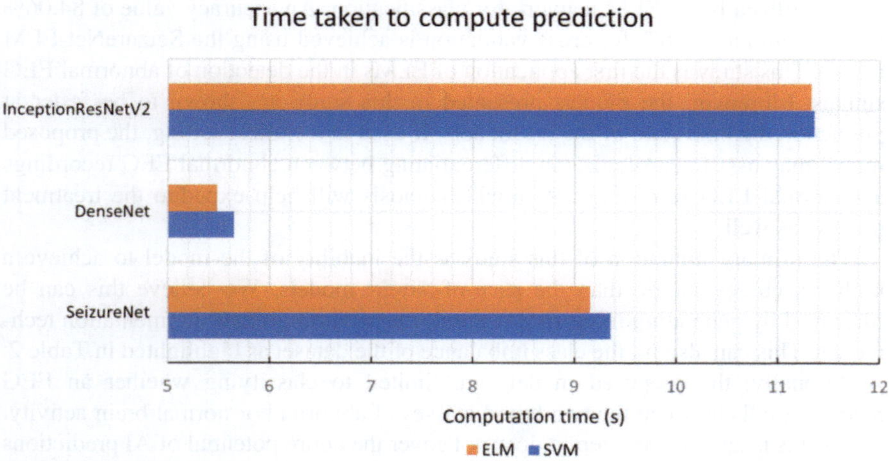

Fig. 11 Computation time for prediction for different decision layers

the SVMs use a softer margin in their decision-making to avoid overfitting to outliers in the data. This helps them generalize better. While the accuracy scores were not better than the state-of-the-art, this approach to abnormal EEG classification has a few other advantages that can give it the edge in a clinical setting.

Figure 11 compares the computation run times of all the CNN-based models along with their decision layers. This graph includes the metrics of the time taken by CNN feature extraction and ELM decision-making from this study and [32]. It doesn't include loading and pre-processing times as the input modalities for both studies are different. From the figure, we can identify DenseNet-ELM as the model with the lowest evaluation time. The models generally found in literature have greater than 100 million parameters as compared to approximately 10 million parameters of SeizureNet and DenseNet. Since the prediction time of a model is directly proportional to the size of the CNN model, we can safely assume that models like SeizureNet-ELM and DenseNet-ELM would also outperform popularly used CNN models in literature like AlexNet in terms of evaluation speed due to their drastically smaller size. In this study, SeizureNet-ELM model took an average detection time of approximately 9.15 s. Additionally, the ELM model also loads extremely quickly as compared to SVMs mentioned in [32].

6 Conclusion

In this study, a CNN-based deep learning system for classifying abnormal EEG was proposed. Each channel of the input signal was converted into a spectrogram image. The spectrograms of each channel were concatenated along the z-axis to create a 3D image, which was fed to the CNN models to extract features. These features were finally utilized by an ELM network for classification. An accuracy value of 84.06% for pre-split and 85.8% for cross-validation is achieved using the SeizureNet-ELM model. This study is the first application of ELMs in the detection of abnormal EEG signals. Moreover, the models presented in this study are shown to be faster at prediction than the state-of-the-art models. If used in a clinical setting, the proposed model may assist neurologists in differentiating between abnormal EEG recordings and normal EEG recordings. A quick diagnosis will help expedite the treatment process as well.

The primary limitation of this study is the inability of the model to achieve a higher accuracy metric than the state-of-the-art models. We believe this can be improved by generating more spectrogram images through data augmentation techniques. This can also fix the class imbalance of the dataset as highlighted in Table 2. Additionally, the proposed models are limited to classifying whether an EEG recording falls between the two broad classes of abnormal or normal brain activity. While this is an essential step, it does not cover the entire potential of AI predictions for EEG. Multiple brain-related disorders can be classified through EEG, and this front is being explored for further studies.

References

1. İnce, R., Adanır, S. S., & Sevmez, F. (2021). The inventor of electroencephalography (EEG): Hans Berger (1873-1941). In Child's nervous system: ChNS : official journal of the International Society for Pediatric Neurosurgery 37(9):2723–2724. doi: https://doi.org/10.1007/s00381-020-04564-z
2. Pathirana, S., Asirvatham, D., & Johar, G. (2018). A Critical Evaluation on Low-Cost Consumer-Grade Electroencephalographic Devices. In: 2nd International Conference on BioSignal Analysis, Processing and Systems (ICBAPS)
3. Prasanth, T., Thomas, J., Yuvaraj, R., Jing, J., Cash, S. S., Chaudhari, R., Leng, T. Y., Rathakrishnan, R., Rohit, S., Saini, V., Westover, B. M., & Dauwels, J. (2020). Deep Learning for Interictal Epileptiform Spike Detection from scalp EEG frequency sub bands. In: Annual International Conference of the IEEE Engineering in Medicine and Biology Society.
4. Yuvaraj, R., Thomas, J., Kluge, T., & Dauwels, J. (2018). A deep Learning Scheme for Automatic Seizure Detection from Long-Term Scalp EEG. Paper presented at the 52nd Asilomar Conference on Signals, Systems, and Computers, USA.
5. Dharamsi, T., Das, P., Pedapati, T., Bramble, G., Muthusamy, V., Samulowitz, H., Varshney, K. R., Rajamanickam, Y., Thomas, J., & Dauwels, J. (2017). Neurology-as-a-Service for the Developing World. ArXiv Preprint ArXiv:1711.06195.
6. Bhurane, A.A., Dhok, S., Sharma, M., Yuvaraj, R., Murugappan, M. and Acharya, U.R. (2019). Diagnosis of Parkinson's disease from electroencephalography signals using linear and self-similarity features. Expert Systems, p.e12472.
7. Al-Nuaimi, F. A., Al-Nuaimi, R. J., Al-Dhaheri, S. S., Ouhbi, S., & Belkacem, A. N. (2020). Mind Drone Chasing Using EEG-based Brain Computer Interface. In: 16th International Conference on Intelligent Environments (IE)
8. Machado, S., Araújo, F., Paes, F., Velasques, B., Cunha, M., Budde, H., Basile, L. F., Anghinah, R., Arias-Carrión, O., Cagy, M., Piedade, R., de Graaf, T. A., Sack, A. T., & Ribeiro, P. (2010). EEG-based brain-computer interfaces: an overview of basic concepts and

clinical applications in neurorehabilitation. Reviews in the Neurosciences, 21(6):451–468. doi: https://doi.org/10.1515/revneuro.2010.21.6.451
9. Grant, A. C., Abdel-Baki, S. G., Weedon, J., Arnedo, V., Chari, G., Koziorynska, E., Lushbough, C., Maus, D., McSween, T., Mortati, K. A., Reznikov, A., & Omurtag, A. (2014). EEG interpretation reliability and interpreter confidence: aa large single-center study. Epilepsy & Behavior: E&B, 32:102–107. doi: https://doi.org/10.1016/j.yebeh.2014.01.011
10. Halford, J. J., Shiau, D., Desrochers, J. A., Kolls, B. J., Dean, B. C., Waters, C. G., Azar, N. J., Haas, K. F., Kutluay, E., Martz, G. U., Sinha, S. R., Kern, R. T., Kelly, K. M., Sackellares, J. C., & LaRoche, S. M. (2015). Inter-rater agreement on identification of electrographic seizures and periodic discharges in ICU EEG recordings. Clinical Neurophysiology: Official Journal of the International Federation of Clinical Neurophysiology, 126(9):1661–1669. doi: https://doi.org/10.1016/j.clinph.2014.11.008
11. López, S., Suarez, G., Jungreis, D., Obeid, I., & Picone, J. (2015). Automated identification of abnormal adult EEGs. In: IEEE Signal Processing in Medicine and Biology Symposium (SPMB)
12. Brogger, J., Eichele, T., Aanestad, E., Olberg, H., Hjelland, I., & Aurlien, H. (2018). Visual EEG reviewing times with SCORE EEG. Clinical Neurophysiology Practice, 3:59–64. doi: https://doi.org/10.1016/j.cnp.2018.03.002
13. Impact of artificial intelligence on radiology: a EuroAIM survey among members of the European Society of Radiology. (2019). Insights into Imaging, 10(1):105. doi: https://doi.org/10.1186/s13244-019-0798-3
14. Justy Antony, Jemin J Webster, Neema Jayadas & Supriya Medapati (2021) Time is Brain: AI helps cut down stroke diagnosis time in the Himalayan foothills. https://blog.qure.ai/notes/ai-cuts-down-stroke-diagnosis-time-himalayas. Accessed 22 Sep 2021
15. Dev, K., Khowaja, S.A., Bist, A.S., Saini, V. and Bhatia, S. (2021) Triage of potential covid-19 patients from chest x-ray images using hierarchical convolutional networks. Neural Computing and Applications, pp.1-16.
16. O'Mahony, N., Campbell, S., Carvalho, A., Harapanahalli, S., Hernandez, G. V., Krpalkova, L., Riordan, D., & Walsh, J. (2020). Deep Learning vs. Traditional Computer Vision BT - Advances in Computer Vision (K. Arai & S. Kapoor (Eds.); pp. 128–144). Springer International Publishing.
17. Sejnowski, T. J. (2020). The unreasonable effectiveness of deep learning in artificial intelligence. Proceedings of the National Academy of Sciences, 117(48):30033–30038. doi: https://doi.org/10.1073/pnas.1907373117
18. Oh, S.L., Hagiwara, Y., Raghavendra, U., Yuvaraj, R., Arunkumar, N., Murugappan, M. and Acharya, U.R., (2018) A deep learning approach for Parkinson's disease diagnosis from EEG signals. Neural Computing and Applications, pp. 1-7. doi: https://doi.org/10.1007/s00521-018-3689-5
19. Basheer, S., Bhatia, S. and Sakri, S.B., 2021. Computational Modeling of Dementia Prediction Using Deep Neural Network: Analysis on OASIS Dataset. IEEE Access, 9, pp.42449-42462.
20. Oh, S. L., Vicnesh, J., Ciaccio, E., Rajamanickam, Y., & Acharya, U. R. (2019). Deep Convolutional Neural Network Model for Automated Diagnosis of Schizophrenia Using EEG Signals. Applied Sciences, 9:2870. doi: https://doi.org/10.3390/app9142870
21. Roy, Y., Banville, H., Albuquerque, I., Gramfort, A., Falk, T. H., & Faubert, J. (2019). Deep learning-based electroencephalography analysis: a systematic review. Journal of Neural Engineering, 16(5):51001. doi: https://doi.org/10.1088/1741-2552/ab260c
22. Obeid, I., & Picone, J. (2016). The Temple University Hospital EEG Data Corpus. Frontiers in Neuroscience, 10:196. doi: https://doi.org/10.3389/fnins.2016.00196
23. López, S., Obeid, I. and Picone, J. (2017) Automated interpretation of abnormal adult electroencephalograms. Dissertation, Temple University.
24. Schirrmeister, R. T., Springenberg, J. T., Fiederer, L. D. J., Glasstetter, M., Eggensperger, K., Tangermann, M., Hutter, F., Burgard, W., & Ball, T. (2017). Deep learning with convolutional

neural networks for EEG decoding and visualization. Human Brain Mapping, 38(11): 5391–5420. doi: https://doi.org/10.1002/hbm.23730
25. Yıldırım, Ö., Baloglu, U. B., & Acharya, U. R. (2020). A deep convolutional neural network model for automated identification of abnormal EEG signals. Neural Computing and Applications, 32(20):15857–15868. doi: https://doi.org/10.1007/s00521-018-3889-z
26. Shukla, U., Saxena, G.J., Shailani, M., Bafila, A.S., Pundir, A. and Singh, S., (2021). An Improved Decision Support System for Identification of Abnormal EEG Signals Using a 1D Convolutional Neural Network and Savitzky-Golay Filtering. IEEE Access
27. Roy S., Kiral-Kornek I., Harrer S. (2019) ChronoNet: A Deep Recurrent Neural Network for Abnormal EEG Identification. In: Riaño D., Wilk S., ten Teije A. (eds) Artificial Intelligence in Medicine. AIME 2019. Lecture Notes in Computer Science, 11526. doi: https://doi.org/10.1007/978-3-030-21642-9_8
28. Amin, S. U., Hossain, M. S., Muhammad, G., Alhussein, M., & Rahman, M. A. (2019). Cognitive Smart Healthcare for Pathology Detection and Monitoring. IEEE Access, 7:10745–10753. doi: https://doi.org/10.1109/ACCESS.2019.2891390
29. Alhussein, M., Muhammad, G., & Hossain, M. S. (2019). EEG Pathology Detection Based on Deep Learning. IEEE Access, 7:27781–27788. doi: https://doi.org/10.1109/ACCESS.2019.2901672
30. Gemein, L. A. W., Schirrmeister, R. T., Chrabąszcz, P., Wilson, D., Boedecker, J., Schulze-Bonhage, A., Hutter, F., & Ball, T. (2020). Machine-learning-based diagnostics of EEG pathology. NeuroImage, 220:117021. doi: https://doi.org/10.1016/j.neuroimage.2020.117021
31. Singh, R., Ahmed, T., Kumar Singh, A., Chanak, P., & Singh, S. K. (2021). SeizSClas: An Efficient and Secure Internet-of-Things-Based EEG Classifier. IEEE Internet of Things Journal, 8(8):6214–6221. doi: https://doi.org/10.1109/JIOT.2020.3030821
32. Bajpai, R., Yuvaraj, R., & Prince, A. A. (2021). Automated EEG pathology detection based on different convolutional neural network models: Deep learning approach. Computers in Biology and Medicine, 133:104434. doi: https://doi.org/10.1016/j.compbiomed.2021.104434
33. Asif U., Roy S., Tang J., Harrer S. (2020) SeizureNet: Multi-Spectral Deep Feature Learning for Seizure Type Classification. In: Kia S.M. et al. (eds) Machine Learning in Clinical Neuroimaging and Radiogenomics in Neuro-oncology. MLCN 2020, RNO-AI 2020. Lecture Notes in Computer Science, 12449. doi: https://doi.org/10.1007/978-3-030-66843-3_8.
34. Acharya, J. N., Hani, A. J., Thirumala, P. D., & Tsuchida, T. N. (2016). American Clinical Neurophysiology Society Guideline 3: A Proposal for Standard Montages to Be Used in Clinical EEG. Journal of Clinical Neurophysiology: Official Publication of the American Electroencephalographic Society, 33(4): 312–316. doi: https://doi.org/10.1097/WNP.0000000000000317
35. Szegedy, C., Liu, W., Jia, Y., Sermanet, P., Reed, S., Anguelov, D., Erhan, D., Vanhoucke, V., & Rabinovich, A. (2015). Going deeper with convolutions. In: IEEE Conference on Computer Vision and Pattern Recognition (CVPR)
36. He, K., Zhang, X., Ren, S., & Sun, J. (2016). Deep Residual Learning for Image Recognition. In: IEEE Conference on Computer Vision and Pattern Recognition (CVPR)
37. Li, C. Y., & Vu, N. T. (2018). Densely Connected Convolutional Networks for Speech Recognition. Speech Communication. In: 13th ITG-Symposium
38. Huang, G.-B., Zhu, Q.-Y., & Siew, C.-K. (2006). Extreme learning machine: Theory and applications. Neurocomputing, 70(1):489–501. doi: https://doi.org/10.1016/j.neucom.2005.12.126

Physical Action Categorization Pertaining to Certain Neurological Disorders Using Machine Learning-Based Signal Analysis

Asad Mansoor Khan, Sajid Gul Khawaja, Muhammad Usman Akram, and Ali Saeed Khan

Abstract The daily life of thousands of individuals around the globe suffers due to physical or neurological disorders that can hinder their limb movement to varying degrees in addition to other effects. Assistive applications can alleviate the effects of the disease leading to improvement of quality of life for such individuals. In such computer-aided application, classification of physical actions is required. Surface electromyography (sEMG) presents a noninvasive mechanism which helps in translating physical movement to signals which can be used in applications for different neurological disorders such as tonic-clonic seizure and epilepsy. In this article, a framework for classification of four physical actions is proposed. Successful classification of these actions can help build systems that can passively monitor epileptic patients and others and provide useful data for improving the quality of life for such patients or providing better treatment. The framework makes use of various features from various signal signatures with contribution from time domain, frequency domain, and inter-channel statistics. Next, we conducted a comparative analysis of SVM, 3-NN, and ensemble learning with accuracy of 97.405%, 95.7%, and 96.5%, respectively. Finally, we have reported the effect which combinations of different signatures have on the classification of physical activities. It was observed that we can achieve an accuracy of 97.26% using SVM classifier via a subset of 48 features. These findings can help design an algorithm for a constrained environment such as real-time processing.

Keywords Physical action classification · sEMG signals · Signal processing and analysis · Feature engineering · Machine learning

A. M. Khan · S. G. Khawaja (✉) · M. U. Akram
Department of Computer and Software Engineering, College of Electrical and Mechanical Engineering, National University of Sciences and Technology, Rawalpindi, Pakistan
e-mail: asad.mansoor@ceme.nust.edu.pk; sajid.gul@ceme.nust.edu.pk; usman.akram@ceme.nust.edu.pk

A. S. Khan
Department of Computer Science, National University of Modern Languages, Islamabad, Pakistan

M. Murugappan, R. Yuvaraj (eds.), *Biomedical Signals Based Computer-Aided Diagnosis for Neurological Disorders*, https://doi.org/10.1007/978-3-030-97845-7_2

1 Introduction

Even with the advent of cutting-edge healthcare, individuals suffering from a range of physical disabilities lack optimum treatment and/or rehabilitation solutions. Broadly, disability can be classified into two groups: physical or mental. Unfortunately, each has various factors that can be the cause, e.g., aging process can cause gait disorder or limb impairment [1] and work-related injuries or trauma caused by accidents all lead to physical disability. Similarly, stroke is a major cause of limb disability in adults [2]. Neurological disorders are not only a cause of mental disorders; rather, they are also a leading cause of accidents [3]. Among the neurological disorders, epilepsy is a major disorder affecting almost 50 million people around the world [4]. Unusual nerve cell activity in the brain causes epilepsy or commonly known as seizure [5]. The unpredictable nature of seizure onset is stressful and disabling for the affected [6] inducing isolation and lowering quality of life. Moreover, unattended patients suffering from generalized tonic-clonic or bilateral seizures constitute primary risk for sudden unexpected death in epilepsy (SUDEP) [7, 8]. In short, both mental and physical disorders contribute to accidents and sudden death.

Recent years have seen rise in the popularity of wearable devices and gadgets, and their use in healthcare applications is rapidly increasing. A large number of wearable devices are available in the market that can measure various vitals and health parameters on the go. In recent times, wearable seizure detection devices have emerged as a useful tool that can help in preventing SUDEP [6–8]. Thus, for a number of neurological disorders, it may be possible to noninvasively monitor a relevant biomarker, which can aid in detection, prevention, or treatment of the said neurological disorder. Surface electromyography (sEMG) is a commonly used biomarker which can noninvasively identify muscle and motor neuronal activity [9]. sEMG is the measure of electrical activity produced as a product of skeletal muscles' movement. Owning to the fact that sEMG can be captured noninvasively and has a unique repeatable property, it can be used in wearable devices allowing for real-time classification applications [10, 11]. Therefore, sEMG seems like a promising modality candidate that can be used for the detection of motor seizures. Seizure activity induces specific pathological changes which can be reflected in the sEMG recordings [12]. Different EMG parameters can be useful in differentiating epileptic and non-epileptic muscle activity [13–15]. We can also make use of sEMG to analyze ailments of the muscular system, clinical or biomedical applications which in turn can help researchers and developers in presenting modern human computer interaction [3]. Similarly, medical abnormalities [16, 17], emotion detection [18], prosthetic arms, and hand and lower limb control [19] can be identified by the use of sEMG signals and their variations. Various approaches have been suggested by different people to analyze EMG signal.

In this paper, we propose a multi-signature-based feature vector for classification of multiple physical actions using sEMG. The initial part of the paper discusses the acquisition mechanism of sEMG signals from participants performing a range of activities. sEMG data collection was achieved by a combination of dry sEMG electrodes and a controller for acquisition, processing, and its transmission. The aim of this device was to monitor the muscle activity from the belly of bicep brachii. A classification algorithm for the collected sEMG signals was then proposed which uses signatures from different domains including time, frequency, and higher-order statistics between channels for action classification.

The structure for the rest of the article is as follows: State-of-the art algorithms and literature survey is discussed in Sect. 2. Acquisition methodology for collection of sEMG signals is discussed in Sect. 3. Section 4 presents the proposed methodology for classification of actions based on sEMG signals, followed by experimentation and results under Sect. 5. Sections 6 and 7 respectively discuss the findings of these experimentations and conclude them.

2 Related Work

A lot of research is focused on recognition of patterns within sEMG signals to differentiate between various limb motions. The complexity of the electromyographic signal and its relationship with the arm movement state enforce the use of machine learning-based schemes. Machine learning-based recognition frameworks consist of multiple phases: pre-processing of data, feature extraction, and classification. Feature extraction and classification are among the important aspects of this framework for sEMG signal analysis. Researchers have actively presented different set of discriminative features consisting of domain knowledge [11, 20–22] while making use of machine learning algorithms for classification [23, 24]. Existing feature set for the classification of sEMG signals can be divided into three different categories [25]: time domain, frequency domain, and time-frequency domain features.

Overall, the literature survey undertaken can be divided into three separate subsections as general sEMG classification, binary aggressive and normal action classification, and multi-class physical action classification based on sEMG signals.

The quality of acquired signals can be critical in development of any decision support system. Akhundov et al. [26] presented a framework for quality assessment of sEMG signal using five different classifiers from both supervised and unsupervised categories. Experimentation showed superiority of unsupervised artificial neural network (ANN) giving over 98% accuracy in comparison to the supervised ANNs. Among various works where sEMG has been used, Mishra et al. presented an improved empirical mode decomposition (EMD)-based approach which used median filtering for the classification of amyotrophic lateral sclerosis (ALS) and normal EMG signals [27]. Samar et al. proposed a technique to provide reference control signals for prosthetic hands based on sEMG signals [28]. The proposed framework used wavelet analysis/decomposition for extraction of features while using ANNs. In [29], Podrug et al. proposed a framework for EMG pattern recognition using multi-scale principal component analysis for de-noising and discrete wavelet transform for features extraction. Duan et al. proposed a framework for multi-class gesture identification using spectrogram of the EMG signal as input to the convolutional neural network (CNN) [30]. CNN's inherent traits of local connection and weight sharing exhibit good translation invariance. Consequently, 94.06% accuracy is obtained by the proposed framework. These works indicate the usability of sEMG signatures in various domains.

Research related to classification of aggressive and normal activities can assist in detecting abnormalities in patients' behavior such as abnormalities of the musculoskeletal system or GTC seizure. In this regard, we have two distinct research avenues in binary and multi-class classification. Sezgin et al. proposed an extreme learning machine (ELM) algorithm based on EMG signal making use of bispectrum and quadratic phase couplings (QPCs) for the classification of aggressive and normal activity [31]. "EMG physical action dataset" from UCI repository, containing aggressive and normal activity actions from four subjects, was used. The authors were able to achieve an impressive accuracy of 99% but only for binary classification. In another approach presented by Jana et al. [32], wavelet decomposition of sEMG signals was used to extract features with the aim of discriminating between aggressive and normal actions. The proposed adaptive neuro-fuzzy inference system (ANFIS) for binary classification achieved an overall accuracy of 98%. Alaskar et al. proposed a novel convolutional neural network-based approach using time-frequency representation of an EMG signal [33]. The authors focused on binary classification by converting EMG signal into spectrogram and scalogram. A pre-trained AlexNet was used for training and testing which achieved an overall training and testing accuracy of 100% and 94.1%.

Classification of normal physical actions using the same UCI repository was proposed in [34]. Authors proposed extraction of QPC and use of multilayer artificial neural network for classification of ten normal activities. The resultant framework achieved an accuracy of 86.5%. The limited feature set used in the paper restricts the accuracy of the system which can be enhanced by exploring a wider range of features and classifiers. Sukumar et al. proposed an improved framework using variational mode decomposition (VMD) for classification of ten normal actions [35]. The framework proposed decomposition of sEMG signal into modes, where statistical features are extracted from each mode. Multi-class least squares support vector machine (MC-LS-SVM) classifier with radial basis function (RBF) was used obtaining an accuracy of 98.17%. Sravani et al. proposed extreme learning machine (ELM) classifier based on flexible analytic wavelet transform features for the identification of ten normal physical actions [24]. Another approach for classification of normal actions is proposed by Demir et al. where time-frequency maps are used as input to pre-trained convolutional neural networks [36]. Authors propose use of SVM classifier and deep feature concatenation utilizing AlexNet and VGG16 variants. The proposed framework attains 99.04% accuracy score for deep feature concatenation. The paper ignores half of the available dataset by discarding aggressive actions. Moreover, the traditional feature extraction mechanism is replaced by an overly complex and computationally intensive mechanism. Recent literature focuses on classification of individual aggressive and normal activities. A 20-class classification framework is proposed in [37] by introducing modified spectral moment-based features and inter-channel correlation features. An ensemble of subspace k-nearest neighbors and probabilistic neural network are used which obtain accuracy of 93.9% and 92.7%, respectively. The most common available physical actions dataset used in research articles make use of eight channels for collection of sEMG signals. The eight channels are further subdivided into four, each for upper and lower limbs, respectively. Further, the physical activities performed for the collection of data are mixed with no particular focus on any particular muscle activity. A system directed as a first step in designing GTC seizures can benefit from focused activities targeted toward upper limb with actions of varying intensity.

In this paper, we have focused on building a model with far fewer number of sEMG channels going all the way down to two. While reducing the number of channels can reduce accuracy, the fewer number of channels can facilitate real-time processing which is a desired attribute in a wearable device. Thus, the acquisition of data has been done using a single electrode band placed on each of the upper limbs. In addition, while the main focus of sEMG acquisition and subsequent detection has mainly been on prosthetic development, the actions included in our dataset target the daily life activities that an epileptic patient may perform on daily basis. The aim of collecting a local dataset is to analyze the muscular activity of contradicting activities showing varying muscle activity. Successful analysis of such activities can enable us to have a working principle for classification of seizure activity in the near future. Therefore, the dataset can be valuable for building models that are specifically geared toward detection of epileptic seizures in an unrestrained environment.

The contributions of our paper are as follows:

- Dataset generation for daily life activities
- Comparison of machine learning models capable of real-time performance
- Signal signature selection with minimal loss of accuracy
- Basic design of a wearable armband for sEMG acquisition

3 Materials and Methods

In this section, we will discuss the details regarding the dataset used in this article and shed light on the associated mechanism for its acquisition. The collection of this dataset has been targeted to enhance the pool of available dataset aiming to present discernible signals which can be used for classification of abnormal muscle intensive activity such as seizures. The generation of dataset follows two steps: (1) acquisition band realization and (2) data acquisition, which are discussed in the subsections below.

3.1 Acquisition Band

In order to acquire the sEMG data from the upper limb, a wearable armband with sEMG sensors was designed. The device contains two sEMG electrodes, a sensor board for pre-amplification and noise removal and a Bluetooth Low Energy (BLE)-enabled MCU for wireless transmission of signals. Figure 1 shows the block diagram view of the acquisition device.

The acquisition band consists of a set of dry electrodes that are placed on the belly of bicep brachii of each arm resulting in a two-channel data. The use of dry electrodes eliminates the need of application of any conductive gel; however, it is to be ensured that each electrode has good contact with the muscle. The dry electrodes used in the acquisition device are of stainless steel having dimension of 22 mm $*$ 35 mm. The acquisition band is accompanied by a custom application for receiving the signals. The band along with the mobile application is shown in Fig. 2a. The placement of the band on the subject's bicep is shown in Fig. 2b.

Researches in [13] captured sEMG signals from deltoid muscles of patients and showed that the frequency ranges from different types of seizures can go a little over 110 Hz with the range and the median frequency for different types of seizures varying slightly. The highest of these observed ranges and median frequencies is for tonic seizures with the range being 63.3 to 113 Hz and the median frequency being 86.2 Hz, respectively. On the other hand, tonic-clonic seizures happen to have the

Fig. 1 An overview of our acquisition band

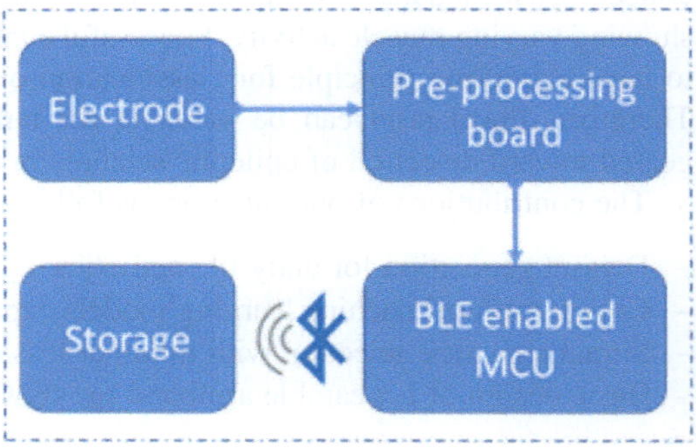

a b

Fig. 2 (**a**) The acquisition band. (**b**) The placement on band on the subject's bicep lowest range and median frequency measuring only 59 to 80.8 Hz and 73.6 Hz, respectively. Similarly, in [38], the team gathered the sEMG data from an assembly of eight electrodes for 20 different daily life actions performed by the cohort. The frequency analysis of the data clearly shows that the maximum frequency lies around 153 Hz and 122 Hz, respectively, for normal and aggressive actions.

The above clearly shows that whether the sEMG data is collected from daily activities or comes from the subjects suffering from seizures, the maximum observed frequency remains well within the 200 Hz range. Therefore, even if the data is sampled at 500 Hz, it can well accommodate frequencies up to 250 Hz keeping in-line with the Nyquist theorem. Keeping this in view, the data has been sampled at a frequency of 500 Hz for this study. Furthermore, to remove the effect of aliasing, an anti-aliasing filter with a cutoff of 250 Hz was applied before sampling of data.

3.2 Data Acquisition

The data collection phase was performed using our designed band discussed in the previous section. With this aim, some daily life activities have been simulated under controlled conditions. Five healthy subjects of varying ages (mean age 32.7 ± 7.4) consented for data collection and to be part of the cohort. The subjects were asked to perform varying daily life activities which have different magnitudes of muscle. Table 1 contains the list of performed activities.

A calibration step was carried out for each subject whenever the band is worn for the first time using the approach described in [39]. The aim of the calibration step is to set a baseline for each subject by measuring the maximum voluntary contractions. Once the calibration is carried out, each subject performed the set of four activities from Table 1 where each activity was performed five times by each subject. In this manner, each subject performed simulation for 20 instances of said activities. The duration of each instance was 70 s. The protocol for data collection was set such that for each activity, the initial and ending 5 s of data is invalid and the remaining 60 s of the data is considered as valid. This wait time acts as a buffer to decrease the possibility of getting amalgamation of noise and signal.

The raw sEMG signals are then subjected to pre-processing on the embedded board before it being sent for storage and further analysis. A fourth-order IIR notch filter at 50 Hz is applied on the raw sEMG signal to reduce the effect of power line noise. Figure 3 shows both filtered and thresholded signals, along with the start and end intervals, for a sample push-up activity.

Table 1 Summary of physical actions

Sr.	Activities
1	Typing
2	Push-ups
3	Lifting heavy objects (LHO)
4	Upper limb voluntary movement (UVLM)

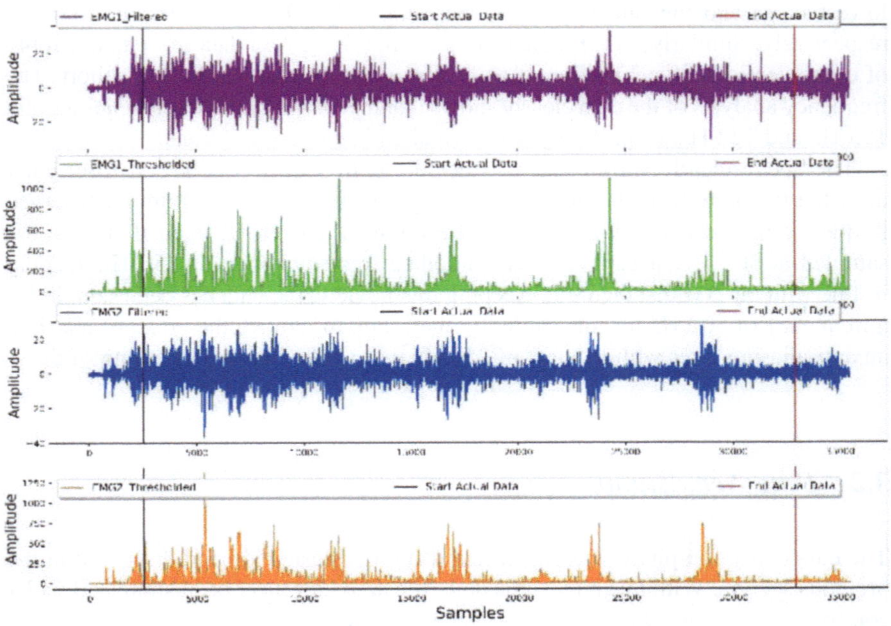

Fig. 3 Sample filtered and thresholded signal for push-up activity

The acquired sEMG signals were eventually transmitted through the BLE connection for storage and analysis at later stage.

4 Methodology

In our paper, we propose a framework for the classification of physical activities using two-channel sEMG signals. The overall classification framework can be broken down into sub-blocks where initially sEMG signals are pre-processed into chunks, and each chunk is then passed to the feature extraction block for feature vector generation, which is fed to a classification engine. The proposed framework is represented as block diagram in Fig. 4.

4.1 Pre-Processing and Segmentation

The first step in our proposed methodology is to pre-process and segment out each channel of sEMG signal. As discussed in Sect. 3, denoising has already been performed on the sEMG before storing the signals; thus, no pre-processing filter is applied here.

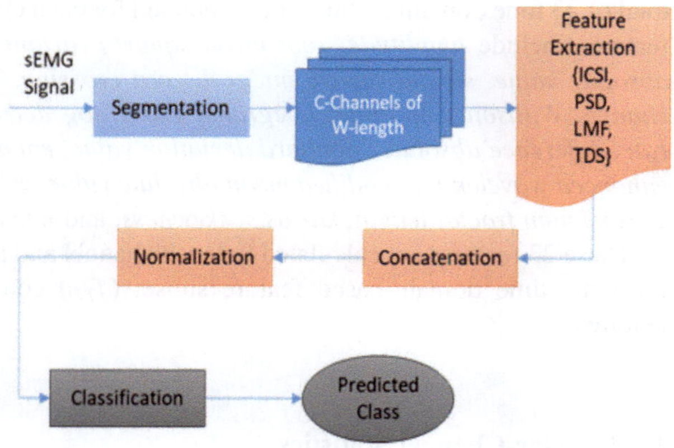

Fig. 4 System level flow diagram of our proposed methodology

Here each pre-processed signal, sch, is divided into W-length segments having an overlap of 25% where "**W**" represents the length of the segment that is controlled by sampling frequency. After segmentation each sEMG signal is now converted into N_f multi-channel (two channels) frames, ς, having length **W**. These segments are generated using sliding hamming window of length **W**.

$$\varsigma = s_{ch} * w_l \tag{1}$$

where

$$w_l(n) = \begin{cases} 0.54 - 0.46\,cos\left(\frac{2\pi n}{N}\right) & 0 \leq n \leq N \\ 0 & \text{otherwise} \end{cases}$$

These segments are fed to the feature vector generation block for further processing.

4.2 *Feature Vector Generation*

The sEMG signals show variations in both time and frequency domains, so in our proposed framework, we make use of different signatures for extracting features. These signatures and their details are discussed in the following subsections.

4.2.1 Time Domain Features

Time domain features are most commonly and frequently used in the analysis of sEMG signals [40, 41, 20]; thus, our first subset of features belong to this domain. A total of 23 time domain features are calculated for each channel of segment, ς_i . These features include *amplitude*, *root mean square*, *variance*, *waveform length*, *mean absolute value*, *simple square integral*, *zero crossing*, *Tsallis entropy*, *slope sign change*, *Willison amplitude*, *integrated EMG*, *log detector*, *myopulse percentage rate*, *difference absolute*, *standard deviation value*, *enhanced mean absolute value*, *enhanced wavelength*, *modified mean absolute value*, *modified mean absolute value 2*, *maximum fractal length*, *kurtosis*, *skewness*, and *negative entropy*.

These 23 features are calculated for each channel and are concatenated together to form the time domain-based feature subset (f_{TD}) containing a total of 23 × 2 features.

4.2.2 Inter-Channel Statistics

The second subgroup of our feature vector makes use of inter-channel statistics, a pair of inter-channel statistics that are calculated between the corresponding channels of each segment ς_i.

Maximum Similarity Index

These statistics are based on maximum cross-correlation [42] among the corresponding segments of two channels $\varsigma_{(i,1)}$ and $\varsigma_{(i,2)}$ of an EMG signal which is defined in Eq. (2).

$$CC_{a,\mathrm{b}} = max\left(\varsigma_{i,\mathrm{a}} \star \varsigma_{i,\mathrm{b}}\right) \tag{2}$$

Equation (2) is used to compute the value of maximum correlation between respective segments of $\varsigma_{i,a}$ and $\varsigma_{i,b}$ channels of EMG signal. And we have only two channels so we can get one value after performing the maximum correlation between corresponding channels.

Maximum Covariance Index

Covariance index belongs to higher-order statistics, which represent the co-variance between respective segments of $\varsigma_{i,1}$ and $\varsigma_{i,2}$ channels of sEMG signal. Eq. (3) shows the mathematical expression for this feature.

$$CV_{a,\mathrm{b}} = max\left(Cov\varsigma_{\mathbf{i,a}}, \varsigma_{\mathbf{i,b}}\right) \tag{3}$$

Thus, a total of two features are extracted for each segment ς_i, using this mechanism.

4.2.3 Power Spectral Density (PSD)

This subset of features is based on spectral energy of EMG signals which is extracted through Burg's transform using Eq. (4). These features have been previously proposed by Khushaba et al. in [43]. For each channel of an EMG pattern, assuming a model order v for calculation of σ_{burg}, the PSD is estimated using Eq. (4).

$$\psi(w_k) = |H(w_k)|^2 \sigma_{burg}^2 \tag{4}$$

Here σ_{burg}^2 represents Burg's method's error variance. After computation of spectrum, it is divided into N_b bands and power in the respective band is computed as per Eq. (5).

$$\eta_b^m = \sum\nolimits_{k \in b^{th}\ band} \psi(w_k) \tag{5}$$

Here η_b^m represents the power in bth band of mth channel where $b \in 0, 1, 2, \ldots, N_b - 1$ and $m \in [0, 1]$. Finally, N_b features are computed against each segment of every channel which leads to a feature vector of $2 * N_b$ for each segment.

4.2.4 Log Moments of Fourier Spectra (LMF)

Ali et al. proposed a new set of features, for EMG classification, based on logarithms of moments and their ratios from the frequency domain [44]. The ith frequency domain moment from the spectrum is defined as [45] where $\psi_m(w_k)$ represents the magnitude response against a segment of mth channel. Using this we can calculate the ith frequency domain moment using following expression.

$$g_j(i) = \sqrt{\sum_{i=1}^{n} k^i \psi_j(w_k)} \tag{6}$$

After calculation of frequency moments, 17 log moments can be calculated using expressions (7)–(14).

$$f_j(1) = \ln g_j(0) \tag{7}$$

$$f_j(2) = \ln g_j(2) \tag{8}$$

$$f_j(3) = \ln g_j(4) \tag{9}$$

$$f_j(4) = \ln g_j(1) - \frac{1}{2} \ln\left(g_j(0) - g_j(2)\right) - \frac{1}{2} \ln\left(g_j(0) - g_j(4)\right) \tag{10}$$

Table 2 Summary of features

Feature type	Feature count
TDS (F1)	46
ICS (F2)	2
LMFS (F3)	34
PSD (F4)	20
Total	**102**

$$f_j(5) = \ln g_j(2) - \frac{1}{2} \ln \left(g_j(0) - g_j(4) \right) \tag{11}$$

$$f_j(6) = \ln g_j(0) - \frac{1}{2} \ln \left(g_j(1) - g_j(3) \right) \tag{12}$$

$$f_j(7) = \ln g_j(0) - \frac{1}{2} \ln \left(g_j(2) - g_j(6) \right) \tag{13}$$

The pair-wise features based on moment products are calculated using Eq. (14) where $n \in \{8, 9, \ldots, 17\}$ and the values of i and j are chosen from the set C2 of ten ordered pairs defined using Eq. (15).

$$f_j(n) = \frac{1}{2} \ln g_j(i).g_j(j)) \tag{14}$$

$$C_2 = \{(\mathrm{i}, j) : i = 1, 2, 3, 4 \text{ and } j = 2, 3, 4, 5 \text{ where } i \neq j\} \tag{15}$$

The above 17 features are calculated for both channels and are grouped together for the LMF feature set which contains a total of 34 features.

This concludes the discussion regarding the feature vector generation. Eventually, the calculation of all these signatures leads to a feature vector of length 102, which is summarized in Table 2.

4.3 Feature Normalization

Once the feature vector is finalized, it is subjected to normalization before classification phase. Normalization of feature has been carried out using z-score normalization so that each feature comes within range of [−1, 1]. Eq. (16) shows the process of normalization employed.

$$x' = \frac{(x - \mu)}{\sigma} \tag{16}$$

Here x represents the feature value and σ and μ represent standard deviation and mean values for feature x.

4.4 *Classifier*

As the final step in our proposed methodology, the feature vector from sEMG signals is passed to a classifier to separate the signals into four physical activities. In the proposed methodology, we make use of support vector machines (SVM) to classify muscle movement. We use quadratic SVM where one vs. all classification scheme is utilized. Other than this, in order to calculate baseline results, we have used 3-NN and ensemble learning classifiers.

5 Experimental Results and Discussion

5.1 *Dataset*

As discussed in Sect. 3, the dataset has been self-generated using an in-house designed armband. The collected dataset contains sEMG signals of five male subjects which were asked to perform four different physical activities.

5.2 *Experimental Results*

In this section, we will discuss the experimental setup and the accompanying validation parameters that have been measured for classification of four physical activities.

As the first step in our proposed framework, each signal is divided into windows of 2 s each with an overlap of half second (25%). This division ensures that each activity has a healthy sample space, which ranges from 600 to 900 depending on the length of original recorded signal. Finally, this sample space is used to extract a feature vector for all signatures as defined in Sect. 4. In order to have basic results, a baseline model using time domain features for SVM, k-NN, and ensemble learning classifiers has been generated. The selection of these classifiers for the initial analysis has been derived from literature. Table 3 shows the accuracies for all four classes using only time domain features while employing fivefold cross-validation.

The above analysis leads to selection of quadratic SVM as the best classifier for our framework. As the next step, our focus has been toward highlighting the contribution of each signal signature toward prediction of physical actions. In order to perform further evaluation at feature level, we have used each signature to calculate the accuracy using SVM as our classifier. Table 4 represents the signature-wise accuracies achieved by SVM.

It can be seen from the table that the time domain features have the highest accuracy among individual signatures. Thus, we have kept it as the baseline and aggregated different signatures for performance evaluation. Inclusion of various

Table 3 Accuracy comparison of base models using only time domain features vs. complete feature set

Classifier	Accuracy (time domain features) (%)	Accuracy (complete features set) (%)
k-NN	94.5	95.7
SVM	95.6	97.405
Ensemble	95	96.5

Table 4 Performance comparison of different signatures

Features	Time domain (F1)	ICS (F2)	LMFs (F3)	PSD (F4)
Accuracy	95.6	72	89.9	90.3

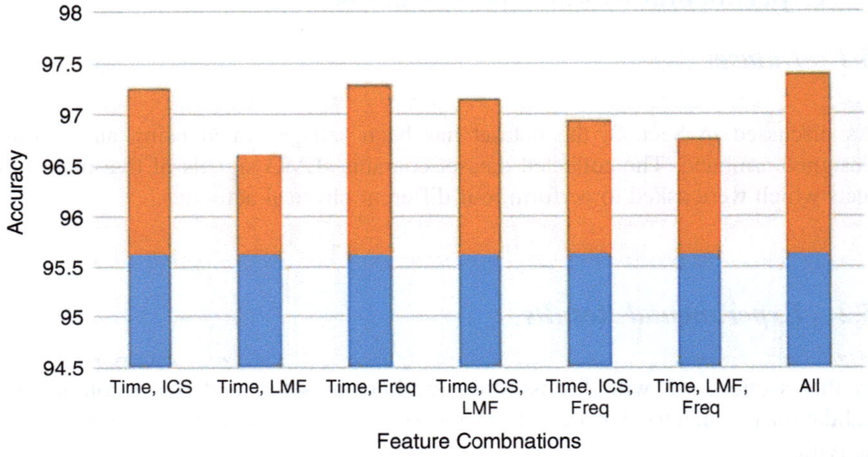

Fig. 5 Performance improvement evaluation in accuracy by inclusion of the different signatures

signatures from the signal shows that performance of the proposed methodology can be enhanced. It is clear from Fig. 5 that the baseline accuracy achieved by time domain features varies by inclusion of ICS, PSD, and LMFs.

Performance evaluation of feature subset space enabled us to assess the possibility of better or similar accuracy with use of a smaller number of features. Moreover, it highlights the contribution of each signal signature in correctly identifying the activity from sEMG signal. In this regard, Fig. 6 depicts comparison of feature vector count and the accompanying accuracy achieved by SVM. Though the highest accuracy is achieved by using the complete feature set, a thorough examination shows that there is not a significant difference in accuracy even with use of fewer features. In this regard, sub-feature vector of time- and ICS-based features gives 97.26% accuracy which shows a decrease of 0.149% while using only 47% of the original features.

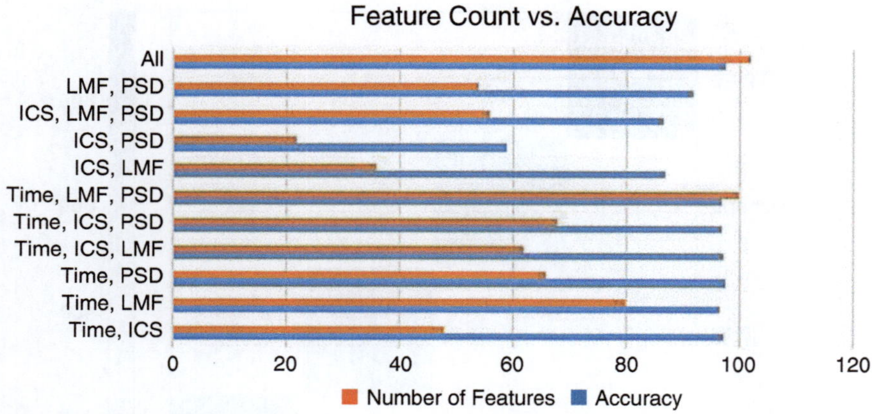

Fig. 6 Feature subset count and their achieved accuracies

Improvement in classification results using ICS-based feature subset further highlights the importance of calculating statistics between different channels of a single recording.

The detailed results of the proposed methodology can be evaluated using confusion matrix. Figure 7 represents confusion matrices against SVM for all features.

We have measured the performance of our proposed algorithm in terms of accuracy, sensitivity (recall), specificity, precision, F-measure, balanced accuracy (BA), and misclassification rate (MR). Specificity and sensitivity represent the true negative rate and true positive rate of our classifier against proposed feature set, whereas precision quantifies the true positive rate that actually belongs to that class. F-measure or F-1 score provides a measure of accuracy via weighted harmonic average of precision and recall. Moreover, balanced accuracy has also been calculated to nullify the imbalance in the dataset. Table 5 summarizes the abovementioned parameters for our complete feature set using SVM classifier.

Furthermore, a comparison between the baseline classifiers is also conducted while highlighting the effectiveness of SVM classifier for the classification of four physical actions. The results are visualized in Fig. 8. Finally, kappa coefficient κ has also been calculated for SVM, 3-NN, and ensemble learning which gives 0.9628, 0.934, and 0.9454 values, respectively.

6 Discussion

sEMG plays a pivotal part in improving the daily life standard of people who suffer from physical or neurological disorders. In this regard, the noninvasive techniques for acquisition of signals, such as sEMG, help the designers to provide real-time wearable solutions that can elevate the lifestyle of people suffering from numerous

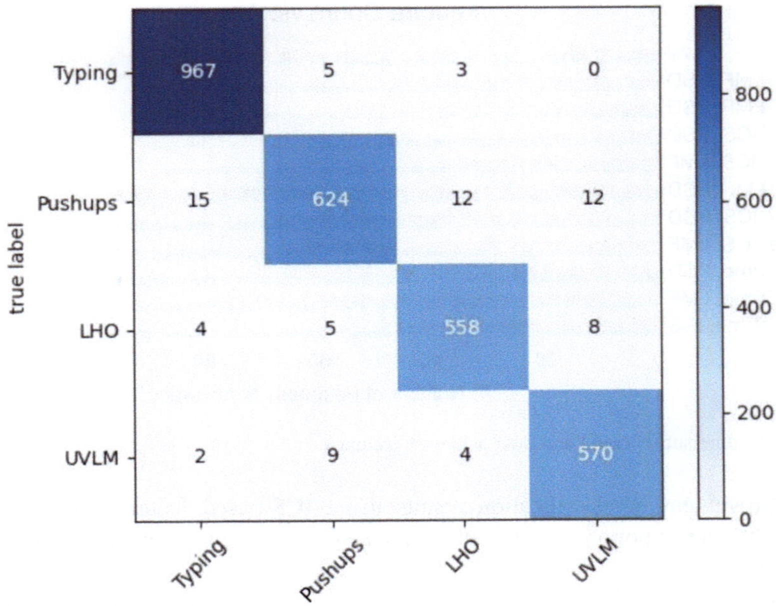

Fig. 7 Confusion matrix for our proposed model using feature subset of time domain and inter-channel statistics

Table 5 Performance parameter summary for proposed framework using feature subset

Classes	Typing	Pushups	LHO	UVLM
Sensitivity	0.992	0.952	0.962	0.974
Specificity	0.990	0.989	0.993	0.991
Precision	0.982	0.963	0.972	0.968
F2-measure	0.987	0.957	0.967	0.971
Balanced accuracy	0.991	0.970	0.977	0.983
Misclassification rate	0.009	0.020	0.014	0.012

physical and neurological diseases. First system in such monitoring systems is to correctly catalog the physical actions of such patients. Furthermore, a system that can distinguish between daily physical activities and an occurrence of a seizure also requires such a dataset. A framework for classification of physical actions has been proposed in this article which can help in designing assistive technologies. To achieve this aim, we have collected sEMG signals for four physical activities. The signals are pre-processed at the collection module to remove baseline and environmental noise. In addition, the protocol that has been employed for collecting the data ensures that noise is minimized by having buffer zones at both the start and end of an activity.

The classification results show that accuracy achieved by analyzing sEMG signal can be improved by inclusion of other signal signatures. The signatures for contrasting activities can provide much stronger differences by use of other

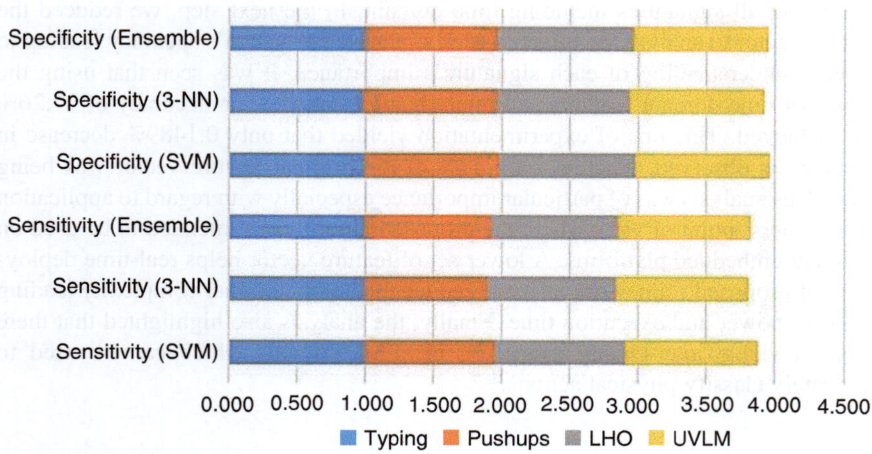

Fig. 8 Performance comparison of parameters for SVM, 3-NN, and ensemble learning

signatures. Furthermore, having made use of both channels for extraction of features also assists in achieving higher accuracy. In this context, analyzing the inter-channel variations is a fruitful activity as well.

The results and discussion lead us to believe that a subset of features can be used to classify sEMG signals without overly affecting the classification accuracy. This is highlighted by Fig. 6 which shows that a feature vector of length 48 gives 94.26% accuracy. Even though the accuracy drops, reducing feature vector can facilitate real-time processing of sEMG signals which is handy for embedded designs especially for monitoring neurological patients. Decreasing the size of feature vector lowers down the tunable parameters of a classifier which in turn lowers the computations which subsequently reduces computational resources, execution time, and the power requirements. Moreover, using a subset of features impedes the classifier's capacity to overfit.

7 Conclusions

In this paper, we have initially discussed about generation of sEMG-based physical activity dataset. A framework for classification of the four activities has been proposed which is based on SVM. The feature vector used for the classification contains signatures from different signatures. We started with a set of 102 features including different signatures such as time domain, frequency domain, and inter-channel statistics. A baseline comparison using traditional time domain features was carried out using SVM, 3-NN, and ensemble learning classification. These achieved an accuracy of 95.6% 94.5%, and 95%, respectively, for the four physical actions. Selected SVM classifier yielded an accuracy of 97.405% using feature vector containing all signatures including time domain. In the next step, we reduced the feature vector to analyze the effect of each signature on system's accuracy leading to a better understanding of each signature's importance. It was seen that using the subset of time domain features and inter-channel statistics, an accuracy of 97.26% was achieved. This line of experimentation yielded that only 0.1489% decrease in accuracy is observed, whereas only 47% of the original feature vector was being used. This analysis was of particular importance especially with regard to application of designers' point of view where the proposed algorithm is intended to be used on different embedded platforms. A lower set of feature vector helps real-time deployment of proposed framework as it can reduce the computational complexity leading to lower power and execution time. Finally, the analysis also highlighted that there exists a viable inter-channel variation in sEMG signals which can be used to accurately classify physical actions.

References

1. Henning Stolze, Stephan Klebe, Christoph Baecker, Christiane Zechlin, Lars Friege, Sabine Pohle, and Günther Deuschl. Prevalence of gait disorders in hospitalized neurological patients. Movement disorders: official journal of the Movement Disorder Society, 20(1):89– 94, 2005.
2. Tommaso Proietti, Vincent Crocher, Agnes Roby-Brami, and Nathanael Jarrasse. Upper- limb robotic exoskeletons for neurorehabilitation: a review on control strategies. IEEE reviews in biomedical engineering, 9:4–14, 2016.
3. Mariska Van Den Broek, Ettore Beghi, and RESt-1 Group. Accidents in patients with epilepsy: types, circumstances, and complications: a European cohort study. Epilepsia, 45(6): 667–672, 2004.
4. World Health Organization. Epilepsy, 2020.
5. Robert S Fisher, Carlos Acevedo, Alexis Arzimanoglou, Alicia Bogacz, J Helen Cross, Christian E Elger, Jerome Engel Jr, Lars Forsgren, Jacqueline A French, Mike Glynn, et al. Ilae official report: a practical clinical definition of epilepsy. Epilepsia, 55(4):475– 482, 2014.
6. Anouk Van de Vel, Kris Cuppens, Bert Bonroy, Milica Milosevic, Katrien Jansen, Sabine Van Huffel, Bart Vanrumste, Lieven Lagae, and Berten Ceulemans. Non-eeg seizure- detection systems and potential sudep prevention: state of the art. Seizure, 22(5):345–355, 2013.
7. Orrin Devinsky, Dale C Hesdorffer, David J Thurman, Samden Lhatoo, and George Richerson. Sudden unexpected death in epilepsy: epidemiology, mechanisms, and prevention. The Lancet Neurology, 15(10):1075–1088, 2016.
8. Torbjorn Tomson, Rainer Surges, Robert Delamont, Serena Haywood, and Dale C Hesdorffer. Who to target in sudden unexpected death in epilepsy prevention and how? risk factors, biomarkers, and intervention study designs. Epilepsia, 57:4–16, 2016.
9. Charles Long and Mary Eleanor Brown. Electromyographic kinesiology of the hand: muscles moving the long finger. JBJS, 46(8):1683–1706, 1964.
10. Marco E Benalcázar, Andrés G Jaramillo, A Zea, Andrés Páez, Víctor Hugo Andaluz, et al. Hand gesture recognition using machine learning and the myo armband. In 2017 25th European Signal Processing Conference (EUSIPCO), pages 1040–1044. IEEE, 2017.
11. Rami N Khushaba, Ali H Al-Timemy, Ahmed Al-Ani, and Adel Al-Jumaily. A frame-work of temporal-spatial descriptors-based feature extraction for improved myoelectric pattern recognition. IEEE Transactions on Neural Systems and Rehabilitation Engineering, 25(10): 1821–1831, 2017.

12. Sándor Beniczky, Isa Conradsen, Ronit Pressler, and Peter Wolf. Quantitative analysis of surface electromyography: biomarkers for convulsive seizures. Clinical Neurophysiology, 127(8):2900–2907, 2016.
13. Isa Conradsen, Peter Wolf, Thomas Sams, Helge BD Sorensen, and Sándor Beniczky. Patterns of muscle activation during generalized tonic and tonic–clonic epileptic seizures. Epilepsia, 52(11):2125–2132, 2011.
14. Sándor Beniczky, Isa Conradsen, Mihai Moldovan, Poul Jennum, Martin Fabricius, Krisztina Benedek, Noémi Andersen, Helle Hjalgrim, and Peter Wolf. Quantitative anal- ysis of surface electromyography during epileptic and nonepileptic convulsive seizures. Epilepsia, 55(7): 1128–1134, 2014.
15. Sándor Beniczky, Isa Conradsen, Mihai Moldovan, Poul Jennum, Martin Fabricius, Krisztina Benedek, Noémi Andersen, Helle Hjalgrim, and Peter Wolf. Automated differentiation between epileptic and nonepileptic convulsive seizures. Annals of neurology, 77(2):348–351, 2015.
16. Varun Bajaj and Anil Kumar. Features based on intrinsic mode functions for classification of emg signals. International Journal of Biomedical Engineering and Technology, 18(2): 156–167, 2015.
17. Rajat Emanuel Singh, Kamran Iqbal, Gannon White, and Jennifer K Holtz. A review of emg techniques for detection of gait disorders. Artificial Intelligence-Applications in Medicine and Biology, 2019.
18. Daniela Girardi, Filippo Lanubile, and Nicole Novielli. Emotion detection using noninva- sive low-cost sensors. In 2017 Seventh International Conference on Affective Computing and Intelligent Interaction (ACII), pages 125–130. IEEE, 2017.
19. Mahmoud Tavakoli, Carlo Benussi, and Joao Luis Lourenco. Single channel surface emg control of advanced prosthetic hands: A simple, low cost and efficient approach. Expert Systems with Applications, 79:322–332, 2017.
20. Angkoon Phinyomark, Pornchai Phukpattaranont, and Chusak Limsakul. Feature re- duction and selection for emg signal classification. Expert systems with applications, 39(8): 7420–7431, 2012.
21. Bernard Hudgins, Philip Parker, and Robert N Scott. A new strategy for multifunction myoelectric control. IEEE transactions on biomedical engineering, 40(1):82–94, 1993.
22. Yi-Chun Du, Chia-Hung Lin, Liang-Yu Shyu, and Tainsong Chen. Portable hand mo- tion classifier for multi-channel surface electromyography recognition using grey relational analysis. Expert Systems with Applications, 37(6):4283–4291, 2010.
23. Ying Sun, Cuiqiao Li, Gongfa Li, Guozhang Jiang, Du Jiang, Honghai Liu, Zhigao Zheng, and Wanneng Shu. Gesture recognition based on kinect and semg signal fusion. Mobile Networks and Applications, 23(4):797–805, 2018.
24. C Sravani, V Bajaj, S Taran, and A Sengur. Flexible analytic wavelet transforms-based features for physical action identification using semg signals. IRBM, 41(1):18–22, 2020.
25. Mohammadreza Asghari Oskoei and Huosheng Hu. Myoelectric control systems—a survey. Biomedical signal processing and control, 2(4):275–294, 2007.
26. Riad Akhundov, David J Saxby, Suzi Edwards, Suzanne Snodgrass, Phil Clausen, and Laura E Diamond. Development of a deep neural network for automated electromyographic pattern classification. Journal of Experimental Biology, 222(5), 2019.
27. Vipin K Mishra, Varun Bajaj, Anil Kumar, Dheeraj Sharma, and GK Singh. An efficient method for analysis of emg signals using improved empirical mode decomposition. AEU-International Journal of Electronics and Communications, 72:200–209, 2017.
28. Samar M Hatem, Geoffroy Saussez, Margaux della Faille, Vincent Prist, Xue Zhang, Del- phine Dispa, and Yannick Bleyenheuft. Rehabilitation of motor function after stroke: a multiple systematic review focused on techniques to stimulate upper extremity recovery. Frontiers in human neuroscience, 10:442, 2016.
29. Ermin Podrug and Abdulhamit Subasi. Surface emg pattern recognition by using dwt feature extraction and svm classifier. In The 1st conference of medical and biological engineering in Bosnia and Herzegovina (CMBEBIH 2015), pages 13–15, 2015.

30. Na Duan, Li-Zheng Liu, Xian-Jia Yu, Qingqing Li, and Shih-Ching Yeh. Classification of multichannel surface-electromyography signals based on convolutional neural networks. Journal of Industrial Information Integration, 15:201–206, 2019.
31. Necmettin Sezgin. Analysis of emg signals in aggressive and normal activities by using higher-order spectra. The Scientific World Journal, 2012, 2012.
32. Gopal Chandra Jana, Aleena Swetapadma, and Prasant Pattnaik. An intelligent method for classification of normal and aggressive actions from electromyography signals. In 2017 1st International Conference on Electronics, Materials Engineering and Nano-Technology (IEMENTech), pages 1–5. IEEE, 2017.
33. Haya Alaskar. Deep learning of emg time-frequency representations for identifying normal and agressive actions. Int. J. Computer Sci. Netw. Security, 18:16–25, 2018.
34. Hana Săhinbegović, Laila Mušić, and Berina Alić. Distinguishing physical actions using an artificial neural network. In 2017 XXVI International Conference on Information, Communication and Automation Technologies (ICAT), pages 1–5. IEEE, 2017
35. Nagineni Sukumar, Sachin Taran, and Varun Bajaj. Physical actions classification of surface emg signals using vmd. In 2018 International Conference on Communication and Signal Processing (ICCSP), pages 0705–0709. IEEE, 2018.
36. Fatih Demir, Varun Bajaj, Melih C Ince, Sachin Taran, and Abdulkadir Şengür Surface emg signals and deep transfer learning-based physical action classification. Neural Computing and Applications, 31(12):8455–8462, 2019.
37. Anish C Turlapaty and Balakrishna Gokaraju. Feature analysis for classification of physical actions using surface emg data. IEEE Sensors Journal, 19(24):12196–12204, 2019.
38. Dheeru Dua and Casey Graff. UCI machine learning repository, 2017.
39. Gabriele Rescio, Alessandro Leone, and Pietro Siciliano. Supervised machine learning scheme for electromyography-based pre-fall detection system. Expert Systems with Applications, 100: 95–105, 2018.
40. Jingwei Too, Abdul Rahim Abdullah, and Norhashimah Mohd Saad. Classification of hand movements based on discrete wavelet transform and enhanced feature extraction. Int. J. Adv. Comput. Sci. Appl, 10(6):83–89, 2019.
41. Angkoon Phinyomark, Pornchai Phukpattaranont, and Chusak Limsakul. Fractal analysis features for weak and single-channel upper-limb emg signals. Expert Systems with Applications, 39(12):11156–11163, 2012.
42. Lauren H Smith and Levi J Hargrove. Comparison of surface and intramuscular emg pattern recognition for simultaneous wrist/hand motion classification. In 2013 35th annual international conference of the IEEE engineering in medicine and biology society (EMBC), pages 4223–4226. IEEE, 2013.

A Comparative Study on EEG Features for Neonatal Seizure Detection

S. Abirami, John Thomas, Rajamanickam Yuvaraj, and Jac Fredo Agastinose Ronickom

Abstract Epileptic seizure is one of the common neurological disorders, and its clinical manifestation is different from that of the adult as the neonate's brain is not yet fully developed. In clinical practice, manual observation of EEG recordings to diagnose and identify epileptic seizures is expensive and time-consuming. Computer-aided diagnosis (CAD) tools will enable clinicians to examine the EEG expeditiously and effectively. In this study, we investigate 37 statistical and time-frequency domain features to discriminate between seizure and non-seizure EEG segments. The analysis is performed on the publicly available Helsinki University database. The significant features were identified by using the Wilcoxon rank sum test and then ranked using manual threshold area under the curve (AUC) value and eXtreme Gradient Boosting (XGBoost) feature importance method. The performance of the features was analyzed using XGBoost and support vector machine (SVM) classifier with fourfold cross-validation. We found that entropy plays a significant role in the discrimination of seizure and non-seizure segments. We achieved an average AUC of 0.84 and 0.76 using XGBoost and SVM classifiers, respectively. This study presents the significance of each extracted feature and will be beneficial to the neurologists for the continuous monitoring and diagnosis of seizures in neonates.

Keywords Epilepsy · Seizure · EEG · Neonate · Machine learning · Computer-aided diagnosis (CAD)

S. Abirami (✉) · J. F. Agastinose Ronickom
School of Biomedical Engineering, Indian Institute of Technology (BHU), Varanasi, Uttar Pradesh, India
e-mail: abiramis.bme20@itbhu.ac.in; jack.bme@iitbhu.ac.in

J. Thomas
Montreal Neurological Institute, McGill University, Montreal, Canada
e-mail: john007@e.ntu.edu.sg

R. Yuvaraj
Nanyang Technological University (NTU), Singapore, Singapore
e-mail: yuvaraj.rajamanickam@nie.edu.sg

M. Murugappan, R. Yuvaraj (eds.), *Biomedical Signals Based Computer-Aided Diagnosis for Neurological Disorders*, https://doi.org/10.1007/978-3-030-97845-7_3

1 Introduction

Epilepsy is a neurological disorder that disrupts the natural electrical activity of the human brain. It is one of the oldest known chronic diseases, with around 70 million people worldwide affected [1]. Seizures in the scalp electroencephalogram (EEG) are deterministic markers of epilepsy [2]. Other electrophysiological biomarkers of epilepsy include high cortical excitability, high-frequency oscillations (HFOs), and spikelike waveforms [3]. The cause of epilepsy is unknown primarily, while the risk factor may increase in the case of patients suffering from a brain tumor, developmental disorders, trauma, and central nervous and cerebrovascular diseases [4]. Individuals with epilepsy suffer from low cognitive performance, anxiety, and depression that make their quality of life poor [5, 6]. The available modalities for diagnosing and managing neurological disorders include EEG (electroencephalogram), MRI (magnetic resonance imaging), fMRI (functional MRI), and other functional imaging techniques. Imaging modalities such as MRI, fMRI, or PET (positron emission tomography) are occasionally used for seizure localization depending upon the need and severity. Among all the available modalities, EEG is the widely used noninvasive method for diagnosing seizures as it is comparatively less complex and provides high precision time measurements.

Seizures are a serious medical condition that leads to psychosocial issues. There are many cases of seizures reported in the neonatal intensive care unit (NICU). In our chapter, we focus on neonatal seizure identification. The probable causes for neonatal seizures are lack of oxygen, stroke, meningitis, hemorrhage, and infection [7]. Seizures are characterized by clinical symptoms such as physical movement or visible indicators [7]. However, in neonatal seizures, a clear manifestation of clinical symptoms is rare. Only 10% of the seizure events detected by the on-site experts are reliable [8]. Further, the automatic seizure detection algorithms built for adult epileptic seizure detection show a high false positive rate when tested with neonatal data [7, 9]. The characteristics of the epileptic waveform are also age-dependent and require an age-specific model for the analysis [10]. Neonatal computer-aided seizure detection has several challenges, such as recording conditions, artifacts, dataset duration, precise labeling, and choice of modality. The fMRI/MRI scanner's micro-environment conditions are not suitable for neonates as their skin cannot withstand the temperature and humidity inside the scanner. It may lead to hypothermia and extreme water loss after the scan [11]. As scalp EEG is noninvasive and safe, it is the most suitable modality for neonatal seizure diagnosis and continuous real-time monitoring. Amplitude-EEG (aEEG) and conventional continuous EEG (cEEG) are two types of scalp EEG used in neonates. It requires only 4–6 electrodes to record using aEEG, while cEEG requires a minimum of 16 electrodes. Although the aEEG is simple to record and interpret, it is less reliable than the cEEG. This study uses the scalp neonatal EEG recording from the Helsinki Hospital [12]. It is a publicly available dataset used for neonatal seizure detection in researches in the past.

The neonatal seizure study in the literature broadly used machine learning (ML), deep learning (DL), and functional brain connectivity methods. The periodic and repetitive characteristics observed in the ictal EEG motivated the researchers to extract the characteristic features from the seizure and compare it with the non-seizure (background) signal for the seizure classification [13]. It involved segmentation of EEG signal into small epochs using a moving window of variable width with 10–50% overlap to extract clinically significant features [9, 13–15]. The time, frequency, wavelet, and energy were widely extracted features in the ML method [7, 16]. Spike correlation and time-frequency correlation-based features were selected for neonatal seizure detection to improve the detection accuracy [13, 17].

Later, several studies on DL methods were used to bypass the feature extraction method and achieved a higher accuracy. The neural networks used by the deep learning methods require an extensive training set compared to the machine learning method. Moreover, the significance of every single feature cannot be identified using the DL methods as it is not interpretable [13]. Besides, functional brain connectivity analysis and clustering methods were also carried out to predict the risk of seizure occurrence using the same dataset [18–23]. Though different studies were performed using feature extraction, the features extracted across various studies are inconsistent based on subjects selected, length of seizure duration considered, epoch width, and EEG neonatal database. Hence, an automated approach with generalized pre-processing steps and evaluation of the features specific to the neonatal dataset is crucial for identifying neonatal seizures. The details of the past studies on seizure classification/detection using the Helsinki dataset are summarized in Table 1.

EEG signals are prone to artifacts and noises. It is essential to pre-process the signal before applying the classification techniques. The power line noise is the most common interference in the EEG recording, and other noises include device interference, muscle movement, and respiration intrusions. The respiration and the body movement artifacts are the primary cause of error in the neonatal EEG. These produce slow rhythmic waves similar to the EEG signal waveform [26]. Different approaches have been followed in the literature to pre-process the noise-prone EEG signal to detect the seizure patterns in neonates. A single pre-processing step may not be able to remove the different types of artifacts in the dataset. A pipeline of pre-processing steps will be helpful to effectively remove the noise, improve the signal-to-noise ratio (SNR), and reduce the misclassification rate [27]. The high-pass filter (HPF), a low-pass filter (LPF), a band-pass filter (BPF) of different cutoff frequencies, and 50 Hz notch filters were employed to remove the noises in the neonatal EEG signal [16, 24, 25, 28]. A generalized pre-processing technique is required to remove the artifacts and improve the EEG data quality. Support vector machine (SVM), artificial neural networks (ANN), k-nearest neighbors (KNN) algorithm, convolutional neural networks (CNN), and wavelet-based combination of methods are used in the literature for the neonatal seizure classification. These methods are still under investigation to be adopted in regular clinical practice.

In this chapter, we used the ML approach to identify the seizure segments automatically. We trained the XGBoost and SVM classifier with 37 time and

Table 1 Comparison of results from the literature

S. no	Ref	Pre-processing	Time window	Features	Classification techniques	Result
1	Açıkoğlu and Tuncer [7]	50 Hz notch filter	–	Multiple features	Top two features, Cubic SVM	Acc: 97.5%
						TPR: 97.5%
						Sen: 100%
						PR: 100%
						F1: 98.7%
					FS algorithm, KNN	Acc: 98.8%
2	Tapani et al. [17]	Adaptive segmentation, edge linking, and time-frequency matched filters	Moving window of 32 s with 4-s shift	Spike correlation, time-frequency correlation	SVM	AUC: 95.5%
	O'Shea et al. [13]	Downsampled to 32 Hz	8-s window with 1-s shift	Without feature extraction	2D-FCN	AUC: 95.6%
3	Ding and Luo [20]	–	–	WPeEn, IncrEn, ApEn, fuzzy entropy, permutation entropy	Entropy-based time domain feature extraction (ETFE)	Kappa value: 0.7–0.8
4	Isaev et al. [14]	Downsampled to 32 Hz	8-s epochs with 50% overlap	–	DL with MultiInstance Learning (MIL)	AUC: 0.925
5	Caliskan and Rencuzogullari [24]	Butterworth HPF ($f_c = 0.5$ Hz), LPF ($f_c = 30$ Hz), Notch filter ($f_c = 50$ Hz)	Moving window of 30 s with 2-s shift	–	P-DCNN (AlexNet, ResNet-18, GoogLeNet, DenseNet, and ResNet-50)	AUC: 0.9917 (DenseNet)
6	Tanveer et al. [15]	Notch filter ($f_c = 50$ Hz)	8-s segments with 50% overlap	–	2D-CNN	Avg AUC: 99.3
7	Frassineti et al. [25]	HPF ($f_c = 0.5$ Hz), LPF ($f_c = 32$ Hz), Notch filter ($f_c = 50$ Hz) and downsampled	8-s window	–	SWT, CNN, FCN	AUC: 81%

8	He et al. [16]	BPF ($f_c = 0.5$–70 Hz), Notch filter ($f_c = 50$ Hz)	–	–	Progressive deep wavelet cascade classification (PDWC) model and random forest	Mean Acc: 0.9914
9	Abbas et al. [18]	SWT, manual artifact removal, seizure duration >5 min only selected	3-s epochs with 1-s overlap	Phase synchrony between EEG electrodes	Circular omega complexity (COC) related algorithms	AUC: 0.5–0.7
10	Khazaei et al. [22]	Seizure duration >2 min only selected, resampled to 100 Hz, BPF ($f_c = 0.3$–30 Hz)	2-min epochs	Microstate metrics: mean duration, occurrence, and coverage	Modified *K*-means clustering	$p < 0.01$
11	Ye et al. [23]	–	Each layer presents 40-s segments	Features in the Riemannian manifolds	Novel sequential-clustering scheme	Clustering Acc: 1
12	Frassineti et al. [21]	BPF ($f_c = 0.5$–30 Hz)	2-s windows with an overlap of 1 s	Stability and phase synchrony indexes	Network analysis using connectivity matrix	$p < 0.001$ Cohen's d: 0.86
13	Ye et al. [19]	–	–	Grassmann manifold features	onlineGCT [kARMA]	Acc: 1
14	Current study	Notch filter ($f_c = 50$ Hz), HPF ($f_c = 1$ Hz), noise statistics	5-s window	Time and frequency features	SVM, XGBoost	AUC: 0.83

Acc Accuracy, *AUC* Area under the curve, *BPF* Band-pass filter, *CNN* Convolutional neural network, *2D-CNN* Two-dimensional CNN, *EEG* Electroencephalogram, *FCN* Fully convolutional network, *DL* Deep learning, f_c Cutoff frequency, *FS* Feature selection, *F*1 Harmonic mean, *HPF* High-pass filter, *Hz* Hertz, *kARMA* Kernel autoregressive-moving-average, *KNN* k-nearest neighbors algorithm, *LPF* Low-pass filter, *P-DCNN* Predictive deep convolutional neural network, *PR* Precision, *SWT* Stationary wavelet transform, *Sen* Sensitivity, *SVM* Support vector machine, *TPR* True positive rate, *XGBoost* eXtreme Gradient Boosting frequency features. Different feature ranking methods were used to identify the best set of features for the neonatal seizure classification. We aim to evaluate and compare the performance of a different set of features to detect seizure and non-seizure patterns using the Helsinki University neonatal EEG database. We pre-processed the EEG data using a generalized pre-processing pipeline to remove the noises and artifacts effectively. We have extracted a set of time domain and frequency domain features and then ranked them using Wilcoxon rank sum test, manual threshold, and XGBoost methods. Finally, the seizure and non-seizure segments were classified using the XGBoost classifier, and the results were compared with the SVM classifier.

The extracted features can be combined and classified as seizure and non-seizure segments using heuristic classifiers. Several ML and DL approaches have been proposed by researchers for the detection of neonatal seizures. These methods are not yet implemented in real time and require additional clinical validation. XGBoost is a decision tree-based gradient boosting algorithm widely used in machine learning models and has recently emerged in the biomedical field. A set of important features identified by the feature ranking algorithm was used to train/validate the XGBoost and SVM classifier for the effective identification of seizure segments. Our study concluded that the top 11 features selected from 37 time and frequency features were sufficient to achieve a higher AUC. The ranking provided by XGBoost was efficient compared to other methods and provided an AUC of 0.84. The SVM classifier provided an AUC of 0.76 for the same top 11 features.

This chapter is organized into various sections such as methodology, results, and discussion. We describe the database used in this study in Sect. 2.1, the pre-processing steps employed in Sect. 2.2, the feature extraction methods in Sect. 2.3, and feature ranking along with classification methods in Sect. 2.4. In Sect. 3, we present our results, followed by the comparison of our chapter with the past studies, limitations, and future directions discussed in Sect. 4, and we summarize the study in Sect. 5.

2 Methodology

The pipeline we followed in this work is summarized in Fig. 1. We used the neonatal EEG data recorded in the Helsinki NICU containing 39 seizure patients. The data was pre-processed, and a set of time and frequency features were extracted. The extracted features were ranked and then used for the classification of seizure and non-seizure segments.

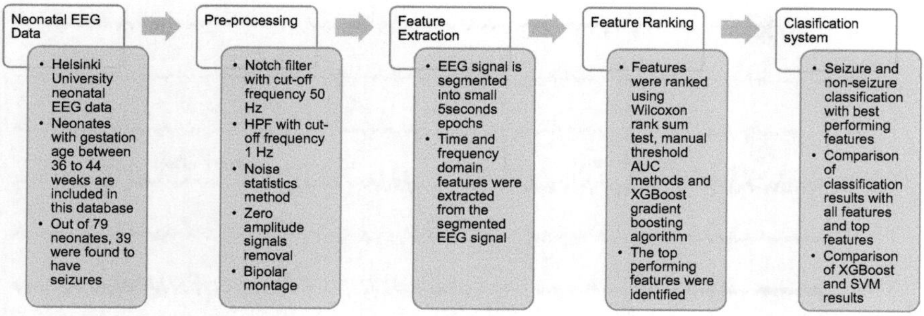

Fig. 1 Process pipeline followed in this study for the classification of the seizure and non-seizure segments. The main steps involved are data preparation, pre-processing, feature extraction, and classification

2.1 Database

The EEG recording from the neonatal intensive care unit (NICU) of the Children's Hospital, Helsinki University Central Hospital, Finland, was used in our study. In total, 79 neonatal multichannel EEG recordings sampled at 256 Hz were recorded from neonates with gestation ages between 36 and 44 weeks. Out of 79 neonates, 39 were diagnosed with seizures, and the others were seizure-free. It is the only publicly available neonatal EEG dataset where three experts have provided their individual visual annotations for marking the seizure. The median recording time of the EEG signals is 74 min approximately. The recordings used 19 electrodes arranged according to the 10–20 international standardized system. These electrodes are connected in a bipolar montage manner [12]. If at least two experts mark a segment as seizure, it is labeled as one; otherwise, it is zero. All the methods proposed in this chapter are trained and tested on this neonatal EEG database. Figure 2 shows the raw EEG signal with seizure onset and end highlighted for the respective channels.

2.2 Pre-processing

The NICU's EEG signal is prone to artifacts like device interference noises, muscle movement, respiration, and power line noises. Pre-processing steps are applied to reject the noises. A notch filter of cutoff frequency 50 Hz and an HPF of cutoff frequency 1 Hz was used to improve the SNR. Artifact removal using the noise statistics method [29] was implemented, in which all the segments with RMS value higher than a threshold (mean + 4 * standard deviation) was first eliminated. By visual inspection of Helsinki EEG data, we found there were few segments with zero amplitude. It is required to remove these segments as they don't contribute to the seizure or non-seizure recordings. The segments with zero amplitude continuously

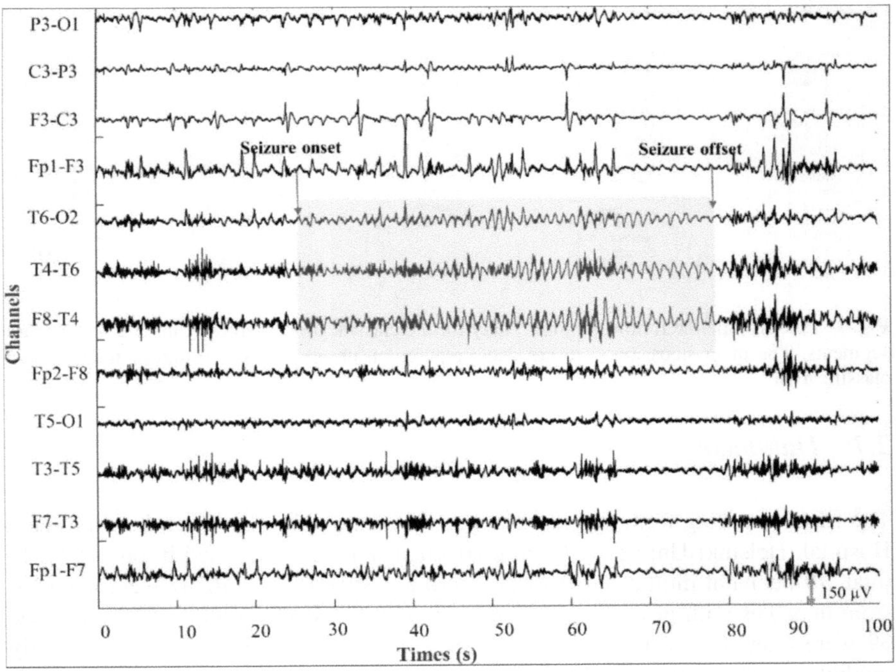

Fig. 2 A raw EEG segment with annotated seizure (onset and end) according to the experts. The seizure events are highlighted

for at least 0.2 s were removed. The bipolar montage was considered to provide the annotations resulting in 18 channels (Fp2-F4, F4-C4, C4-P4, P4-02, Fp1-F3, F3-C3, C3-P3, P3-O1, Fp2-F8, F8-T4, T4-T6, T6-O2, Fp1-F7, F7-T3, T3-T5, T5-O1, FZ-CZ, CZ-PZ). Finally, all the signals present in ±20 seconds of seizure marking were excluded from the dataset. It was performed to eliminate the errors in the labeling of onset and end of the seizure. The pre-processed signal is divided as seizure and background with the help of the annotations and segmented into 5-second epoch. These steps are fixed as a single pipeline for extracting all the features.

2.3 Feature Extraction

The past researches conclude that a reliable seizure detection system can be built using a robust classifier trained with a set of relatively simple hand-engineered features [13]. On this basis, we have extracted and combined 37 simple features from the segmented EEG signal. The features extracted from each of the EEG channel (18 channels) are considered as separate segments and combined patient-wise. Each 5-second segment contains 1280 sample points. Features are extracted for each of these small segments for training the classifier. The time- and frequency-related features from various past researches were extracted and analyzed. The extracted features are total power, sub-bands (0–2 Hz, 1–3 Hz, ... 10–12 Hz), normalized power in same sub-bands, root mean square (RMS), Hjorth parameters (mobility and complexity), zero-crossing rate (ZCR), skewness, kurtosis, variance, Shannon's entropy, curve length, mean power, mean teager energy, mean, standard deviation, entropy, and peak-to-peak value [7, 13, 17, 26, 30]. The details of the features are provided in Table 2.

Table 2 The details of the features extracted are given in the table

Type	Feature	Equation/frequency range
Time domain	Zero-crossing rate	$\frac{1}{T}\sum_{t=1}^{T} \mid x(t) - x(t-1) \mid$
	Entropy	$-\sum (p. * \log_2 p)$
	Mean (μ)	$\frac{\sum x(t)}{N}$
	Standard deviation (σ)	$\sqrt{\sum \frac{(x(t)-\mu)^2}{N}}$
	RMS	$\sqrt{\frac{1}{T}\int_0^T x(t)^2 \mathrm{d}x}$
	Mobility	$\sqrt{\frac{\mathrm{var}\left(\frac{\mathrm{d}x(t)}{\mathrm{d}t}\right)}{\mathrm{var}\, x(t)}}$
	Skewness	$\sum_{i}^{N} \frac{(x(i)-\mu)^3}{(N-1)*\sigma^3}$
	Peak to peak	$\mathrm{RMS} * 2\sqrt{2}$
	Maximum value	$\mathrm{Max}(x(t))$
	Mean line length	$\frac{1}{N}\sum_{i=1}^{N-1} \mid x(i) - x(i-1) \mid$
	Mean teager energy	$\frac{1}{N}\sum_{i=1}^{N-1} \mid x(i)^2 - x(i-1)x(i+1) \mid$
	Shannon's entropy	$-\log p(x)$
	Variance	$\sum \frac{(x(t)-\mu)^2}{N-1}$
	Mean power	$\frac{1}{N}\sum_{\mathrm{i}}^{N-1} x(i)^2$
	Complexity	$\frac{\mathrm{Mobility}\left(\frac{\mathrm{d}x(t)}{\mathrm{d}t}\right)}{\mathrm{Mobility}(x(t))}$
	Kurtosis	$\frac{1}{N}\sum \left(\frac{x(t)-\mu}{\sigma}\right)^4$
Frequency domain	Total bandpower	0–12 Hz
	Bandpower 0–2 Hz	0–2 Hz
	Bandpower 2–4 Hz	2–4 Hz
	Bandpower 3–5 Hz	3–5 Hz
	Bandpower 4–6 Hz	4–6 Hz

(continued)

Table 2 (continued)

Type	Feature	Equation/frequency range
	Bandpower 5–7 Hz	5–7 Hz
	Bandpower 6–8 Hz	6–8 Hz
	Bandpower 7–9 Hz	7–9 Hz
	Bandpower 8–10 Hz	8–10 Hz
	Bandpower 9–11 Hz	9–11 Hz
	Bandpower 10–12 Hz	10–11 Hz
	Bandpower 6–8 Hz	6–8 Hz
	Normalized bp 0–2 Hz	0–2 Hz
	Normalized bp 2–4 Hz	2–4 Hz
	Normalized bp 3–5 Hz	3–5 Hz
	Normalized bp 4–6 Hz	4–6 Hz
	Normalized bp 5–7 Hz	5–7 Hz
	Normalized bp 6–8 Hz	6–8 Hz
	Normalized bp 7–9 Hz	7–9 Hz
	Normalized bp 8–10 Hz	8–10 Hz
	Normalized bp 9–11 Hz	9–11 Hz

RMS Root mean square, *bp* Bandpower, $x(t)$ Signal, N Number of samples, p Probability, T Time interval, *var* Variance, *Max* Maximum

2.4 Feature Ranking and Classification

Feature ranking is the method of selecting the best performing features from a set of features. In order to apply hypothesis testing (parametric or nonparametric test), the normality of the feature set was checked using the one-sample Kolmogorov-Smirnov test (KS test) [31, 32]. It was found that the features were not normally distributed from KS test, and hence the statistical significance was validated using the Wilcoxon rank sum test (nonparametric test). The features that had a maximum number of patients with a p-value <0.05 were considered statistically significant. The weightage of every single feature can be examined only using the individual AUC values. Therefore, the individual feature's area under the curve (AUC) of the receiver operating characteristic (ROC) curve was calculated by manual threshold method to analyze each feature's significance. The ROC plot shows the diagnostic ability of the binary classifier for various thresholds. It is a graphical plot drawn between the true positive rate (TPR) and the false positive rate (FPR). AUC gives the probability that the classifier will rank a random positive class higher than a random negative class [33]. The extracted features from the seizure and the non-seizure segments were combined, and the AUC value was calculated for each feature using the manual threshold method. The threshold values were set in the range of 0.001–1. In the decreasing order of the AUC, the features were ranked.

All 37 features were used to train and validate the XGBoost classifier with fourfold cross-validation for the classification. We considered all the channels for every patient to obtain the classification result. The training of the classifier was done with balanced class data (equal number of training and testing data). The fourfold cross-validation was done by splitting the seizure segments of all the patients into approximately four equal segments. Each fold was divided as patient-specific, at the same time making sure that it contained almost the same number of segments. The patient data used in onefold was not considered in any of the other remaining folds. Average AUC was calculated from the fourfold validation. The Wilcoxon rank sum test and manual threshold method were initially used to understand the significance and weightage of each feature. The best features and the AUC obtained from the Wilcoxon rank sum test, manual threshold method, and XGBoost algorithm were compared. XGBoost algorithm resulted in higher classification accuracy with the least number of features compared to the other feature ranking methods. The XGBoost feature ranking algorithm works in such a way that it neglects the correlated features. The feature importance function of the XGBoost classifier was used to obtain the final feature ranks. The result of the feature ranking obtained using the feature importance function of XGBoost was calculated for each validation set. Finally, the average scores of each validation are considered for obtaining the final ranking of the features. The XGBoost uses a binary decision algorithm to identify the best set of features. The best feature set obtained from XGBoost was fed to SVM, and their performances were compared.

3 Results

The time and frequency features extracted from the seizure and non-seizure segments were analyzed using the Wilcoxon rank sum test, manual threshold, and XGBoost feature importance methods. The features with more number of subjects having a $p < 0.05$ were considered as significant features. According to the Wilcoxon rank sum test, the most significant features were standard deviation, variance, RMS, mean power, entropy, bandpower between 2 and 4 Hz, and Shannon's entropy, with 37 out of 39 patients passing the test. Figure 3 shows the ranking of the Wilcoxon rank sum test result subject-wise.

The performance of individual features to discriminate seizure and non-seizure segments was analyzed using the manual threshold method, and the AUC was calculated by setting 1000 small threshold steps. The feature ranking with respect to the manual threshold AUC method is shown in Fig. 4. According to this method, the top three features were ZCR, entropy, and standard deviation.

The feature ranking results of the best features obtained from the XGBoost feature importance algorithm are shown in Fig. 5. According to this method, ZCR, entropy, and total bandpower were the top three features. We found that the best features according to the manual threshold and XGBoost were not the same. The gradient boosting algorithm selects only one best feature out of all the correlated features, while the manual threshold gives each feature's particular importance.

The best performing features obtained from the Wilcoxon rank sum test, manual threshold, and XGBoost feature importance methods were used for training and

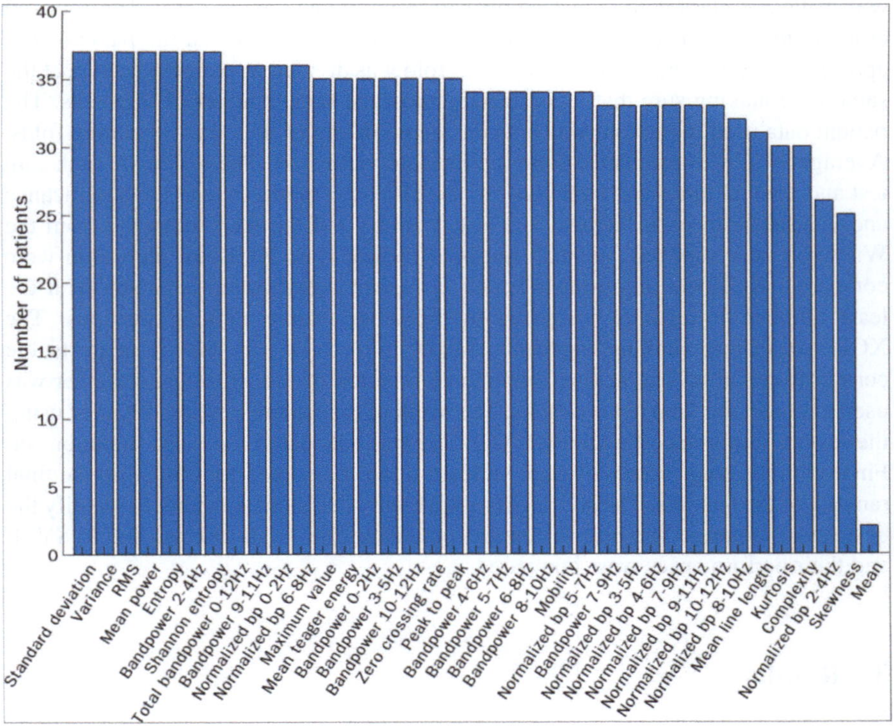

Fig. 3 Feature ranking using Wilcoxon rank sum test. It can be inferred that standard deviation, variance, RMS, mean power, entropy, bandpower between 2 and 4 Hz, and Shannon's entropy features have a $p < 0.05$ for 35 patients; bp, bandpower

testing the XGBoost classifier. In order to measure the performance of the system, the area under the ROC was calculated. The AUC was calculated using the top 2 to 37 features in the same order of the ranking obtained from each method. The AUC values attained on each trial are shown in Fig. 6. The features ranked based upon the Wilcoxon rank sum test's result are fed to the XGBoost classifier, and the corresponding AUC values obtained for each feature set are shown in Fig. 6a. It can be observed that the AUC value increased with the number of features included. We achieved a peak AUC value of 0.84 with a large number of features (37 features). Figure 6b shows the AUC values calculated using the XGBoost classifier based on the manual threshold method's feature ranking. We achieved the highest AUC value of 0.84 with the feature set that consists of 29 features. Figure 6c shows the AUC values obtained by the XGBoost classifier using the features ranked based on the XGBoost feature importance method. The average AUC increased from 76% (top 2 features) to 84% (top 11 features). It can be inferred that the XGBoost feature ranking method distinguishes the best features and gives a better performance with fewer features. The top 11 features of the XGBoost ranking resulted in an AUC of

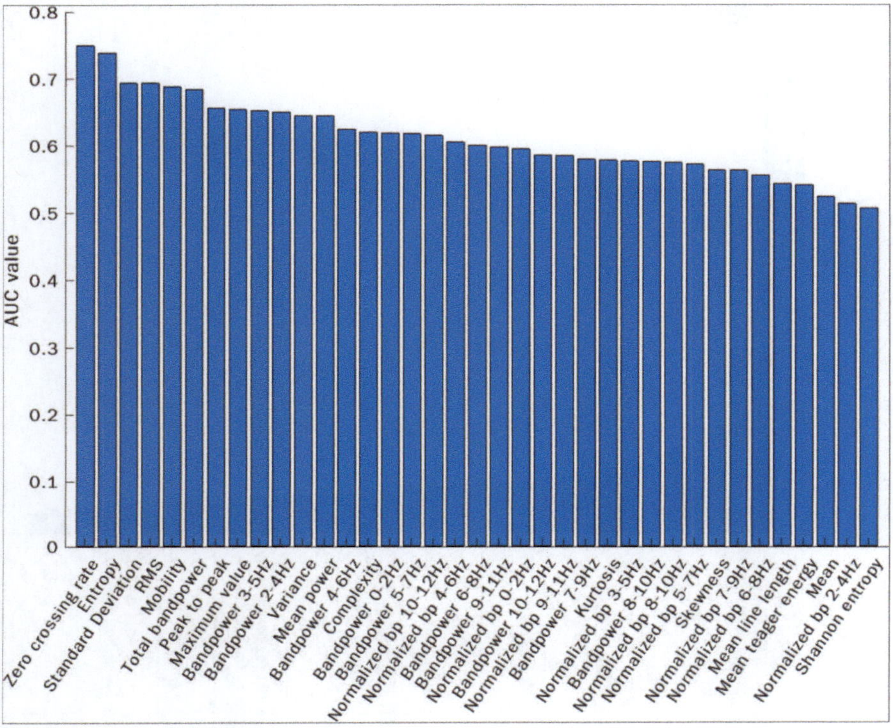

Fig. 4 Feature ranking using manual threshold method. Individual feature's importance's result calculated using the AUC value of single features by manual threshold method; bp, bandpower

0.84, higher than all the other ranking method's top 11 feature's AUC. We can observe in Fig. 6c that the AUC value was high with the top 11 features, and similar AUC values were obtained for the rest of the feature set except the feature set with all the 37 features.

The best 11 features obtained from the XGBoost algorithm that resulted in a higher AUC value (0.84) are shown in Fig. 7. For the best 11 features, the model achieved a sensitivity of 0.87 and specificity of 0.5. We can observe that ZCR, entropy, and total bandpower influenced the classifier to achieve higher AUC.

We compared the results of the XGBoost classifier with SVM for the top 11 features obtained using XGBoost feature importance. We achieved an average AUC of 0.76 using the SVM classifier. Out of these two classifiers, the random forest-based XGBoost algorithm, resulted in superior performance. We compared the AUC values obtained when the classifier was trained with the different sets of best features. The average AUC of the fourfold cross-validation of the single features, all features, and best features of XGBoost are shown in Fig. 8a. The best features obtained by the XGBoost method are used to train the SVM classifier for comparing the performance. The SVM hyper-parameters were tuned for each fold

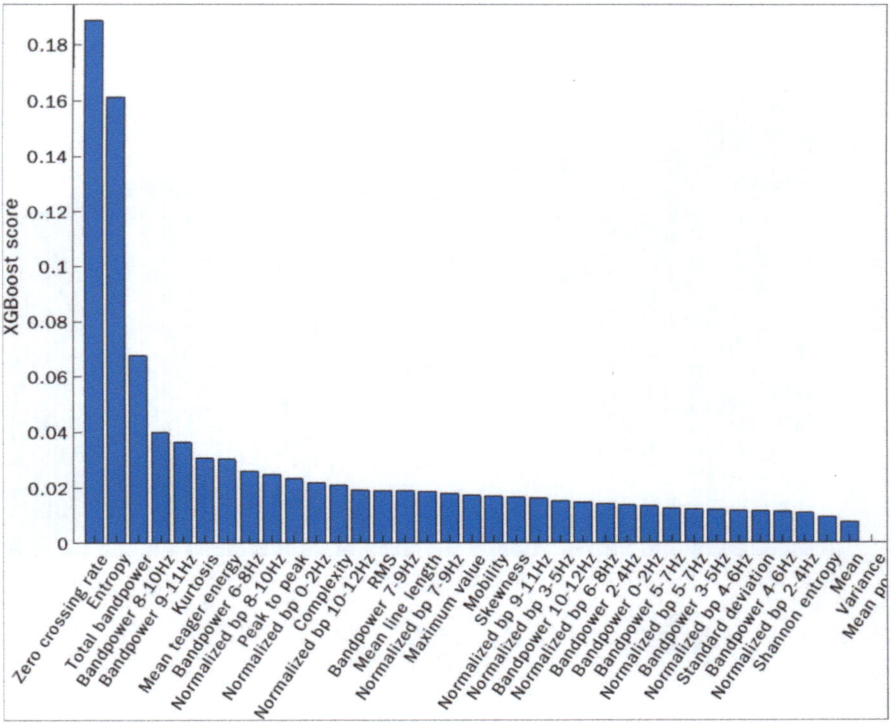

Fig. 5 Feature ranking using XGBoost feature importance method. ZCR, entropy, and total bandpower were the well-distinguished features according to this method

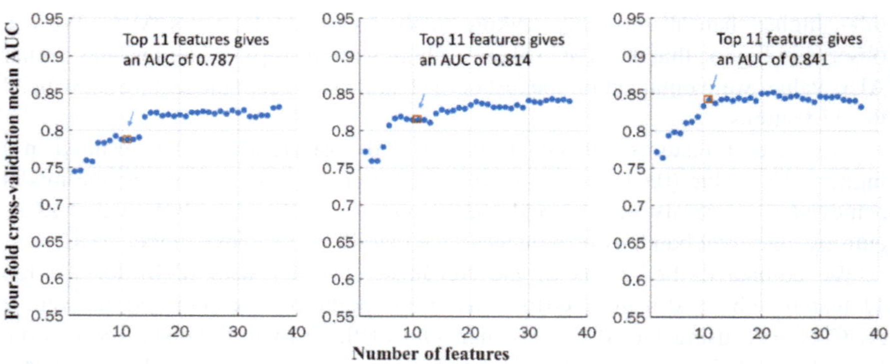

Fig. 6 (**a**) The AUC value obtained using the features in the ranking order obtained from the Wilcoxon rank sum test; (**b**) AUC value obtained using the features in the ranking order obtained from the manual threshold result; (**c**) AUC value obtained using a different number of best features used with XGBoost classifier. It can be observed that the AUC value increased with the number of features included. The AUC obtained using top 11 features ranked by XGBoost gives a comparatively higher AUC than other ranking methods

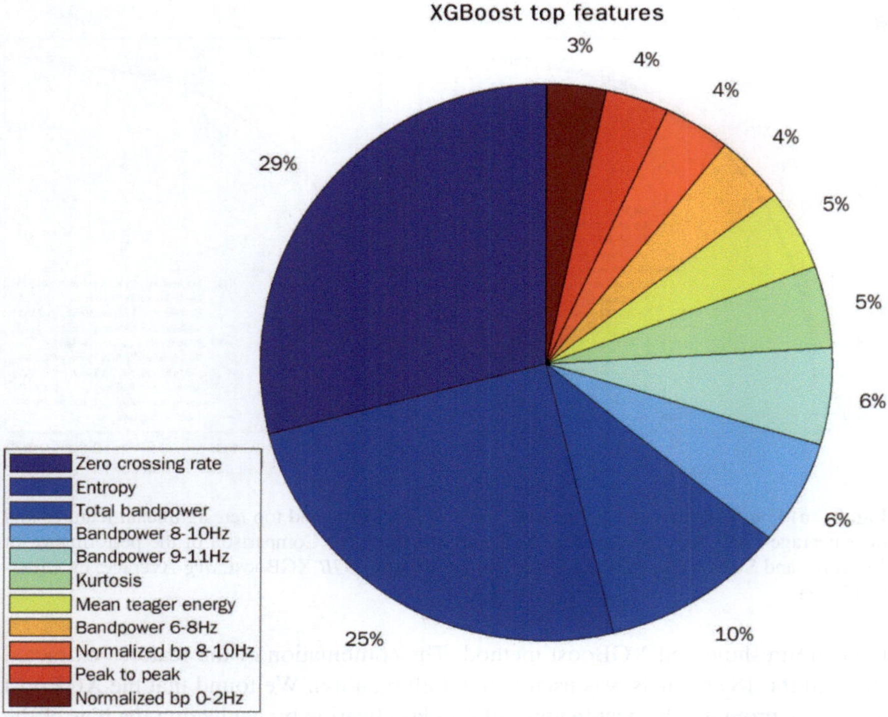

Fig. 7 Feature ranking acquired from XGBoost feature importance algorithm and the percentage of their importance is depicted in the figure. These features were able to improve the classifier's performance by improving the AUC

using "gridsearch" method to optimize the results. We concluded from Fig. 8b that the XGBoost classifier (AUC: 0.84) outperforms the nonlinear SVM classifier (AUC: 0.76) widely used in neonatal seizure detection.

4 Discussions

In this chapter, we employed a set of time and frequency domain features extracted from the publicly available neonatal EEG database (Helsinki University database) for seizure classification. Our study included all the seizure patient's EEG data available in the database and bipolar EEG channels. We used a typical pre-processing pipeline to remove the artifacts, including HPF, notch filter, and noise statistics methods. Unlike most of the studies mentioned in Table 1, we avoided downsampling and band-pass filtering to prevent data loss. Further, we segmented the seizure and non-seizure segments and extracted each segment's time and frequency features. We evaluated the weightage of each feature using the

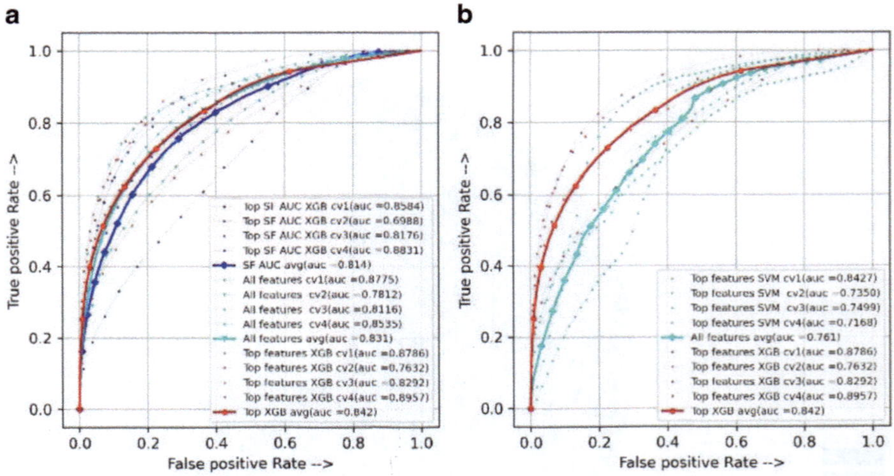

Fig. 8 (**a**) Fourfold cross-validation ROC for a single feature and top ten significant features and their average (avg) AUC obtained by XGBoost classifier; (**b**) Comparison of the performance of XGBoost and SVM using ROC curve; *SF* Single feature, *XGB* XGBoost, *avg* Average, cv Cross-validation

manual threshold and XGBoost method. The combination of the features extracted from all the 18 channels was used without elimination. We found that the XGBoost algorithm provided the best feature set for classification by neglecting the bias of the correlated features. The best features' performance was evaluated using the XGBoost classifier, and the results were compared with the SVM. We obtained an AUC of 0.84 using the XGBoost and 0.76 with the SVM classifier for the best 11 features. The proposed system was designed in Python (version 3.7.1) and MATLAB software using Intel(R) Core(TM) i3-5010U CPU @ 2.10 GHz processor.

4.1 *Effect of Number of Channels*

Our study extracted the features from all the available 18 channels and achieved higher classification accuracy. Most of the studies that used the Helsinki database used all the available channels in the EEG data [13, 14, 16, 19, 21, 23]. However, a few studies used the channel selection method to reduce the computational complexity and improve the classification accuracy [7, 13, 24, 25]. Seizure tends to spread to the adjacent channels after its onset. Hence, a single channel cannot be fixed for all the datasets, and each time, different channel selection techniques might be required to localize the seizure and achieve higher accuracy. It may also lead to the missing of true positives in the other channels. We treated each channel separately for the classification. We understand that considering each channel separately might lead to a slight skewness in classification results as a single seizure is considered multiple segments. However, this wouldn't affect the result as we have performed a comparative study for identifying the best features on the same dataset.

4.2 *Effect of Feature Extraction Methods*

We found that using the manual threshold method, ZCR, entropy, and standard deviation provides higher AUC. Figure 9 shows the box plot of these features extracted from the seizure and background segments. ZCR was able to distinguish between seizure and non-seizure segments. We can observe from Fig. 9a that the background segments have higher ZCR compared to the seizure segments. This suggests that the rate of signal change from either positive to negative or negative to positive in background segments is higher. The background signal has higher randomness than the seizure signal, as shown in Fig. 9b. This concludes that the seizure signals have more uniformity than the non-seizure signal. We can observe

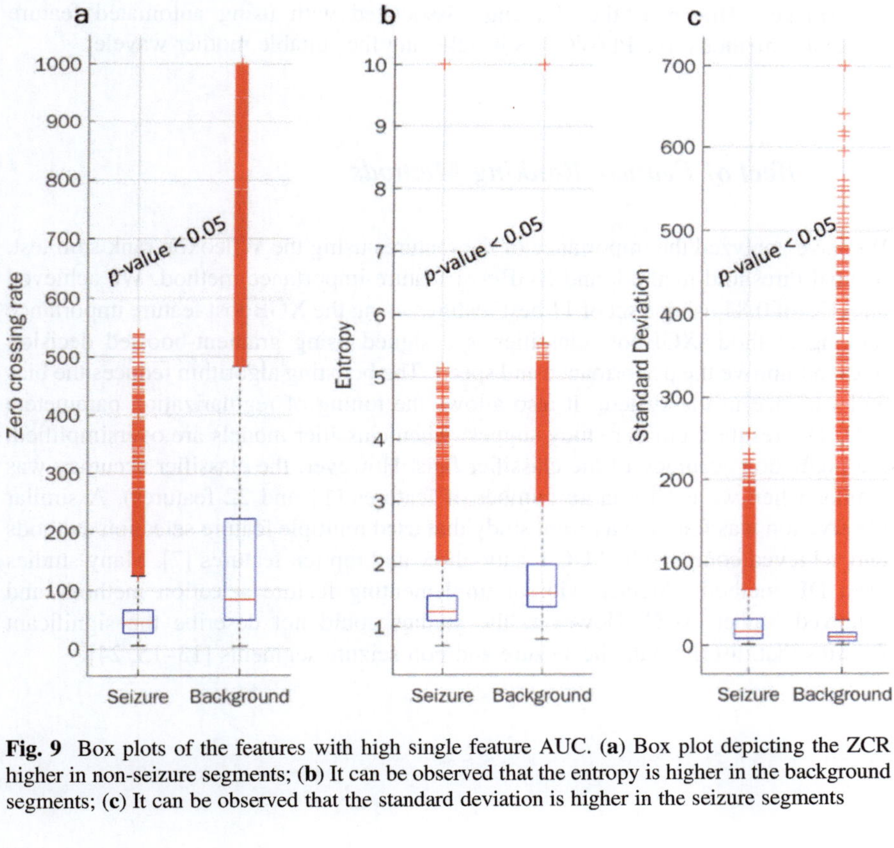

Fig. 9 Box plots of the features with high single feature AUC. (**a**) Box plot depicting the ZCR higher in non-seizure segments; (**b**) It can be observed that the entropy is higher in the background segments; (**c**) It can be observed that the standard deviation is higher in the seizure segments

from the result that the periodicity pattern of the seizure signal distinguishes seizure from the background. The same observation was found in Açıkoğlu and Tuncer [7] using multiple feature selection methods. The standard deviation was found to be higher for the seizure segments than the background, shown in Fig. 9c.

In our study, XGBoost supplied ZCD, entropy, and total bandpower as the top three features. The manual threshold and XGBoost feature weightage methods conclude "entropy" and "zero-crossing rate" as one of the best features. We also found that the normalized bandpower in the lower frequency was higher in the background signal compared to the seizure, resulting in normalized bandpower between 0 and 2 Hz as an important feature according to XGBoost.

We extracted 37 time and frequency features and obtained an AUC of 0.84 using the 11 best features. The features considered in our study outperformed the wavelet and brain connectivity features [18, 25]. However, few studies used brain connectivity features to cluster neonatal seizure segments and detect brain synchronicity [19, 21–23]. Spike and time-frequency correlation-related features were used for neonatal seizure and non-seizure classification in the literature and obtained a higher accuracy compared to our results [13, 17]. However, these features are more complex to implement and require high computational efforts. Some studies used automated feature extraction [13–15] to reduce the complexity and improve the performance. However, the challenge associated with using automated feature extraction methods like PDWC lies in selecting the suitable mother wavelet.

4.3 *Effect of Feature Ranking Methods*

We have analyzed the importance of the features using the Wilcoxon rank sum test, manual threshold method, and XGBoost feature importance method. We achieved an AUC of 0.84 using a set of 11 best features using the XGBoost feature importance ranking method. XGBoost classifier is designed using gradient-boosted decision trees to improve the performance and speed. The boosting algorithm reduces the bias and variance in the system. It also allows the tuning of regularization parameters [34]. The results from our study suggest when classifier models are oversimplified, the prediction accuracy of the classifier falls. However, the classifier accuracy was similar when we used a large number of features (11 and 22 features). A similar observation was found in a recent study that used multiple feature selection methods and achieved comparable AUC for top three and top ten features [7]. Many studies used DL methods directly without implementing feature selection methods and achieved higher AUC. However, the studies could not describe the significant features that discriminate the seizure and non-seizure segments [13–15, 24].

4.4 Effect of Classifiers

We compared the results of the XGBoost classifier with SVM and found that XGBoost performs well compared to SVM. Our study achieved an average AUC of 0.84 with the XGBoost classifier and 0.76 with the SVM classifier. Few studies reported smaller classification performance [18, 25], and many reported higher classification accuracy compared to our study [7, 13–16, 24]. The higher accuracy resulted in the other studies may be due to the usage of a selective duration of the seizure, a limited number of samples [22], selected channels [7, 13], and complex features [17].

Higher classification performance can be achieved by using high dimensionality deep learning methods [13, 15, 24]. However, DL-based neural network methods have a drawback of optimizing hyper-parameters, complexity in implementing higher dimensionality classifiers, and a requirement of a large training dataset and require advanced computational setup. With the limited availability of neonatal EEG data, it is difficult to efficiently train the DL classifier [35]. Over-fitting is also a problem often seen in DL classifiers validated with limited data. CNN models were developed based upon three independent individual expert's labeling, and the ensemble model generated out of the three models gave better results [15]. However, the availability of several expert annotations for the same dataset is uncommon, and the implementation of such an ensemble model might have other dependencies.

4.5 Effect of Performance Metrics

We evaluated the performance of the XGBoost and SVM classifier using AUC. However, studies used different metrics like precision, specificity, F1 score (harmonic mean between precision and sensitivity), sensitivity [7], AUC [13–15, 18, 24, 25], accuracy [16, 19, 23], and kappa value [20] to validate their model. The performance of the supervised [21] and unsupervised [22] classification was analyzed using *p*-values. Hence, it is difficult to compare the performance of the classifier models between the studies. Refer to Table 1 for the respective metrics and their values achieved in each of the studies.

4.6 Limitations and Future Directions

Our study focused on comparing a set of time and frequency domain features to rank them according to their effectiveness in classifying neonatal seizures. We utilized a publicly available database that contains multiple expert annotations. However, this is the only database we used to analyze our features. The number of seizures available in this database is less than the adult EEG databases. Extending the study with a huge neonatal database would help in conducting a robust comparison of features. Moreover, the inclusion of features such as spike and frequency correlations would detect neonatal seizures more precisely and improve the system's performance. Our study also has a few common limitations seen in other studies, such as lack of clinical trials, reliable elimination of eye movement and other artifacts, etc. It is also possible to explore more on a few DL methods as proposed in other studies to improve the accuracy.

5 Conclusion

In this study, we evaluated different time and frequency features and identified a set of significant features. The identified best features were able to classify the seizure and non-seizure segments of the neonatal EEG data with an effective classifier like XGBoost. Our study can also infer that the extraction of a large number of features will only lead to an increase in the computational time and data redundancy instead of improving the model performance. A set of feature selection based on prior knowledge helped in boosting the accuracy of the model. Hence, we used the manual threshold to analyze each feature and the XGBoost algorithm for feature selection and classification. The XGBoost algorithm is much simple to implement, and it also effectively eliminates the correlated features. The processing pipeline followed in our study supplied an average AUC of 0.84. The usage of channel and patient selection, complicated spike-correlation features, hybrid wavelet, and deep learning methods, or elimination of particular width seizure segments in a few past studies could be the reason for the higher performance of these models. This study will help the neurologists in long-time monitoring and diagnosing epileptic seizures in NICU. It also will facilitate the understanding of the significance of different features concerning the neonatal database.

References

1. Thijs, R. D. *et al.* (2019) 'Epilepsy in adults', *The Lancet*, 393(10172), pp. 689–701. doi: https://doi.org/10.1016/S0140-6736(18)32596-0.
2. Hauser, W. A. and Kurland, L. T. (1975) 'The Epidemiology of Epilepsy in Rochester, Minnesota, 1935 Through 1967', *Epilepsia*, 16(1), pp. 1–66. doi: https://doi.org/10.1111/J.1528-1157.1975.TB04721.X.
3. Sueri, C. *et al.* (2018) 'Diagnostic Biomarkers of Epilepsy', *Current Pharmaceutical Biotechnology*, 19(6), pp. 440–450. doi: https://doi.org/10.2174/1389201019666180713095251.
4. Annegers, J. F. and Rocca, W. A. (1996) 'Causes of Epilepsy: Contributions of the Rochester Epidemiology Project', *Mayo Clinic Proceedings*, 71(6), pp. 570–575. doi: https://doi.org/10.4065/71.6.570.
5. Holmes, G. L. (2015) 'Cognitive impairment in Epilepsy: The Role of Network Abnormalities', *Epileptic disorders : international epilepsy journal with videotape*, 17(2), p. 101. doi: https://doi.org/10.1684/EPD.2015.0739.

6. Kwon, O. Y. and Park, S. P. (2014) 'Depression and Anxiety in People with Epilepsy', *Journal of Clinical Neurology (Seoul, Korea)*, 10(3), p. 175. doi: https://doi.org/10.3988/JCN.2014.10.3.175.
7. Açıkoğlu, M. and Tuncer, S. A. (2019) 'Incorporating feature selection methods into a machine learning-based neonatal seizure diagnosis'. doi: https://doi.org/10.1016/j.mehy.2019.109464.
8. Murray, D. M. *et al.* (2008) 'Defining the gap between electrographic seizure burden, clinical expression and staff recognition of neonatal seizures', *Archives of Disease in Childhood: Fetal and Neonatal Edition*, 93(3). doi: https://doi.org/10.1136/adc.2005.086314.
9. Aarabi, A. *et al.* (2007) 'A multistage knowledge-based system for EEG seizure detection in newborn infants', *Elsevier*. Available at: https://www.sciencedirect.com/science/article/pii/S1388245707004294.
10. Thomas, J. *et al.* (2021) 'Automated Adult Epilepsy Diagnostic Tool Based on Interictal Scalp Electroencephalogram Characteristics: A Six-Center Study', *International Journal of Neural Systems*, 31(5). doi: https://doi.org/10.1142/S0129065720500744.
11. Erberich, S. G. *et al.* (2003) 'Functional MRI in neonates using neonatal head coil and MR compatible incubator', *NeuroImage*, 20(2), pp. 683–692. doi: https://doi.org/10.1016/S1053-8119(03)00370-7.
12. Stevenson, N. J. *et al.* (2019) 'Data Descriptor: A dataset of neonatal EEG recordings with seizure annotations', *Nature Publishing Group*. doi: https://doi.org/10.1038/sdata.2019.39.
13. O'Shea, A. *et al.* (2020) 'Neonatal seizure detection from raw multi-channel EEG using a fully convolutional architecture', *Neural Networks*, 123, pp. 12–25. doi: https://doi.org/10.1016/J.NEUNET.2019.11.023.
14. Isaev, D. *et al.* (2020) 'Attention-Based Network for Weak Labels in Neonatal Seizure Detection', http://ncbi.nlm.nih.gov. Available at: https://www.ncbi.nlm.nih.gov/pmc/articles/PMC7521836/.
15. Tanveer, M. A. *et al.* (2021) 'Convolutional neural networks ensemble model for neonatal seizure detection', *Journal of Neuroscience Methods*, 358.
16. He, H. *et al.* (2021) 'A progressive deep wavelet cascade classification model for epilepsy detection', *Elsevier*. Available at: https://www.sciencedirect.com/science/article/pii/S093336572100110X.
17. Tapani, K. T., Vanhatalo, S. and Stevenson, N. J. (2019) 'Time-Varying EEG Correlations Improve Automated Neonatal Seizure Detection', *International Journal of Neural Systems*, 29(4). doi: https://doi.org/10.1142/S0129065718500302.
18. Abbas, A. *et al.* (2021) 'An EEG-based methodology for the estimation of functional brain connectivity networks: Application to the analysis of newborn EEG seizure', *Elsevier*. Available at: https://www.sciencedirect.com/science/article/pii/S1746809420303591 (Accessed: 24 August 2021).
19. C. Ye, K. Slavakis, J. Nakuci, S. F. M. and J. M. (2021a) *Online Classification of Dynamic Multilayer-Network Time Series in Riemannian Manifolds, ICASSP 2021 - 2021 IEEE International Conference on Acoustics, Speech and Signal Processing (ICASSP)*. Available at: https://ieeexplore.ieee.org/abstract/document/9413560/ (Accessed: 25 August 2021).
20. Ding, Fengqian and Luo, C. (2019) 'The entropy-based time domain feature extraction for online concept drift detection', *Entropy*. doi: https://doi.org/10.3390/e21121187.
21. Frassineti, L. *et al.* (2021) 'Multiparametric EEG analysis of brain network dynamics during neonatal seizures', *Elsevier*. Available at: https://www.sciencedirect.com/science/article/pii/S016502702030426X.
22. Khazaei, M. *et al.* (2021) 'Microstate Alterations Associated with Seizure Events in Full-term Neonates', https://anyscreeninc.com. doi: https://doi.org/10.1016/j.neuroimage.2017.11.062.
23. Ye, C. *et al.* (2021b) 'Fast Sequential Clustering in Riemannian Manifolds for Dynamic and Time-Series-Annotated Multilayer Networks', *IEEE Open Journal of Signal Processing*, pp. 1–1. doi: https://doi.org/10.1109/ojsp.2021.3051453.

24. Caliskan, A. and Rencuzogullari, S. (2021) 'Transfer learning to detect neonatal seizure from electroencephalography signals', *Neural Computing and Applications*. doi: https://doi.org/10.1007/S00521-021-05878-Y.
25. Frassineti, L. *et al.* (2020) 'Neonatal Seizures Detection using Stationary Wavelet Transform and Deep Neural Networks: Preliminary Results', https://ieeexplore.ieee.org. Available at: https://ieeexplore.ieee.org/abstract/document/9140713/.
26. Temko, A. *et al.* (2012) 'Instantaneous measure of EEG channel importance for improved patient-adaptive neonatal seizure detection', *IEEE Transactions on Biomedical Engineering*, 59. Available at: https://ieeexplore.ieee.org/abstract/document/6096393/.
27. Yuvaraj, R. *et al.* (2019) 'A deep Learning Scheme for Automatic Seizure Detection from Long-Term Scalp EEG', *Conference Record - Asilomar Conference on Signals, Systems and Computers*, 2018-October, pp. 368–372. doi: https://doi.org/10.1109/ACSSC.2018.8645301.
28. Temko, A. *et al.* (2011) 'Performance assessment for EEG-based neonatal seizure detectors', *Elsevier*. Available at: https://www.sciencedirect.com/science/article/pii/S1388245710006048.
29. Thomas, J. *et al.* (2020) 'Automated Detection of Interictal Epileptiform Discharges from Scalp Electroencephalograms by Convolutional Neural Networks', *International journal of neural systems*, 30(11), p. 2050030. doi: https://doi.org/10.1142/S0129065720500306.
30. Ansari, A. H. *et al.* (2016) 'Improved multi-stage neonatal seizure detection using a heuristic classifier and a data-driven post-processor', *Clinical Neurophysiology*, 127(9), pp. 3014–3024. doi: https://doi.org/10.1016/J.CLINPH.2016.06.018.
31. Dickinson Gibbons, J., Chakraborti, S. and Kk, D. E. (2020) *Nonparametric statistical inference*. Available at: https://books.google.com/books?hl=en&lr=&id=nWwNEAAAQBAJ&oi=fnd&pg=PP12&dq=%5B1%5D+Gibbons,+J.+D.,+and+S.+Chakraborti.+Nonparametric+Statistical+Inference,+5th+Ed.,+Boca+Raton,+FL:+Chapman+%26+Hall/CRC+Press,+Taylor+%26+Francis+Group,+2011.&ots=LAooLpOcHK&sig=QhLZd8_RH8lyNNIADTGIRBJqzBA.
32. Hollander, M., and D. A. Wolfe. (1999) 'Nonparametric Statistical Methods', *Hoboken, NJ: John Wiley & Sons, Inc.,*.
33. Manzouri, F. *et al.* (2018) 'A Comparison of Machine Learning Classifiers for Energy-Efficient Implementation of Seizure Detection', *Frontiers in Systems Neuroscience*, 0, p. 43. doi: https://doi.org/10.3389/FNSYS.2018.00043.
34. Dhaliwal, S. S., Nahid, A.-A. and Abbas, R. (2018) 'Effective intrusion detection system using XGBoost' Information 2018, Vol. 9, Page 149, 9(7), p. 149. https://doi.org/10.3390/INFO9070149.
35. Shoeibi, A. *et al.* (2021) 'Epileptic seizures detection using deep learning techniques: A review', *International Journal of Environmental Research and Public Health*, 18(11). doi: https://doi.org/10.3390/ijerph18115780.

Hilbert Huang Transform (HHT) Analysis of Heart Rate Variability (HRV) in Recognition of Emotion in Children with Autism Spectrum Disorder (ASD)

B. Anandhi and S. Jerritta

Abstract Emotion recognition has become a boon to society since it aids in the early management of uncontrollable sudden emotional outbursts in children with ASD prior to their onset. As a result, developing an alarm system assists caregivers, doctors and therapists in preventing such aggressive behaviours before they occur. Only a few studies have studied the physiological markers in ASD children and adults. The Pan-Tompkins equation was used to determine HRV from the RR index. The Hilbert-Huang transform (HHT) of heart rate variability (HRV) data was used to extract emotional characteristics corresponding to two emotions (positive and negative valence). The k-nearest neighbour (KNN) and ensemble classifiers were used to identify only the statistically important features. The HHT analysis of the HRV data achieved good classification rate with a mean accuracy of 74.5% and 73.5% in children with ASD and in TD children using the ensemble classifier. Further feature reduction was done using the principal component analysis (PCA), and it was seen that the accuracy improved to 77% and 78.3% in children with ASD and TD children, respectively using the ensemble classifier. Though the children with ASD are deficit in emotion recognition, we found no significant impairment in identifying the basic emotions. However, they were deficit in correctly predicting the positive valance state than the TD children

Keywords Autism spectrum disorder (ASD) · Heart rate variability (HRV) · Hilbert transform · Principal component analysis (PCA)

B. Anandhi (✉)
Department of Biomedical Engineering, Agni College of Technology (ACT), Chennai, Tamil Nadu, India

S. Jerritta
Department of Electronics and Communication Engineering, Vels Institute of Science Technology and Advanced Studies (VISTAS), Chennai, Tamil Nadu, India
e-mail: jerritta.se@velsuniv.ac.in

M. Murugappan, R. Yuvaraj (eds.), *Biomedical Signals Based Computer-Aided Diagnosis for Neurological Disorders*, https://doi.org/10.1007/978-3-030-97845-7_4

1 Introduction

ASD is widespread over the world, and its prevalence has risen to 10% since the previous surveillance in 2018.As per the latest prevalence figures from the US Centers for Disease Control and Prevention (CDC), ASD affects roughly 1 out of every 54 people. Boys are four times more likely than girls to be diagnosed, which is consistent with prior data [1]. In general, ASD is a neurodevelopmental disorder which typically has a developmental delay, impairment in communication, emotion deficit and social interaction due to lack of theory of mind (ToM) and has difficulty in understanding other's thoughts, intentions and feelings. In addition to this, the individual with ASD shows stereotyped behaviours and has varied sensory issues [2]. No two individuals with ASD experience the same difficulties but exhibit specific behaviours such as hand flapping, echolalia, screaming, walking on toes, head rolling and extreme behaviours. Persons with ASD rarely start speech and frequently fail to respond to any questions or do not express affection towards others [3]. It has been assumed that the children with ASD have difficulty in recognizing the emotions when compared to their neurotypical peers [4, 5]. Few other works by researchers have proved that children and adults with ASD can recognize the basic emotions (anger, fear, sadness and happiness) accurately as the TD groups [6, 7]. To bring the contradictory findings in recognition of basic emotions, researchers have hypothesized that the ASD population are capable of doing simple emotion recognition tasks [5]. Many researchers have worked on extraction of emotion in individuals with ASD and control population through facial expressions, gestures and behavioural measures [8–11]. However, studies suggest that the individuals with ASD have trouble in emotion recognition than the controls [12, 13]. In recent years, emotion recognition mostly focusses on the physiological signals such as electrocardiogram (ECG), electroencephalogram (EEG), electrooculogram (EOG), electromyogram (EMG), respiration (RSP) and skin temperature (SKT) both in control and ASD groups [14–18]. Many works on emotion recognition were done on individuals with Parkinson's disease [19]. Most of the studies were mainly focussed on brain and eye tracking [20, 21]. But only a very few researchers have worked on the ECG signals to determine the underlying emotions of the individuals with ASD [16, 17, 22–24]. The research in emotion recognition through physiological signals are increasing in recent years as it is very evident that the autonomous nervous system (ANS) which is a complex system that connects the central nervous system(CNS) [25] helps to detect the true underlying emotions. The parasympathetic nervous systems is very active during the resting state and can be evaluated from the high-frequency bands of the HR. This leads to suppress the sinus arterial nodes which in turn decreases the heart rate and increases the HRV. HRV is the most promising electrophysiological indicator for identifying different heart disorders [26]. In contradictory the parasympathetic nervous system has increased heart rate which helps in regulating the heart [27]. In few studies done to 2- to 6-year-old toddlers, it was found that there were no substantial differences in heart rate and HRV among them [16, 28].

In this work, we mainly focus on the valence state detection in children with ASD and TD children using ECG signals using analyses of the derived HRV data using HHT which is deeply discussed in the below section.

2 Methodology

The various stages involved from designing of elicitation protocol through emotion classification employed in emotion recognition study are shown in Fig. 1. This study was carried on 30 children (15 children with ASD and 15 TD children) between 5 and 11 years old taking part in the pre-trials and the real-time experiments with all the exclusion criteria for serious hearing loss and vision impairment. The children with ASD were screened for their developmental quotient (DQ) and the Indian Scale for Assessment of Autism (ISAA) by the paediatrician and the psychologist. The children in the ASD group belonging to the ASD severity level of mild (ISAA-70–106) and moderate (ISAA-107–153) autism were considered in this work. Children with TD were included who are the school goers if they had no history of parent's concern for their child's development.

3 Protocol Design

Triggering emotion in the individuals often plays a major role in the emotion recognition task. There are various approaches through which the emotions are induced such as visual, audio, audiovisual, imagery, recall and games. This study uses audio and audiovisual cues to elicit the emotions. Since the study involves young children with ASD as well as the control children, designing the protocol becomes quite challenging. After a thorough observation of the children with ASD in their classrooms and interaction with the parents, therapists and psychologists, emotion elicitation protocol was designed. Figure 2 shows the design protocol of the experiment.

As per the protocol design, the baseline data for 2 min was recorded when the child is at relaxed state followed by the positive and negative valence emotional cues comprising of audio- and audiovisual clips of length that lasted for 0.16 s to 1 min

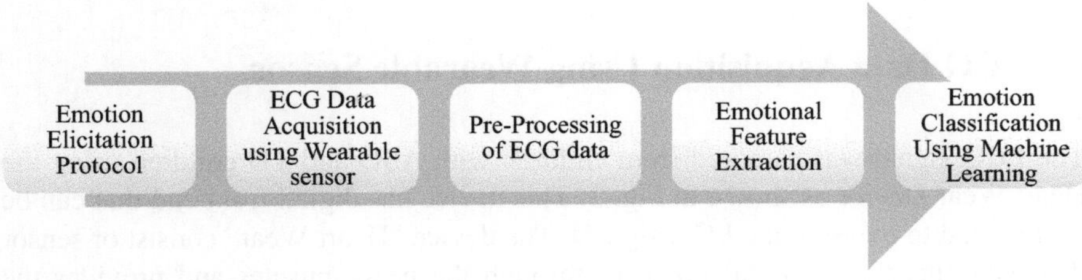

Fig. 1 Methods involved in emotion recognition

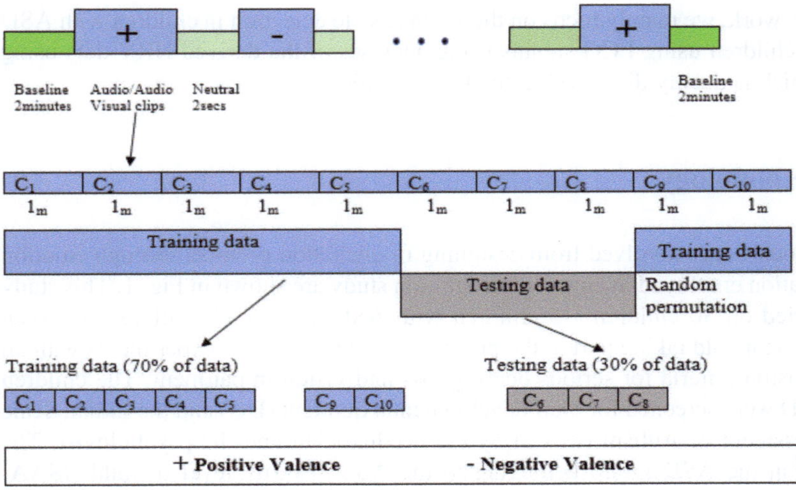

Fig. 2 Protocol for emotion elicitation and its timing paradigm

which were taken from various sources from the Internet. A neutral image for 2 s was played in-between the opposite valence cues. There were three trials each lasted for 30–40 min with a break of 10 min in-between each trial. The emotional ECG data were collected from the child using a wearable shimmer ECG device along with chest belt. An emotional assessment questionnaire was rated by the child's parent to scale the felt emotion by their child while subjected to view the emotional cues. The emotion elicitation protocol design of the study was approved by the ethics board of National Institute for Empowerment of Persons with Multiple Disabilities (NIEPMD). The children with ASD were identified from NIEPMD's clinic and school, Chennai. The parent or the guardian of the participating child provided the informed consent. Each child was observed for a week for their behavioural issues and had a parental interview session to know their "likes" and "dislikes" regarding child's daily routine, favourite cartoon, film songs, advertisements and dramas and their aversions towards various audios and videos. Based on the parent's information, a personalized protocol was designed for each child using the audio and audiovisual cues.

4 ECG Data Acquisition Using Wearable Sensor

The ECG signal was captured from children with ASD and TD children using the Heart Wear Device as shown in Fig. 3. This device has digital front end that can be customized to measure the ECG signals. The device "Heart Wear" consist of sensor that measures the electrical impulses through the heart muscles and provides the morphological information to the heart's response to emotional instances, during

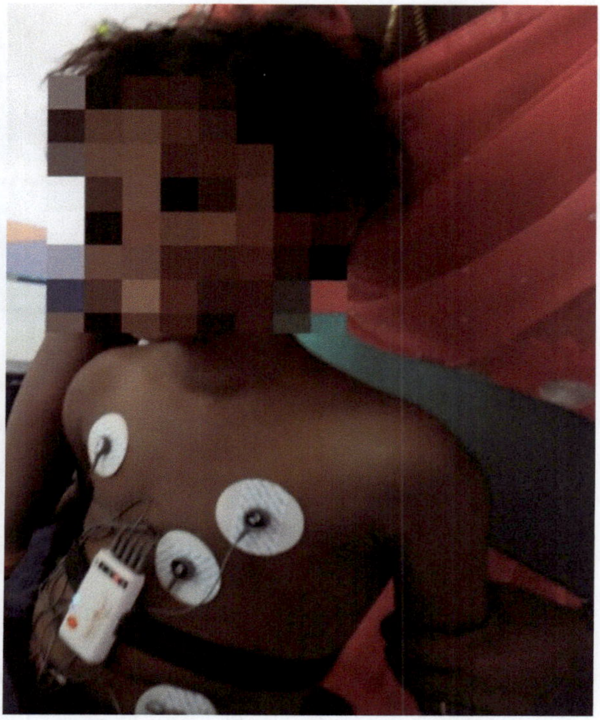

Fig. 3 ECG wearable sensor for ECG data acquisition

exercise, ambulatory subjects and to detect various cardiac arrhythmias. It has five wires and four lead solutions for measuring bipolar limb lead and choice to measure v1 to v6 chest positions. With the help of the ConsenSys Pro software, all the signals are streamed via Bluetooth radio RN-42 in real time for visualization, and also it can be stored in an integrated 8 GB SD card and can be accessed for post-processing. It consists of MSP430 microcontroller (24 KHz) as the central processing unit, has a long lasting 450 mAh rechargeable Li-on battery and assisted with a waist strap to hold the device firmly.

This emotion recognition experiment was conducted in a confined room with a 42-inch LCD television with built-in speakers as shown in Fig. 4. The emotion elicitation protocol was shown on the TV screen using a laptop linked to the TV via HDM1 cable and another laptop streaming ECG signals using the ConsenSys software.

5 Data Pre-processing

The acquired ECG data carried various noises such as baseline wandering, powerline interferences and muscle artefacts due to the stereotyped and repetitive behaviour of the child and also due to respiration which alters the morphological information of

Fig. 4 Experimental setup and streaming of ECG signal from child with ASD

the ECG signal. The baseline wandering of the ECG signal was efficiently removed in this work by subtracting the eighth level of decomposition from the original raw ECG signal using Daubechies (db4) mother wavelet, and a sixth-order low-pass Butterworth filter with a cut-off frequency of 45 Hz was employed to remove the additional high frequency disturbances.

6 Feature Extraction

The filtered ECG signal as such does not convey any emotional information, and hence the main QRS amplitude of ECG signal and the widely practiced Pan-Tompkins algorithm was used to calculate heart rate variability (HRV) from its RR index. The HRV signal is then transformed to its time-frequency domain using the Hilbert-Huang transform (HHT), and the features such as statistical, higher-order statistics (HOS), frequency domain and energy-based features are extracted and are used for emotional classification. Figure 5 depicts the methods used to extract the emotional features from HRV using the HHT algorithm. This method employs descriptive analysis for the two group (like and dislike) differences analysis using the student's t-test using SPSS tool.

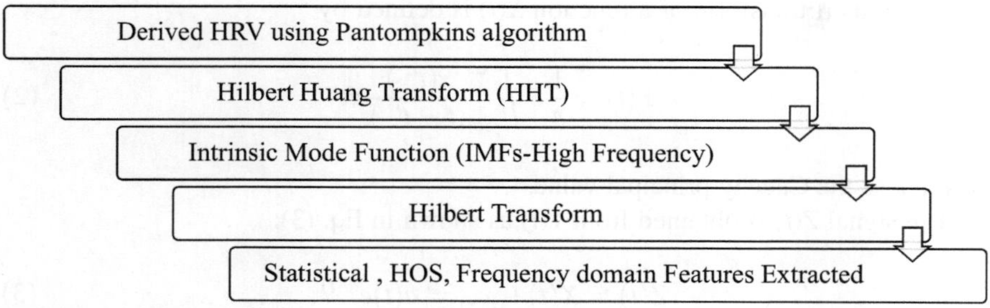

Fig. 5 ECG wearable sensor for ECG data acquisition

7 Hilbert-Huang Transform (HHT) Algorithm

The non-linear and non-stationary data are analysed using the HHT. It was proposed by Huang et al. [29]. The empirical mode decomposition (EMD), in which the signal is decomposed iteratively into a series of intrinsic mode functions (IMFs), is an important feature of HHT. There are an equal number of extrema and zero crossings in each IMF. Furthermore, the envelope's mean defined by local maxima and minima is zero. The algorithm for the sifting process is shown below.

- The signal's upper and lower envelopes, $X_{\max}(t)$ and $X_{\min}(t)$, are created by interpolating the signal's local maxima and minima from the derived HRV data $X(t)$.
- Calculate the upper and lower envelopes' mean $\widehat{M}(t) = (X_{\max}(t) + X_{\min}(t))/2$ and subtract it from the actual input signal $h1(t) = x(t) - \widehat{M}(t)$.
- The residue $r(t) = x(t) - h1(t)$ is evaluated when the $h1(t)$ satisfies the IMF's properties and the sifting process continued through the residue signal or else the sifting process is continued on $h1(t)$ component. Figure 3 shows the HRV data is decomposed to various IMFs and its corresponding HF and LF components are calculated as follows.

When the standard deviation (SD) of two consecutive iterations, calculated using Eq. (1), is between 0.2 and 0.3, or the residue is a monotonous function, the sifting process should be stopped:

$$\text{Standard Deviation} = \sum_{t=0}^{T} \frac{|h_{n-1}(t) - h_n(t)|^2}{h_{n-1}^2(t)} \tag{1}$$

The sifting component is represented by $h_n(t)$, where k is the number of siftings that occur during the extraction of an IMF. The IMFs are monotonous components and have no meaning on their own. The Hilbert transform, as indicated in Eqs. (2) through (4), can be used to acquire information on the local attributes of the IMFs, such as instantaneous amplitude and instantaneous frequency.

The Hilbert transform of a function *X*(*t*) is defined by

$$Y(t) = \frac{1}{\pi} P \int_{-\infty}^{\infty} \frac{x(t'\ )}{(t - t'\ )} dt' \tag{2}$$

where *P* is the Cauchy principal value.

The signal *Z*(*t*) is obtained from *Y*(*t*) as shown in Eq. (3):

$$Z(t) = X(t) +_{iY(\mathrm{y})} = a(t)e^{i\theta(t)} \tag{3}$$

a(*t*) is the instantaneous amplitude and the *θ*(*t*) is the instantaneous phase of the signal.

The spectral power is calculated using the following equation:

$$\text{Spectral Power } P = \frac{1}{N} \sum\nolimits_{i=1}^{N} {a_i}^2 \tag{4}$$

The HF and LF power are calculated from the first *N*. IMFs are considered to contain more oscillations and to have high frequency, and the rest of the IMFs other than the residue is considered to be the low-frequency oscillations [30]

8 Feature Classification

The obtained attributes are then categorized using the KNN and ensemble classifier into two emotional states (positive and negative valence). The feature data were permuted at random, with almost 70% of the data being used to train the model and 30% being used to test it. The accuracy was calculated from the confusion matrix obtained from the classification model. Feature classification is done using k-nearest neighbour (KNN) and ensemble classifier and 70 percentage of data set for training and 30 percentage of data set for testing. Individual significant features and combination of all the significant features were performed, and its accuracy was analysed.

The calculation of accuracy is as follows [31]:

$$\text{Accuracy}_{\text{valence}} = \frac{\text{TP} + \text{TN}}{\text{TP} + \text{TN} + \text{FP} + \text{FN}} \tag{5}$$

9 Results

9.1 Pre-processing

Figure 6 shows the pre-processed output of the raw ECG signal. It is seen from the figure that baseline wander present in the raw ECG signal is eliminated using the wavelet transform method applying Daubechies (db4) as the mother wavelet. And further the signal is passed through the low-pass Butterworth filter using a cut-off frequency of 45 Hz for the removal of high-frequency noises.

9.2 Extraction of Features

Apart from the basic six statistical features derived [10], higher-order statistical (HOS), other linear, nonlinear and frequency domain features are extracted. The HRV data was utilized to produce five IMFs and a residue signal, as shown in Fig. 7. The derived IMF residue has no message signal and should not be used in any analysis. The number of IMFs is determined by the length and extrema of HRV data. It also shows that the lower-order IMFs have very dense oscillations, and it gradually decreases as the order of IMF increases. In this study, we considered the high-frequency IMFs of the HRV signal.

The student's *P*-test was carried to statistically validate all the extracted features derived from the HHT of the HRV data are listed in Table 1. It shows that 18 features were selected due to its significance where its *P*-Value is <0.05 that supports to select

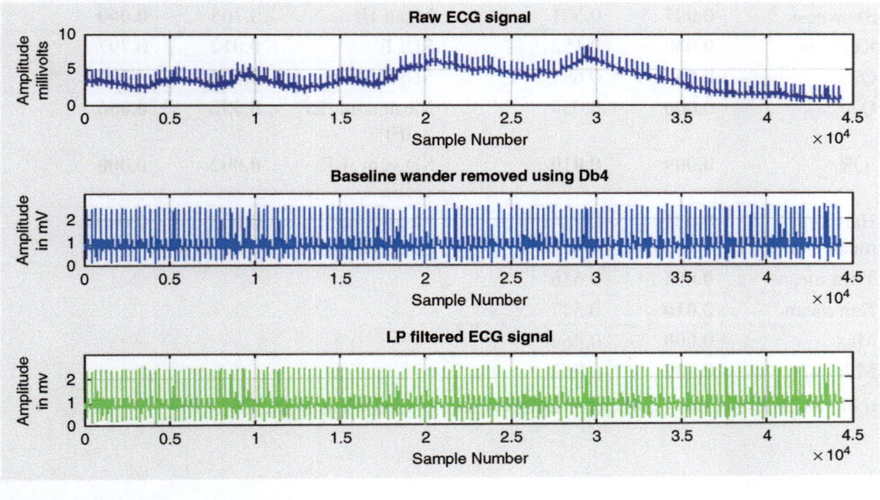

Fig. 6 ECG signal processing

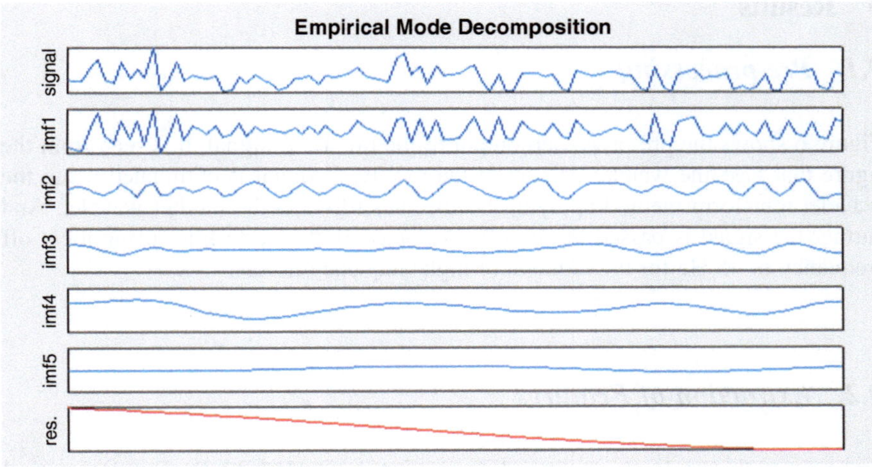

Fig. 7 Intrinsic mode function

Table 1 Statistical analysis of features derived from HRV using HHT of HRV data

Emotional features	TD children	Children with ASD	Emotional features	TD children	Children with ASD
	***P*-Value**			***P*-Value**	
Mean	**0.010**	0.547	MIN	**0.122**	0.953
Median	**0.003**	0.681	PSE	**0.040**	0.211
Mean I	**0.000**	0.387	Power	**0.017**	0.602
Mean II	**0.001**	0.767	Mean PSD	**0.026**	0.284
Mean III	**0.000**	0.438	Std PSD	0.534	0.640
Mean IV	**0.006**	0.971	Mean LF	**0.001**	0.361
Skewness	**0.007**	0.531	Mean HF	0.765	**0.000**
Kurtosis	**0.000**	0.152	SDLF	**0.012**	0.797
Q2	**0.003**	0.681	SDHF	**0.004**	**0.000**
Q3	**0.000**	0.057	Mean sum (LF +HF)	**0.052**	**0.006**
IQR	**0.009**	**0.019**	SD sum (LF +HF)	**0.002**	**0.000**
Harmonic mean	**0.031**	0.934	Mean ratio (LF/HF)	**0.005**	**0.000**
Trim mean	**0.005**	0.616			
Nan mean	**0.010**	0.547			
Max	**0.000**	0.469			
Min	**0.122**	0.953			

Bold values shows the significant features whose p-values less than 0.05

the features and provides better results in classification. Though the IMFs were generated to analyse the HRV data in low- and high-frequency components, it does not show much significance in *P*-test analyses as the frequency components of HHT are not very distinct in nature as in DWT. Hence, there is an overlap of frequency components in the order of IMFs which hinders it from giving better classification results.

The emotional features derived from the HHT of the HRV data were classified and the recognition rates are tabulated in Table 2. In this all the significant features were fused together which resulted in mean accuracy of 74.5% and 72% in children with ASD and TD using ensemble classifier which again proved to perform better than the KNN classifier. Trim mean has higher classification accuracy than all other

Table 2 Summary of positive and negative valence classification accuracy of TD children and children with ASD using HHT analysis of HRV data using KNN and ensemble classifiers

	Emotional features	Classifiers	Accuracy					
			typically developed children			Children with ASD		
			Like %	Dislike %	Average %	Like %	Dislike %	Average %
	All significant	KNN (1)	62	79	70.5	65	75	70
		Ensemble	**73**	**74**	**73.5**	**74**	**75**	**74.5**
Time domain statistical features	Mean	KNN (1)	60	80	70	55	74	64.5
		Ensemble	60	78	69	56	74	65
	Median	KNN (1)	74	57	65.5	55	73	64
		Ensemble	74	57	65.5	55	73	64
	Mean I	KNN (1)	62	76	69	59	78	68.5
		Ensemble	62	76	69	60	75	67.5
	Mean II	KNN (1)	68	77	72.5	59	71	65
		Ensemble	68	77	72.5	61	69	65
	Mean III	KNN (1)	60	73	66.5	64	72	68
		Ensemble	61	74	67.5	64	72	68
	Mean IV	KNN (1)	64	75	69.5	61	66	63.5
		Ensemble	64	75	69.5	61	66	63.5
	Skewness	KNN (1)	**72**	**78**	**75**	53	71	62
		Ensemble	**72**	**78**	**75**	56	71	63.5
	Kurtosis	KNN (1)	63	75	69	66	69	67.5
		Ensemble	63	75	69	66	70	68
	Q2	KNN (1)	65	77	71	55	73	64
		Ensemble	65	77	71	55	73	64
	Q3	KNN (1)	63	76	69.5	63	74	68.5
		Ensemble	63	76	69.5	63	70	66.5
	IQR	KNN (1)	60	75	67.5	**67**	**71**	**69**
		Ensemble	60	75	67.5	**67**	**71**	**69**
		KNN (1)	71	73	72	60	76	68

(continued)

Table 2 (continued)

	Emotional features	Classifiers	Accuracy					
			typically developed children			Children with ASD		
			Like %	Dislike %	Average %	Like %	Dislike %	Average %
	Harmonic mean	Ensemble	71	73	72	61	71	66
	Trim mean	KNN (1)	66	77	71.5	48	69	58.5
		Ensemble	66	77	71.5	48	69	58.5
	Nan mean	KNN (1)	60	80	70	53	70	61.5
		Ensemble	61	79	70	55	74	64.5
	Max	KNN (1)	52	76	64	54	72	63
		Ensemble	52	76	64	54	70	62
	Min	KNN (1)	68	79	73.5	59	65	62
		Ensemble	69	77	73	62	61	61.5
Non-linear features	PSE	KNN (1)	**69**	**77**	**73**	57	61	59
		Ensemble	69	75	72	62	59	60.5
Energy features	Power	KNN (1)	**62**	**84**	**73**	62	**76**	**69**
		Ensemble	71	77	74	63	74	68.5
Frequency domain features	Mean PSD	KNN (1)	63	78	70.5	57	72	64.5
		Ensemble	63	78	70.5	57	72	64.5
	Mean LF	KNN (1)	64	82	73	61	77	69
		Ensemble	66	81	73.5	61	78	69.5
	Mean HF	KNN (1)	66	77	71.5	68	70	69
		Ensemble	66	77	71.5	67	70	68.5
	SDLF	KNN (1)	**78**	**74**	**76**	67	70	68.5
		Ensemble	**78**	**74**	**76**	62	68	65
	SDHF	KNN (1)	72	74	73	61	77	69
		Ensemble	72	74	73	61	75	68
	Mean sum (LF+HF)	KNN (1)	60	70	65	56	79	67.5
		Ensemble	60	70	65	56	75	65.5
	SD sum (LF+HF)	KNN (1)	63	75	69	64	70	67
		Ensemble	63	75	69	64	70	67
	Mean ratio (LF/HF)	KNN (1)	63	78	70.5	**70**	**78**	**74**
		Ensemble	63	78	70.5	**70**	**78**	**74**
Principal component analysis (PCA)	All significant features	KNN (1)	**76**	**66**	**71.4**	**65**	**81**	**73.4**
		Ensemble	**78**	**76**	**77**	**75**	**81**	**78.3**

Bold values shows the significant features that achieved maximum mean accuracy

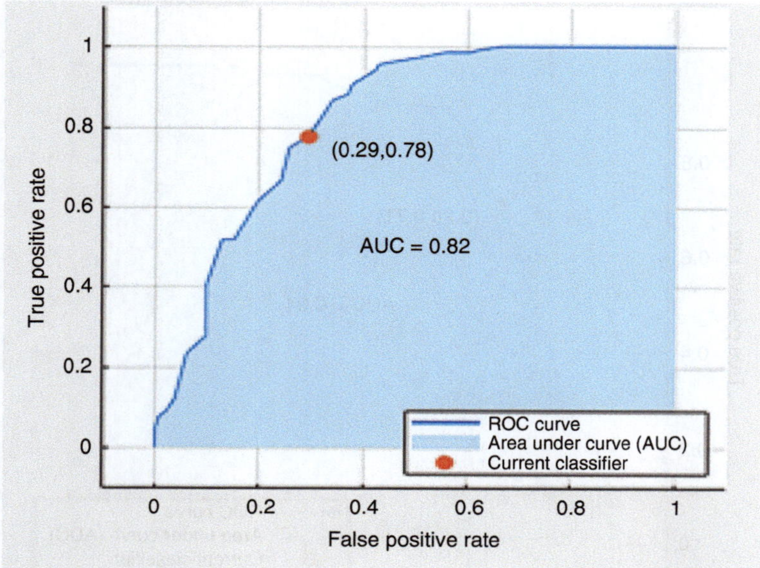

Fig. 8 AUC for children with ASD

features in TD children, whereas mean I and minimum have higher classification accuracies in children with ASD of 66.5% using KNN and ensemble classifiers.

Figure 8 shows that the area under the curve is 0.82 for children with ASD which shows a better performance of the ensemble classifier. It has achieved F1 score of 0.7315, sensitivity of 70.67% and specificity of 77.61% with an error rate of 0.2584. The computation time is observed as 2.4884 s, 11000 obs/s.

Figure 9 shows that the area under the curve is 0.84 for TD children which is proven to have achieved a better performance accuracy using ensemble classifier. It has achieved F1 score of 0.7650, sensitivity of 79.54% and specificity of 71.26% with an error rate of 0. 2457. The computation time is measured as 2.7562 s, 6900 obs/s.

It is observed from Fig. 10 that the ensemble classifier outperforms the KNN classifier by achieving an overall accuracy by combining all the significant features of 73.5% and 74.5% in TD children and children with ASD, respectively. The accuracy was improved to 77% and 78.3% in TD children and children with ASD after the application of PCA.

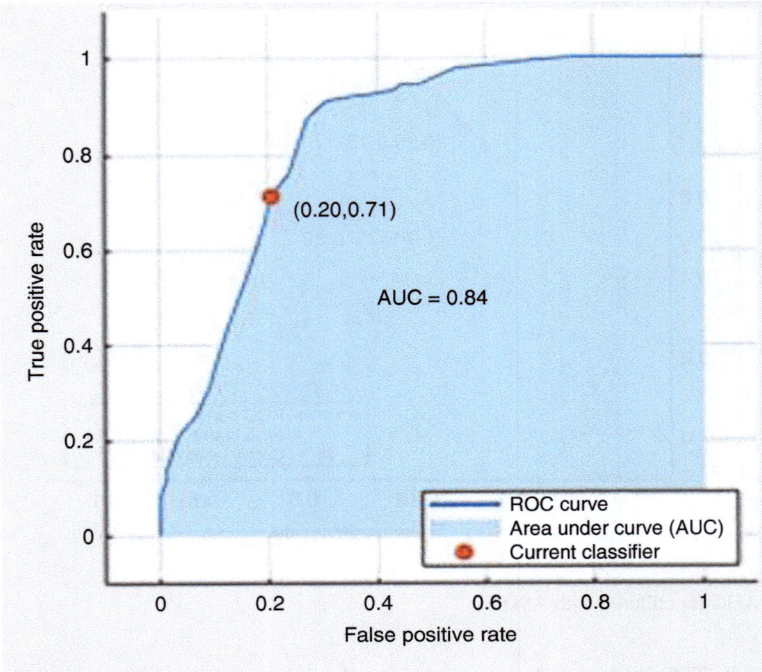

Fig. 9 AUC for TD children

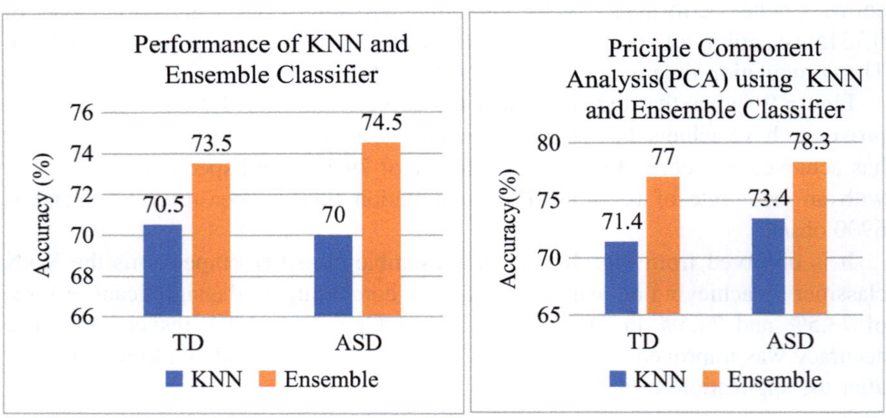

Fig. 10 Comparison of performance of KNN and ensemble classifiers using HHT analysis of HRV data

10 Discussion and Conclusion

In this paper, we discussed the detailed insights of the emotion elicitation protocol for children with ASD using the audio and video emotional cues, emotional ECG data collection, methodology and analysis of the data for devising the smart band for detecting the sudden emotional outburst in children with ASD using ECG signals that helps the caretakers, parents and doctors to prevent such extreme behaviours from happening in advance. The two basic emotional states (positive and negative) were considered for emotion detection in children with ASD. The raw emotional ECG data was acquired from 30 children (15 ASD and 15 TD) which were pre-processed using the digital filters. The emotional HRV was derived using the well-known Pan-Tompkins algorithm. The HRV was analysed using the time-frequency domain using HHT analyses, and it has been proven that the HHT analysis of the HRV data captured the valence information using the ratio of high- and low-frequency information, power of HRV, skewness, IQR and PSE which has achieved a good classification rate with an overall mean accuracy of 74.5 % and 73.5% in children with ASD and in TD children using the ensemble classifier. Further feature reduction was done using the principal component analysis (PCA), and it was seen that the accuracy improved to 77% and 78.3% in children with ASD and TD children, respectively, using the ensemble classifier. The results show that the children with ASD and TD have no much difference in emotion recognition of the basic emotions using the HHT of HRV data. Though the children with ASD are deficit in emotion recognition, we found no significant impairment in identifying the basic emotions. However, they were deficit in correctly predicting the positive valance state than the TD children, and individuals with ASD's emotion detection abilities may improve with age, even if they may not appear to reach the level of typically developing individuals.

Compliance of Ethical Standards

Conflicts of Interest
There are no conflicts of interest declared by the authors.

Ethical Consideration
Prior to conducting the trials, the protocol design and data acquisition process were approved by the Ethics Committee Board of the National Institute for Empowerment of Persons with Multiple Disabilities (NIEPMD), Chennai..

Ethical Approval ID: SE: -0101/2018

References

1. Baio, J., Wiggins, L., Christensen, D. L., Maenner, M. J., Daniels, J., Warren, Z., . . . Dowling, N. F. (2018). Prevalence of autism spectrum disorder among children aged 8 Years - Autism and developmental disabilities monitoring network, 11 Sites, United States, 2014. *MMWR Surveillance Summaries*, *67*(6), 1–23. doi: https://doi.org/10.15585/mmwr.ss6706a1
2. Lord, C., Cook, E. H., Leventhal, B. L., & Amaral, D. G. (2000). Autism spectrum disorders. *Neuron*, *28*, 355–363. doi: https://doi.org/10.4103/0253-7613.40480
3. Duffy, C., & Healy, O. (2011). Spontaneous communication in autism spectrum disorder: A review of topographies and interventions. *Research in Autism Spectrum Disorders*, *5*(3), 977–983. doi: https://doi.org/10.1016/j.rasd.2010.12.005
4. Grandgirard, J., Poinsot, D., Krespi, L., Nénon, J. P., & Cortesero, A. M. (2002). Costs of secondary parasitism in the facultative hyperparasitoid Pachycrepoideus dubius: Does host size matter? *Entomologia Experimentalis et Applicata*, *103*(3), 239–248. doi: https://doi.org/10.1023/A:1021193329749
5. Tracy, J. L., Robins, R. W., & Schriber, R. A. (n.d.). *in press,*.
6. Ozonoff, S., Pennington, B. F., & Rogers, S. J. (1990). Are there Emotion Perception Deficits in Young Autistic Children? *Journal of Child Psychology and Psychiatry*, *31*(3), 343–361. doi: https://doi.org/10.1111/j.1469-7610.1990.tb01574.x
7. Piggot, J., Kwon, H., Mobbs, D., Blasey, C., Lotspeich, L., Menon, V., . . . Reiss, A. L. (2004). Emotional attribution in high-functioning individuals with autistic spectrum disorder: A functional imaging study. *Journal of the American Academy of Child and Adolescent Psychiatry*, *43*(4), 473–480. doi: https://doi.org/10.1097/00004583-200404000-00014
8. Enticott, P. G., Kennedy, H.A., Johnston, P. J., Rinehart, N. J., Tonge, B. J., Taffe, J. R., & Fitzgerald, P. B. (2013). Emotion recognition of static and dynamic faces in autism spectrum disorder. *Cognition & Emotion*, *00*(February), 1–9. doi: https://doi.org/10.1080/02699931.2013.867832
9. Hess, U., & Blairy, S. (2001). *Facial mimicry and emotional contagion to dynamic emotional facial expressions and their influence on decoding accuracy*. 129–141.
10. Kim, J., & André, E. (2008). Emotion recognition based on physiological changes in music listening. *IEEE Transactions on Pattern Analysis and Machine Intelligence*, *30*(12), 2067–2083. doi: https://doi.org/10.1109/TPAMI.2008.26
11. Oberman, L. M., Winkielman, P., & Ramachandran, V. S. (2009). Slow echo: Facial EMG evidence for the delay of spontaneous, but not voluntary, emotional mimicry in children with autism spectrum disorders. *Developmental Science*, *12*(4), 510–520. doi: https://doi.org/10.1111/j.1467-7687.2008.00796.x
12. Clark, T. F., Winkielman, P., & Mcintosh, D. N. (n.d.). *Autism and the Extraction of Emotion From Briefly Presented Facial Expressions: Stumbling at the First Step of Empathy*. doi: https://doi.org/10.1037/a0014124
13. Kuusikko, S., Haapsamo, H., Jansson-Verkasalo, E., Hurtig, T., Mattila, M. L., Ebeling, H., . . . Moilanen, I. (2009). Emotion recognition in children and adolescents with autism spectrum disorders. *Journal of Autism and Developmental Disorders*, *39*(6), 938–945. doi: https://doi.org/10.1007/s10803-009-0700-0
14. Murugappan, M., Juhari, M. R. B. M., Nagarajan, R., & Yaacob, S. (2009). An investigation on visual and audiovisual stimulus based emotion recognition using EEG. *International Journal of Medical Engineering and Informatics*, *1*(3), 342. doi: https://doi.org/10.1504/IJMEI.2009.022645
15. Quintana, D. S., Guastella, A. J., Outhred, T., Hickie, I. B., & Kemp, A. H. (2012). Heart rate variability is associated with emotion recognition: Direct evidence for a relationship between the autonomic nervous system and social cognition. *International Journal of Psychophysiology*, *86*(2), 168–172. doi: https://doi.org/10.1016/j.ijpsycho.2012.08.012
16. Rijn van, S., Zantinge, G., Stockmann, L., & Swaab, H. (2017). Psychophysiological Responses to Emotions of Others in Young Children with Autism Spectrum Disorders : Correlates of

Social Functioning. *International Society for Autism Research*, *10*(9), 1499–1509. doi: https://doi.org/10.1002/aur.1794
17. Sasikumar, K., George Priya Doss C, & Adalarasu, K. (2015). Analysis of physiological signal variation between autism and control group in south Indian population. *Biomedical Research*, *26*(3), 525–529.
18. Selvaraj, J., Murugappan, M., Wan, K., & Yaacob, S. (2013a). Classification of emotional states from electrocardiogram signals: a non-linear approach based on Hurst. *Biomedical Engineering Online*, *12*(1), 44. doi: https://doi.org/10.1186/1475-925X-12-44
19. Murugappan, M., Alshuaib, W. B., Bourisly, A., Sruthi, S., Khairunizam, W., Shalini, B., & Yean, W. (2020). Emotion Classification in Parkinson's Disease EEG using RQA and ELM. *Proceedings - 2020 16th IEEE International Colloquium on Signal Processing and Its Applications, CSPA 2020*, (February), 290–295. doi: https://doi.org/10.1109/CSPA48992.2020.9068709
20. Bal, E., Harden, E., Lamb, D., Van Hecke, A. V., Denver, J. W., & Porges, S. W. (2010). Emotion recognition in children with autism spectrum disorders: Relations to eye gaze and autonomic state. *Journal of Autism and Developmental Disorders*, *40*(3), 358–370. doi: https://doi.org/10.1007/s10803-009-0884-3
21. Bedford, R., Gliga, T., Shephard, E., Elsabbagh, M., Pickles, A., Charman, T., & Johnson, M. H. (2017). Neurocognitive and observational markers: Prediction of autism spectrum disorder from infancy to mid-childhood. *Molecular Autism*, *8*(1), 1–10. doi: https://doi.org/10.1186/s13229-017-0167-3
22. Goshvarpour, A., Abbasi, A., & Goshvarpour, A. (2017). ScienceDirect An accurate emotion recognition system using ECG and GSR signals and matching pursuit method. *Biomedical Journal*, *40*(6), 1–14. doi: https://doi.org/10.1016/j.bj.2017.11.001
23. Krupa, N., Anantharam, K., Sanker, M., Datta, S., & Sagar, J. V. (2016). Recognition of emotions in autistic children using physiological signals. *Health and Technology*, *6*(2), 137–147. doi: https://doi.org/10.1007/s12553-016-0129-3
24. Palma, S. Di, Tonacci, A., Domenici, C., Pioggia, G., Muratori, F., & Billeci, L. (2017). Monitoring of autonomic response to sociocognitive tasks during treatment in children with Autism Spectrum Disorders by wearable technologies : a feasibility study. *Computers in Biology and Medicine*, *85*, 143–152. doi: https://doi.org/10.1016/j.compbiomed.2016.04.001
25. Chambers, A. S., & Allen, J. J. B. (2007). Cardiac vagal control, emotion, psychopathology, and health. *Biological Psychology*, *74*(2), 113–115. doi: https://doi.org/10.1016/j.biopsycho.2006.09.004
26. Murukesan, L., Murugappan, M., Iqbal, M., & Saravanan, K. (2014). Machine learning approach for sudden cardiac arrest prediction based on optimal heart rate variability features. *Journal of Medical Imaging and Health Informatics*, *4*(4), 521–532. doi: https://doi.org/10.1166/jmihi.2014.1287
27. Bazelmans, T., Jones, E. J. H., Ghods, S., Corrigan, S., Toth, K., Charman, T., & Webb, S. J. (2019). Heart rate mean and variability as a biomarker for phenotypic variation in preschoolers with autism spectrum disorder. *Autism Research*, *12*(1), 39–52. doi: https://doi.org/10.1002/aur.1982
28. Sheinkopf, S. J., Neal-Beevers, A. R., Levine, T. P., Miller-Loncar, C., & Lester, B. (2013). Parasympathetic response profiles related to social functioning in young children with autistic disorder. *Autism Research and Treatment*, *2013*(September 2013), 868396. doi: https://doi.org/10.1155/2013/868396
29. N. Huang, Z. Shen, S. Long, M. Wu, H. H. Shih, N. C. Yen, C. Tung, and H. L. (1998). The empirical mode decomposition and the Hilbert spectrum for nonlinear and non-stationary time series analysis. *Proceedings of Royal Society, London*, *454*, 903–995.
30. Selvaraj, J., Murugappan, M., Wan, K., & Yaacob, S. (2013b). Electrocardiogram-based emotion recognition system using empirical mode decomposition and discrete Fourier transform. *Expert Systems*, *31*(2), 110–120. doi: https://doi.org/10.1111/exsy.12014
31. Murugappan, M., Murugesan, L., Jerritta, S., & Adeli, H. (2021). Sudden Cardiac Arrest (SCA) Prediction Using ECG Morphological Features. *Arabian Journal for Science and Engineering*, *46*(2), 947–961. doi: https://doi.org/10.1007/s13369-020-04765-3

Detection of Tonic-Clonic Seizures Using Scalp EEG of Spectral Moments

Joseph Mathew, N. Sivakumaran, P. A. Karthick, and Toshihisa Tanaka

Abstract Epilepsy is a brain disorder characterized by recurring excessive neuronal discharge called seizure. These seizures are exhibited by transient episodes of motor, sensory, or psychic dysfunction. Tonic-clonic seizure (TCSZ), a type of electroclinical seizure, has a high risk of injury due to the convulsions. Video EEG is an essential tool, employed in the hospitals for the identification of those seizures. This is not practical outside the clinical settings. Therefore, this study uses scalp EEG of TUH (Temple University Hospital) to develop the machine learning models for the identification of TCSZ. The onset channels of seizures with clinical manifestation (electroclinical) and without clinical manifestations (electrographic), having a period of 4 s, are selected from 28 subjects. The lower- and higher-order spectral moments such as centroid, spread, skewness, and kurtosis are estimated from the distribution of the power spectrum. Different window functions (Hamming, Hann, Rectangular, and Flattop) are also employed to extract these features. The statistical test was conducted and these features were found to be significantly distinct ($p < 0.05$). RBF (radial basis function)-based SVM (support vector machine) is employed to discriminate the electroclinical seizure from electrographic seizure. The model is validated using tenfold cross-validation. The results show that the proposed model detects TCSZ with an accuracy of 88.89%, specificity of 84.33%,

J. Mathew (✉)
Department of Instrumentation and Control, National Institute of Technology, Tiruchirappalli, India

Department of Applied Electronics and Instrumentaion, Rajagiri School of Engineering Technology, Kochi, Kerela, India

N. Sivakumaran · P. A. Karthick
Department of Instrumentation and Control, National Institute of Technology, Tiruchirappalli, India

T. Tanaka
Department of Electrical and Electronic Engineering, Tokyo University of Agriculture and Technology, Fuchu, Japan

M. Murugappan, R. Yuvaraj (eds.), *Biomedical Signals Based Computer-Aided Diagnosis for Neurological Disorders*, https://doi.org/10.1007/978-3-030-97845-7_5

and sensitivity of 90.67%. These approaches hold a great potential and could be an aid for identifying the electroclinical seizures.

Keywords Epilepsy · Machine learning · Seizure

1 Introduction

Epilepsy is a neurological decease, characterized by an enduring predisposition to generate epileptic seizures [1]. This affects individuals of any age and ethnicity. According to the World Health Organization (WHO), approximately 50 million people worldwide suffer from epilepsy [2]. An epileptic seizure is a brief occurrence of signs and/or symptoms caused by abnormally high or synchronous neuronal activity in the brain. [3]. ILAE (International League Against Epilepsy) has categorized seizures into many types based on the localization and manifestations. These seizures can be focal, generalized, or unknown (based on localization) and motor or non-motor (based on manifestations). Seizures with clinical manifestation are generally known as electroclinical seizures, and seizures without clinical manifestation are known as electrographic seizures. Most of the electroclinical seizures are considered as severe seizures because of the risks and injuries associated with them.

Tonic-clonic seizure is the best example of electroclinical seizures. These seizures have a tonic phase, which is characterized by an increase in muscle tone, followed by a clonic phase, which is characterized by bilateral symmetric jerking of the extremities [4]. Normally, these seizure manifestations markedly last 1–3 minutes [3]. During this time, the patients are at the highest risk of seizure-related injuries and even premature deaths. However, electrographic seizures such as focal and generalized non-specified seizures (defined by Temple University Hospital data) do not exhibit any clinical manifestations, and they are not considered as urgent seizures. From the literature, it is clear that 30% of patients are considered as surgical candidates. They are unresponsive to antiepileptic medication [5], and hence they suffer from a number of physical and physiological complications, which worsen when uncontrolled seizures occur [6]. Such circumstances will have a significant impact on the patient's safety and quality of life. Therefore, it is necessary to detect the electroclinical seizures from electrographic seizures.

Seizure identification and reliable assessments are critical for epilepsy diagnosis, syndrome evaluation, treatment selection, and prognosis [7]. Nowadays, there are different modalities that are available for the diagnosis of seizures. Electroencephalography (EEG) and neuroimaging, primarily magnetic resonance imaging (MRI), are the two prominent tools used for the identification of seizures as well as other neurological disorders [8]. Epilepsy is diagnosed primarily on clinical ground, and video EEG is also used for this purpose [9]. Although video-based EEG is potentially viable to use in the clinical settings, it is not practical outside the clinical setup.

As a solution, researchers have developed different warning and alert systems to prevent the complication of electroclinical seizures. To some extent, these devices have assisted the patient in increasing his or her sense of security and independence, reducing anxiety, and avoiding seizure-related injuries [10]. Furthermore, early seizure prediction would allow for interventions such as fast-acting antiepileptic drugs, warning, vagus nerve stimulation, focal cooling, and deep brain stimulation [11]. In this regard, video-based alert systems are also being developed. To develop automated seizure detection systems, researchers focused primarily on motion-strength, motion-trajectory [7], or average differential luminance signals in video segments. However, performance was sometimes hampered by algorithms' inability to generalize and provide good results in the presence of changing luminance or patient occlusion, such as patients covered by a blanket [7]. Other modalities, such as accelerometry [12], electrodermal activity and electromyogram, magnetic sensors, piezoelectric sensors, and smart watches, have also been investigated for this purpose [13–16].

For the past decades, several researches have been carried out for the detection of seizures. The author [17] analyzed and proposed a model for effective use of EEG signals for automatic seizure detection using various computational and statistical techniques. Furthermore, researchers investigated various linear and nonlinear methods for automatic seizure detection. Examples include time domain, frequency domain, time-frequency domain [18], nonlinear methods, and empirical mode decomposition (EMD) [19]. They extracted numerous temporal and spectral features from scalp and intracerebral EEG (iEEG) signals to distinguish between ictal and interictal states and predict impending seizures [20]. Approaches based on machine learning and deep learning [21] have been used for feature extraction as well as classification [22]. However, only a few studies have specifically focused on the detection of electroclinical seizures [5, 23, 24].

Therefore, this study aims to distinguish the electroclinical seizures from electrographic seizures, utilizing the statistical measures of scalp EEG such as lower- and higher-order moments in the spectral domain. Our work is inspired by the potential benefits of assisting at-risk patients in both hospital and home settings.

2 Methods

The scalp EEG signals used in this study were obtained from the Temple University Hospital (TUH) database, which is open to the public. It is one of the largest databases and contains a wide range of seizure types. Focal non-specified seizure (FNSZ), generalized non-specified seizure (GNSZ), and tonic-clonic seizure (TCSZ) were chosen in order from a total of 28 patients. FNSZ is found in the hemispheric or focal part of the brain and has been linked to a wide range of seizure etiologies. The morphology of EEG signals is the primary indicator of a seizure event, and the seizure will be characterized by a spike and slow wave or poly spike and slow wave complex [25]. GNSZ have the same morphology, evolution, and frequency descriptors as FNSZ, but they cover a larger area of the skull. As a result, for the analysis, we combined FNSZ and GNSZ and referred to them as electrographic seizures (EGSZ). Tonic-clonic seizures are a type of seizure that is distinguished by muscle tension and stiffening followed by violent convulsions and jerking [26]. As a result, such seizures cannot be identified in the absence of clinical manifestations. TCP montage was used to optimize and convert the monopolar channels to bipolar

[27]. In this study, only seizures recorded at a sampling rate of 400 Hz were considered. This criterion was met by 64 seizures.

2.1 Feature Extraction

Spectral analysis is one of the important methods employed for the EEG analysis. Power spectral density (PSD) is an important information that can be retrieved from EEG signal via spectral analysis [28]. This describes the signal's power as a function of frequency, per unit frequency. PSD of the EEG signal is considered as a reliable and meaningful information to understand the underlying function of the brain activity [29]. Many methods are used to compute the spectral properties of the signal. Fourier transform, short-time Fourier transform (STFT), and wavelet approaches [30] are used for this purpose. Different window functions are designed to extract the spectral features from the signals. For this, signals are divided into different segments and from each segment, spectral features can be extracted. But the researchers confront a series of choices of experimental factors such as windowing function types and frequency resolution. In this study, four types of windowing functions, namely, Hann, Hamming, Rectangular, and Flattop with different parameter settings, are utilized to compute the power density from the signals.

2.2 Rectangular Window

A Rectangular window is a function that is constant within the interval and zero elsewhere. The normal Rectangular time window expression is defined as [31]

$$W(n) = R_N(n)\begin{cases} 1, & 0 \leq n \leq N-1 \\ 0, & \text{otherwise} \end{cases} \tag{1}$$

2.3 Hanning Window

Because of its shape and form, the Hanning window is also known as a raised-cosine window. It is named after Julius von Hann and is also known as Hann. The Hanning window function has the advantage of reducing the side lobes. The merit of using Hanning window function is that it has narrow side lobes. It outperforms other window methods such as the Rectangular window in terms of frequency resolution. The Hann window takes the form as [32]

$$W(n) = 0.5\left[1 - \cos\left[2\pi\frac{n}{N+1}\right]\right]R_N(n), \quad n = 1, 2, \ldots N \tag{2}$$

2.4 Hamming Window

The Hamming window was proposed by Richard W Hamming. The window is used to optimize to nearest side lobes. For this purpose, its height is adjusted to be about one-fifth that of Hann window [25]. The expression is given by

$$W(n) = 0.54 - 0.46\ \cos\left(\frac{2\pi n}{N+1}\right), n = 1, 2, \ldots N \tag{3}$$

2.5 Flattop Window

Flattop windows are partially negative valued windows with low scalloping loss in the frequency domain. In spectrum analyzers, such windows are available for measuring the amplitudes of sinusoidal frequency components [32]:

$$W(n) = 0.215 - 0.416\ \cos\left(\frac{2\pi n}{N}\right) + 0.136\cos\left(\frac{4\pi n}{N}\right) - 0.01\ \cos\left(\frac{6\pi n}{N}\right), \quad n = 1, 2, \ldots N \tag{4}$$

Figure 1 illustrates the time and frequency domain window functions with a size of 0.5 s.

2.6 Spectral Moments

The spectral moments can also be viewed as a way to use statistics to analyze the power spectrum [33]. Moments are a set of statistical parameters used to describe the distribution. Mean, variance, skew, and kurtosis are four important measures of the distribution. The mean and variance are considered as first- and second-order moments, and skew and kurtosis are regarded as higher-order moments, namely, third- and fourth-order moments. The mathematical equations of statistical moments are

$$\text{Mean}(\mu) = \int xP(x)dx \tag{5}$$

$$\text{Variance}(\sigma^2) = \int (x-\mu)^2 P(x)dx \tag{6}$$

$$\text{Skew}(\varsigma) = \frac{1}{\sigma^3}\int (x-\mu)^3 P(x)dx \tag{7}$$

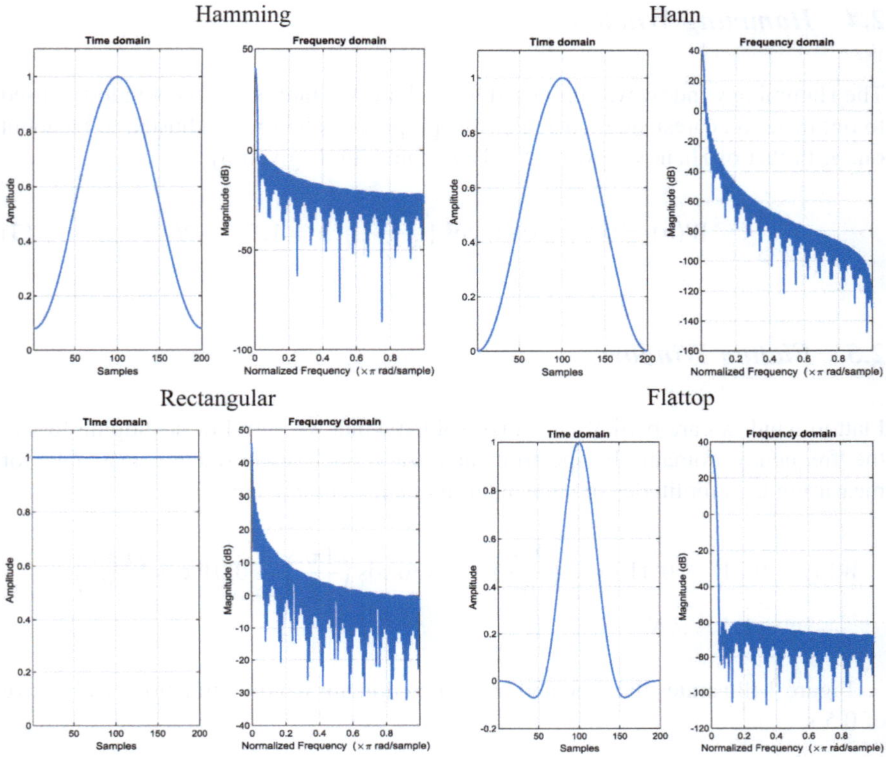

Fig. 1 Window functions in time and frequency domain with a size of 0.5 s

$$\text{Kurtosis}(\varsigma) = \frac{1}{\sigma^4} \int (x - \mu)^4 P(x) dx \tag{8}$$

The mean is the average value of x, and the variance is the spread around the mean. The skew is a symmetry metric. If the value of skew is large in magnitude, it implies that the density has an asymmetric spread around its mean. Suppose any distribution is said to be symmetric, then the skew value will be zero. Kurtosis is a measure of peakedness [34]. Similarly, spectral moments can be defined based on the PSD of the signal. The mean frequency of the signal in the defined time interval is interpreted as the first-order spectral moment. To obtain higher-order shape information, the central moments must be defined. The variance of the power spectrum, or the squared bandwidth, can be interpreted as the second-order spectral moment. The spectral skewness is the third order, and the spectral kurtosis is the fourth order.

Spectral centroid is the barycenter of the spectrum. It is computed by considering the spectrum as a distribution. In this study, spectral centroid is calculated using the bins of the power spectrum created by the window function. The expression is defined as [35]

$$\text{Spectral centroid} = \frac{\sum_{k=b_1}^{b_2} f_k s_k}{\sum_{k=b_1}^{b_2} s_k} \tag{9}$$

The spread of the spectrum around its mean value is denoted by spectral spread and it is defined by the equation [28]:

$$\text{Spectral spread} = \sqrt{\frac{\sum_{k=b_1}^{b_2} (f_k - \mu_1)^2 s_k}{\sum_{k=b_1}^{b_2} s_k}} \tag{10}$$

Spectral skewness is a measure of a distribution's asymmetry around its mean. This is computed using the equation [35]:

$$\text{Spectral Skewness} = \frac{\sum_{k=b_1}^{b_2} (f_k - \mu_1)^3 s_k}{(\mu_2)^3 \sum_{k=b_1}^{b_2} s_k} \tag{11}$$

Spectral kurtosis is a measure of the distribution's flatness around its mean value. The equation is given by [35]

$$\text{Spectral Kurtosis} = \frac{\sum_{k=b_1}^{b_2} (f_k - \mu_1)^4 s_k}{(\mu_2)^4 \sum_{k=b_1}^{b_2} s_k} \tag{12}$$

f_k is the frequency in Hz correspondint to bin k, s_k is the spectral value at bin k, b_1 and b_2 are the band edges in bins, μ_1 is the spectral centroid, and μ_2 is the spectral spread.

2.7 Classification

Support vector machine (SVM) is a supervised linear classification algorithm that could mainly be used for the classification and regression tasks. This finds an optimal decision boundary with the help of support vectors where the data can be separated. In order to classify nonlinear features [36], SVM uses kernel-based approaches. In this study, radial basis function (RBF) is used as kernel. The mathematical expression of RBF is described as

$$k(x,y) = \exp\left(-\|x-y\|^2\right) \tag{13}$$

where x and y represent data.

The sensitivity, specificity, accuracy, and F-score are used to evaluate performance of the model:

$$\text{Sensitivity(SN)} = \text{TP}/(\text{TP}+\text{FN}) \tag{14}$$

$$\text{Specificity(SP)} = \text{TN}/(\text{TN}+\text{FP}) \tag{15}$$

$$\text{Precision (PPV)} = \text{TP}/(\text{TP}+\text{FP}) \tag{16}$$

$$\text{Accuracy(AC)} = \frac{\text{TP}+\text{TN}}{(\text{TP}+\text{TN}+\text{FP}+\text{FN})} \tag{17}$$

$$F-\text{Score(FS)} = \frac{\text{TP}}{\text{TP}+1/2(\text{FP}+\text{FN})} \tag{18}$$

where TP = true positive, TN = true negative, FP = false positive, and FN = false negative.

3 Results and Discussion

There are 476 onset channels involved in 64 seizures. Among these, 83 channels are associated with TCSZ, while the remaining channels are associated with electrographic seizures, specifically FNSZ and GNSZ. The representatives of sample signal with 4-s length during the ictal period from each class are shown in Fig. 2. From this, it is evident that the magnitude of TCSZ (~200 μV) is much higher than that of EGSZ (~30 μV). The reason could be the excessive, hyper-synchronous discharge of neurons in the brain which cause bilateral symmetric jerking of the extremities during TCSZ. That is why TCSZ is considered as clinically urgent seizures.

Moments are a set of statistical parameters used to measure a distribution. The theory of moments has been extensively used in several fields [33]. This has been

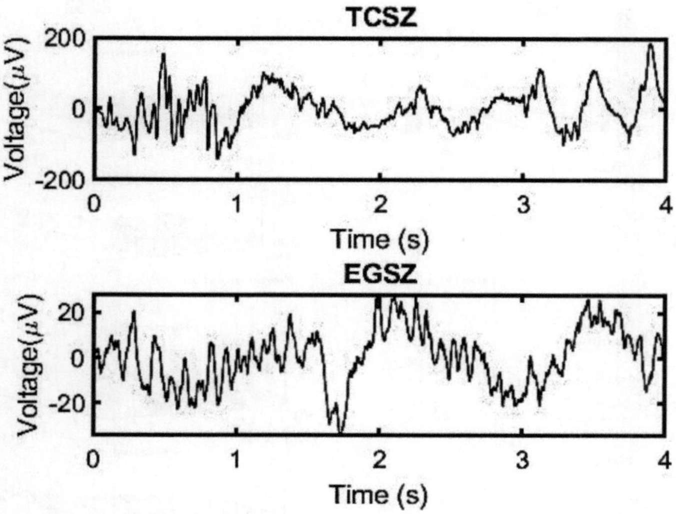

Fig. 2 Representative signals of TCSZ and EGSZ

considered as an effective approach for the feature extraction. Spectral moments are extracted the same way as the moment computed. The basic difference is that while moments are computed in time domain, the spectral moments are computed in temporal-spectral domain. In this study, spectral parameter such as power is utilized to extract spectral moments. Both lower- and higher-order spectral moments, namely, spectral centroid, spectral spread, spectral skewness, and spectral kurtosis, are extracted from the signals for the classification task. For this purpose, PSD is computed from the samples of 4-s duration. Different windowing functions, namely, Hamming, Hann, Rectangular, and Flattop window functions, are utilized to compute the PSD of the signal. The spectrogram obtained from the repetitive signals of TCSZ and EGSZ is presented in Figs. 2, 3, 4, 5, and 6.

Figure 2 presents the power distribution of signals as mentioned in Fig. 1. The window size used here is 1 s and overlapping length is zero percentage. In this, the Rectangular window exhibits more sub-band energy than other window functions. In both TCSZ and TCSZ, the Rectangular window function extracted more energy than other window functions. The percentage of difference in power of TCSZ is 16% (Hann), 15% (Hamming), and 24% (Flattop) with respect to Rectangular window. The average power of EGSZ obtained by the Rectangular window function from 0 to 100 Hz of TCSZ is −21.23 dB for Hann, −21.1 dB for Hamming, −18.30 dB for Rectangular, and −22.71 dB for Flattop. But most of the window function draws maximum energy from lower-frequency bands of TCSZ and EGSZ. It is also found that the energy of EGSZ is less than that of TCSZ. Rectangular function of EGSZ and TCSZ has drawn highest energy in the range of −110 dB.

In Fig. 3 the overlapping length is varied to 25% by keeping window length as 1 s. Then total power in TCSZ and EGSZ using Rectangular window is increased to

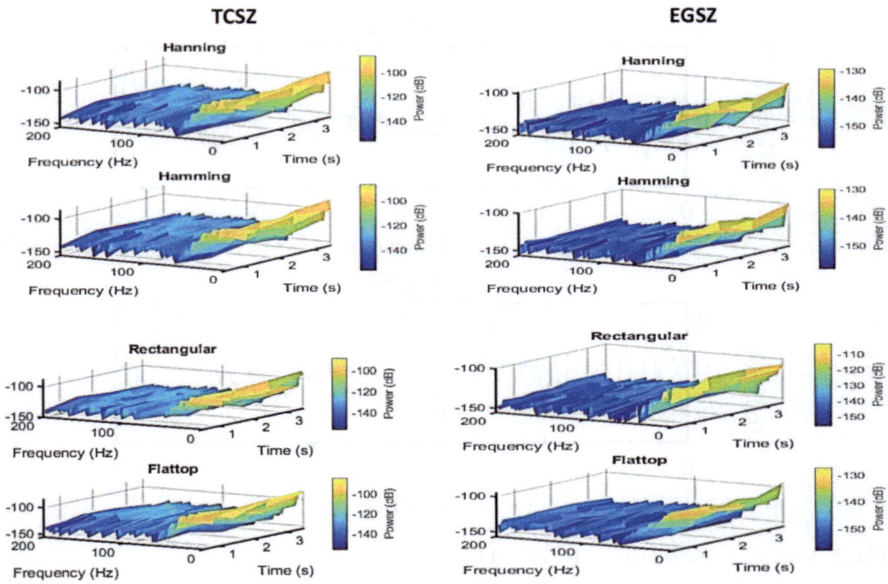

Fig. 3 Spectrogram of representative signals (winsize: 1 s, no overlap)

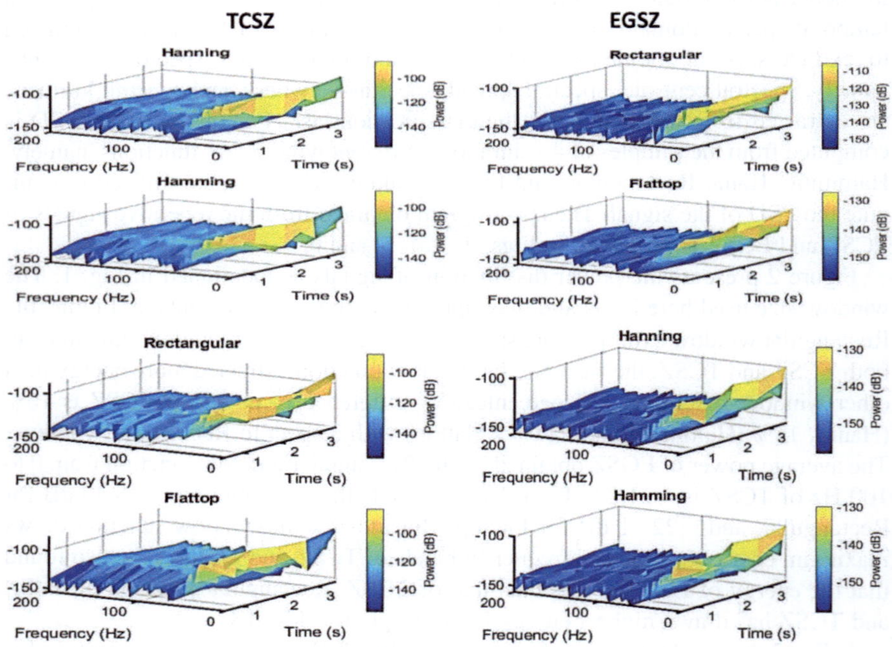

Fig. 4 Spectrogram of representative signals (winsize, 1 s; overlap, 25%)

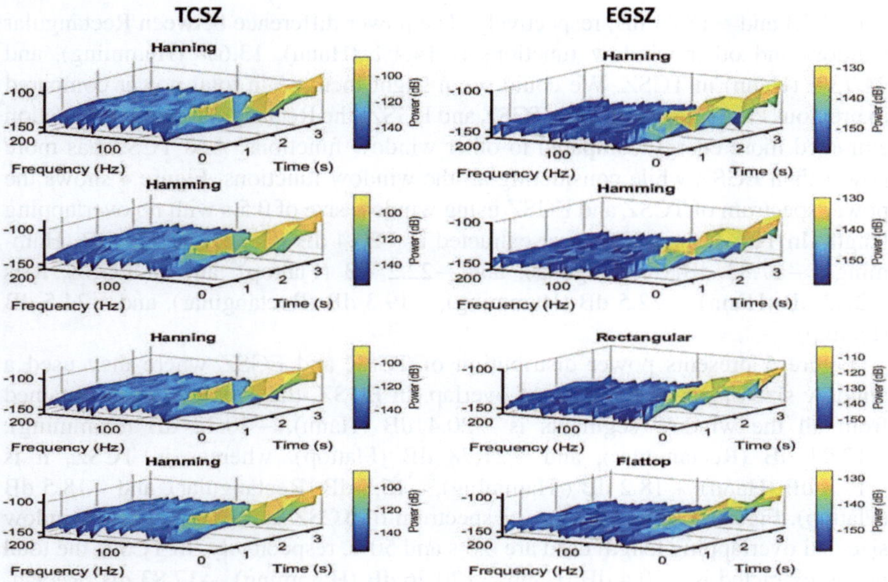

Fig. 5 Spectrogram of representative signals (winsize :1 s, overlap:50%)

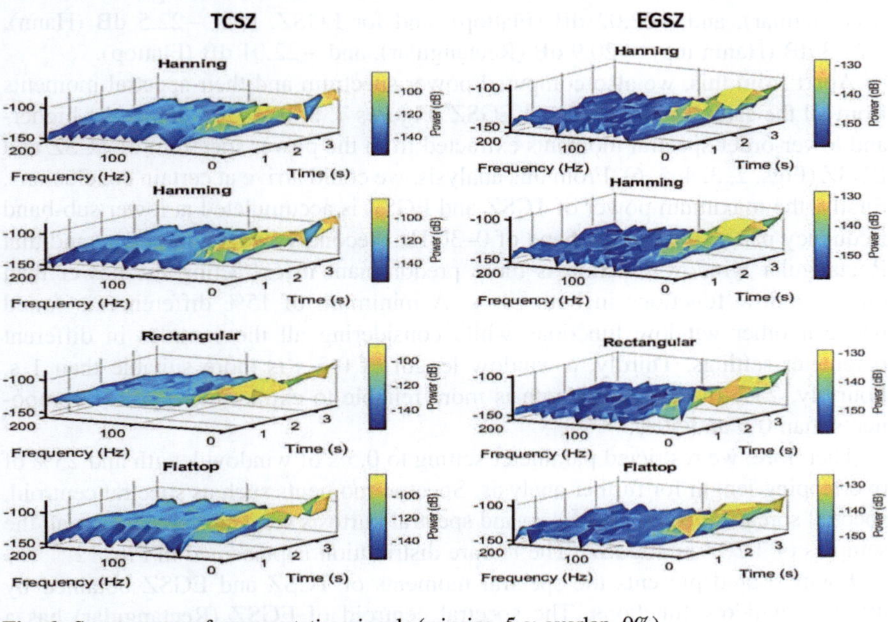

Fig. 6 Spectrogram of representative signals (winsize, 5 s; overlap, 0%)

−17.8 dB and −18.01 dB, respectively. The power difference between Rectangular window and other window functions is 14.4% (Hann), 13.6% (Hamming), and 24.73% (Hann) in TCSZ. We could see a slight increase in total power compared to previous case (Fig. 2). In both TCSZ and EGSZ, the Rectangular window function extracted more energy compared to other window functions. And TCSZ has more power than EGSZ while considering all the window functions. Figure 4 shows the power spectrum of TCSZ and EGSZ using window size of 0.5 s with no overlapping length. In TCSZ, the total power extracted is −20.4 dB (Hann), −20.26 dB (Hamming), −17.83 dB (Rectangular), and −22.2 dB (Flattop), and in EGSZ, it is −22.3 dB (Hann), −22.5 dB (Hamming), −19.3 dB (Rectangular), and −24.5 dB (Flattop).

Figure 5 presents power distribution of TCSZ and EGSZ where they used a window size of 0.5 s with 25% of overlap. In EGSZ, the average power obtained from all the window segments is −20.4 dB (Hann), −20.32 dB (Hamming), −17.93 dB (Rectangular), and −21.78 dB (Flattop), whereas in TCSZ, it is −17.8 dB (Hann), −18.2 dB (Hamming), −15.5 dB (Rectangular), and −18.5 dB (Flattop). Figure 5 shows the power spectrum of TCSZ and EGSZ where window size and overlapping length used are 0.5 s and 50%, respectively. In TCSZ, the total power extracted is −20.4 dB (Hann), −20.26 dB (Hamming), −17.83 dB (Rectangular), and −22.2 dB (Flattop), and in EGSZ, it is −22.3 dB (Hann), −22.5 dB (Hamming), −19.3 dB (Rectangular), and −24.5 dB (Flattop). Figure 6 describes the total power of TCSZ is −20.4 dB (Hann), −20.29 dB (Hamming), −17.92 dB (Rectangular), and −22.02 dB (Flattop), and for EGSZ, it is −22.5 dB (Hann), −22.3 dB (Hamming), −20.9 dB (Rectangular), and −22.01 dB (Flattop).

Apart from this, we also computed power spectrum and their spectral moments from all the samples of TCSZ and EGSZ. Figures 7, 8, 9, 10, 11 depict the higher- and lower-order spectral moments extracted from the power spectrum of TCSZ and EGSZ (Figs. 2, 3, 4, 5, 6). From this analysis, we could arrive at certain conclusions. Firstly, the maximum power of TCSZ and EGSZ is accumulated at lower sub-band frequency particularly in the band of 0–30 Hz. Secondly, we could understand that Rectangular window function is more predominant in extracting the power than other window functions in both cases. A minimum of 15% difference is found between other window functions while considering all the samples in different parameter settings. Thirdly, a window length of 0.5 s is more suitable than 1 s. Fourthly, 25% overlapping length is more reliable to extract more power components than 0% and 50%.

Therefore, we restricted parameter setting to 0.5 s of window length and 25% of overlapping length for further analysis. Spectral moments such as spectral centroid, spectral spread, spectral skewness, and spectral kurtosis were computed from all the samples of TCSZ and EGSZ. The feature distribution is presented in Fig. 12.

Figure 12a–d presents the spectral moments of TCSZ and EGSZ obtained by different window functions. The spectral centroid of EGSZ (Rectangular) has a median of 10.1 (range: 4.7–91.1) and TCSZ has a median of 11.4 (range: 4.1–98.2). However, the spectral centroid of Hann and Hamming windows have almost the same median and range (median, 9.3; range, 3.5–90.7 for EGSZ and

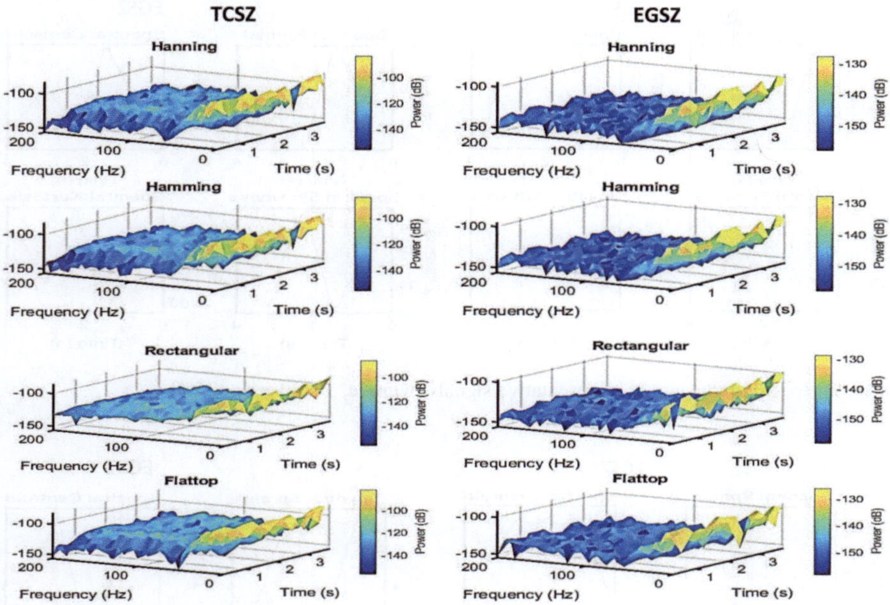

Fig. 7 Spectrogram of representative signals (winsize, 5 s; overlap, 25%)

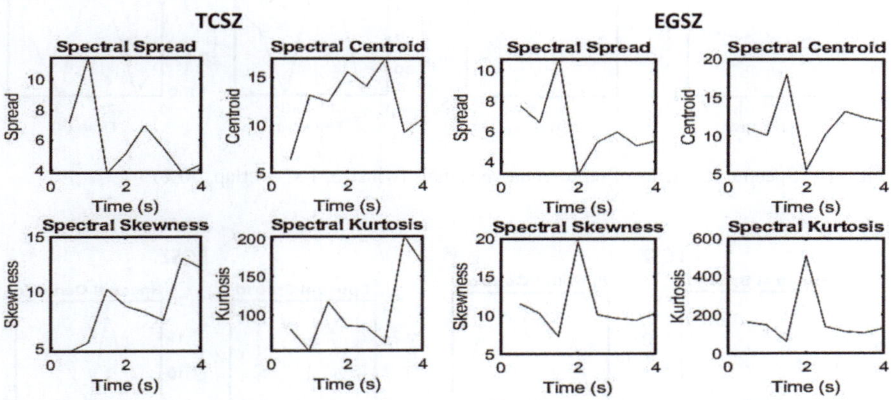

Fig. 8 Spectral moments of representative signals (winsize: 1 s, no overlap)

median, 10.3; range, 2.6–97.5 for TCSZ). The spectral centroid of Flattop has the highest median of 10.7 (range: 4.2–90.2) for EGSZ and 11.7 (range: 3.9–94.2) for TCSZ compared to other window functions. With respect to spectral spread, the highest median values are found in Rectangular window function (median, 17.1; range, 7.6–60.4 for EGSZ and 20.3 (9.9–60.4)). The statistical parameters of spectral moments (median and range) remained as higher for TCSZ than EGSZ for all the window functions. The boxplot diagrams of feature distributions are shown in Figure 13. The details of statistical parameters are presented in Table 1.

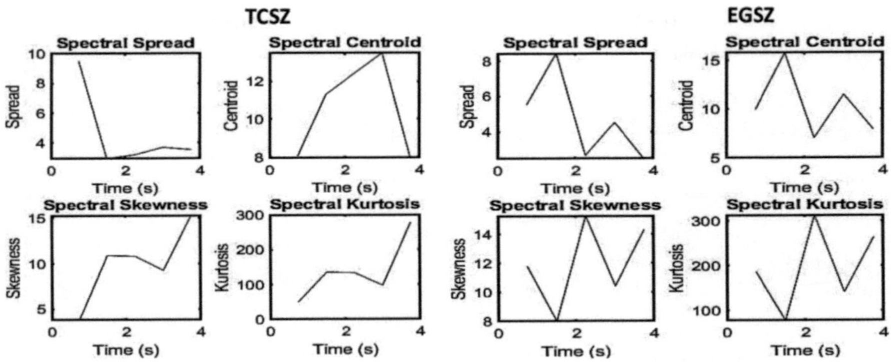

Fig. 9 Spectral moments of representative signals (winsize, 1 s; overlap, 25%)

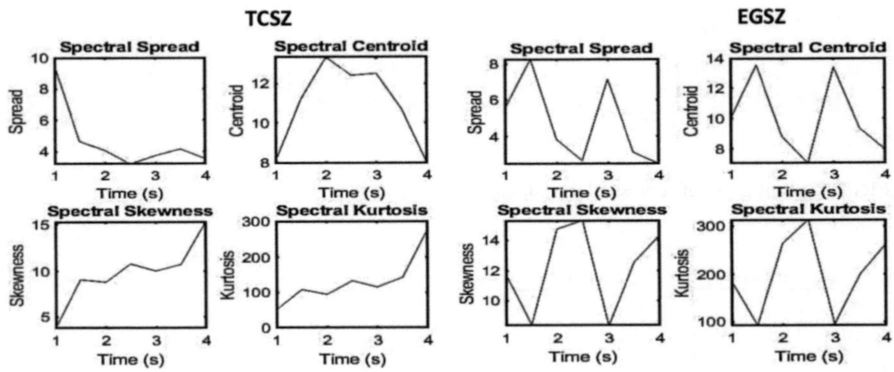

Fig. 10 Spectral moments of representative signals (winsize, 1 s; overlap, 50%)

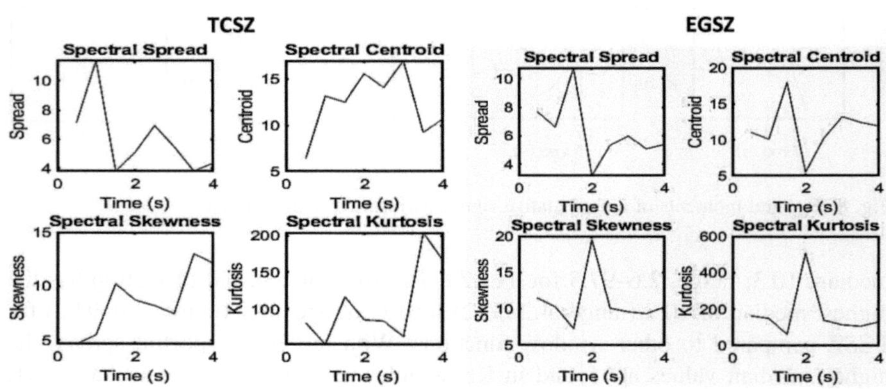

Fig. 11 Spectral moments of representative signals (winsize, 5 s; overlap, 0%)

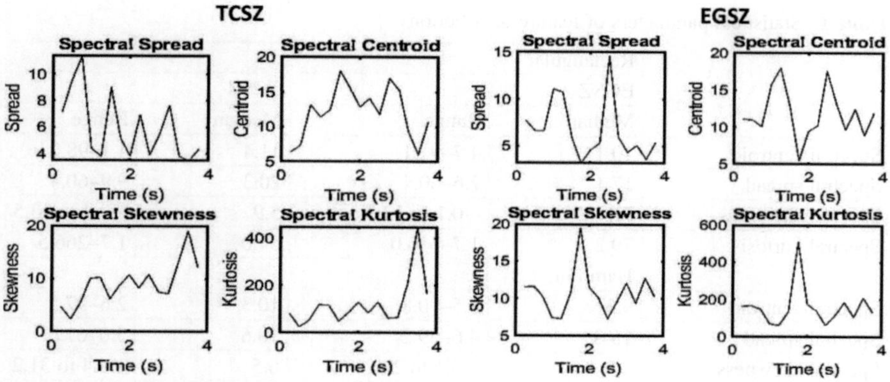

Fig. 12 Spectral moments of representative signals (winsize, 5 s; overlap, 0%)

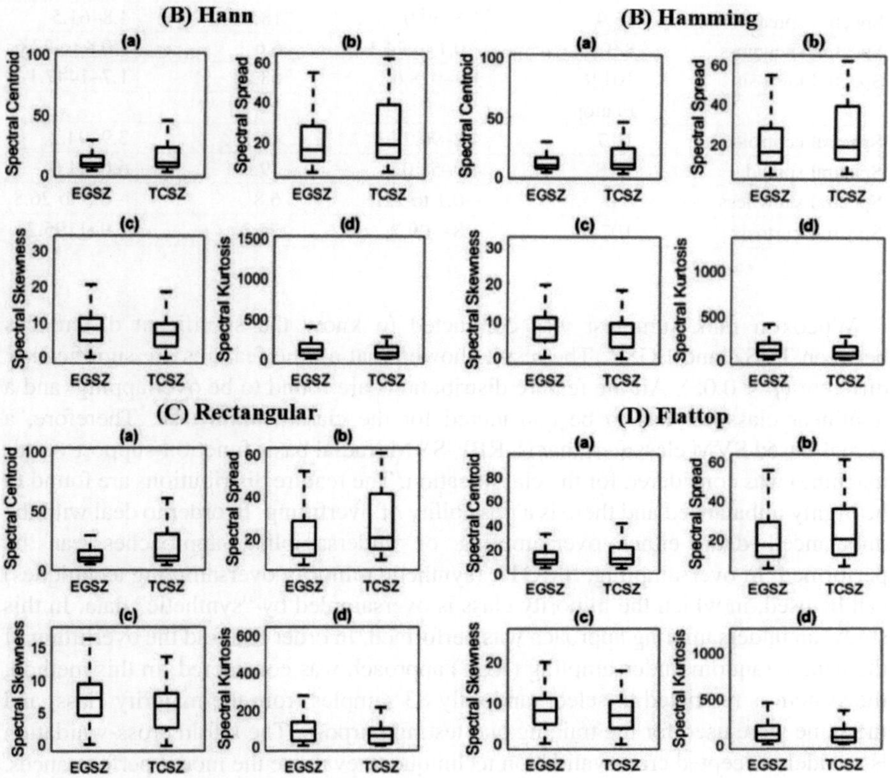

Fig. 13 Feature distributions of different window functions

Table 1 Statistical parameters of feature distributions

	Rectangular			
	EGSZ		TCSZ	
	Median	Range	Median	Range
Spectral centroid	10.1	4.7–91.1	11.4	4.1–98.2
Spectral spread	17.1	7.6–60.4	20.3	9.9–60.4
Spectral skewness	7.1	−0.1 to 17.3	5.9	−0.4 to 13.5
Spectral kurtosis	79.2	1.7–616.0	54.0	1.7–266.5
	Hamming			
Spectral centroid	9.3	3.5–90.8	10.3	2.6–97.5
Spectral spread	15.8	4.6–59.2	18.6	5.0–61.6
Spectral skewness	7.9	−0.1 to 23.2	6.5	−0.4 to 31.2
Spectral kurtosis	100.9	1.8–902.4	67.8	1.7–1316.8
	Hann			
Spectral centroid	9.3	3.5–90.7	10.1	2.7–97.2
Spectral spread	15.9	4.5–59.0	18.6	4.8–61.5
Spectral skewness	8.0	−0.1 to 24.1	6.6	−0.4 to 32.6
Spectral kurtosis	101.9	1.8–928.6	68.2	1.7–1447.1
	Flattop			
Spectral centroid	10.7	4.2–90.2	11.7	3.9–94.2
Spectral spread	16.8	4.9–58.0	19.9	6.0–61.0
Spectral skewness	8.0	−0.1 to 22.7	6.8	−0.3 to 26.8
Spectral kurtosis	107.0	1.8–809.7	74.3	1.9–1195.2

Wilcoxon rank sum test was conducted to know the significant differences between TCSZ and EGSZ. The result showed that all the features are significantly different ($p < 0.05$). All the feature distributions are found to be overlapping, and a nonlinear classifier has to be considered for the classification task. Therefore, a kernel-based SVM classifier, that is, RBF-SVM (radial basis function-support vector machine) was considered for the classification. The feature distributions are found to be highly unbalanced and there is a possibility of overfitting. In order to deal with the unbalanced data, either oversampling or undersampling approaches can be performed. In oversampling, SMOTE (synthetic minority oversampling techniques) can be used, in which the minority class is oversampled by "synthetic" data. In this study, an undersampling approach was performed. In order to avoid the overfitting of the data, a random undersampling (RUS) approach was considered. In this method, the system is restricted to select randomly 83 samples from the majority class, and the same were used for the training and testing purpose. The k-fold cross-validation is a widely accepted cross-validation technique to evaluate the model performances. For this purpose, the data is split, in random, into k number of folds of equal size. The $k-1$ folds are used for training and the remaining fold for the validation. Increasing the value of k improves stability by averaging over more models. Therefore, in this study, we use tenfold cross-validation technique. Tenfold cross-validation approach was used to validate the classification task of the model. In order to confirm the efficacy of the model, this was repeated for 100 times.

Table 2 Performance measures of the model using tenfold cross-validation

Window Functions	ACC	SEN	SPE	PRE	F-ME
Hann	83.33	86.67	66.67	92.86	89.66
Hamming	83.33	90.00	75.00	81.82	85.71
Rectangular	88.89	90.67	84.33	91.67	91.37
Flattop	77.78	80.00	66.67	92.31	85.71

Fig. 14 ROC curve

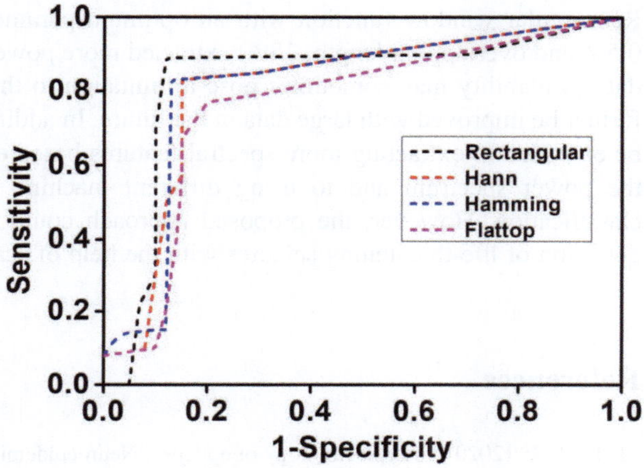

After the 100 times iteration of training and testing, the average performance metrics of the models were computed from the confusion metrics. The details of performance measures are presented in Table 2. Rectangular model yielded highest accuracy of 88.89%, a sensitivity of 90.67%, and precision of 91.67% compared to other models. This model correctly identified an average of 73 channels of the majority class and 75 of the minority class. The models such as Hann and Hamming achieved an accuracy of 83.33%. The lowest accuracy is yielded by Flattop model (ACC, 77.78%, and SEN, 80%), while Hann achieves a sensitivity of 86.67% and precision of 92.86%, specificity drops to 66.67%. F-measure of different models is 89.66% (Hann), 85.71% (Hamming), 91.37% (Rectangular), and 85.71% (Flattop). Figure 14 depicts the ROC curve of different models.

4 Conclusion

This study was aimed to detect electroclinical seizures from electrographic seizures. For this purpose, the spectral moment-based detection system was proposed, and the results imply that this approach is feasible and efficacious, with good performance (mean accuracy, 88.89%, and mean *F*-measure, 91.37%) of Rectangular function model. This study also investigated the influence of parameter settings of the various window functions with respect to TCSZ and EGSZ, and the results showed that the Rectangular window function with an optimum parameter settings (window size, 0.5 s, and overlapping length, 25%) extracted more power from the signals. Limited data availability may sometimes pose a limitation to this approach and this could further be improved with large data in the future. In addition to that, this study could be extended to extracting more spectral features based on orientation and shape of the power spectrum and to using different machine learning approach for the classification. However, the proposed approach could be helpful for the reliable detection of life-threatening seizures with the help of scalp EEG.

References

1. Beghi E. (2020) The epidemiology of epilepsy. Neuroepidemiology. 54(2):185-91
2. Epilepsy: A public health imperative. (2019) Geneva: World Health Organization; https://www.ilae.org/about-ilae/policy-and-advocacy/international-public-policy-activities/global-epilepsy-report-2019.
3. Fisher R.S et al (2019) 2017 International League Against Epilepsy classifications of seizures and epilepsy are steps in the right direction. Epilepsia 60(6):1040-4.
4. Sinha N et al (2021) Focal to bilateral tonic–clonic seizures are associated with widespread network abnormality in temporal lobe epilepsy. Epilepsia 62(3):729-41
5. Karthick PA, Tanaka H, Khoo HM, Gotman J (2018) Prediction of secondary generalization from a focal onset seizure in intracerebral EEG. Clin Neurophysiol 129(5):1030-40
6. Sheikh SR et al (2020) Quantifying the burden of generalized tonic-clonic seizures in patients with drug-resistant epilepsy. Epilepsia 61(8):1627-37
7. Yang Y, Sarkis R, El Atrache R, Loddenkemper T et al (2021)Video-based Detection of Generalized Tonic-clonic Seizures Using Deep Learning. IEEE J. Biomed. Health Inform
8. Murugappan, M. et al (2020) Tunable Q wavelet transform based emotion classification in Parkinson's disease using Electroencephalography. Plos one 15(11):0242014.
9. Catarina Franco A, Parreira S, Bentes C, Pimentel J (2021) Management of a first unprovoked epileptic seizure in adolescence and adulthood. Epileptic Disord 23(4):537-51
10. Mathew, Joseph (2021), Detection of Tonic-Clonic Seizures using Wavelet Entropy of Scalp EEG. In 43rd Annual International Conference of the IEEE Engineering in Medicine & Biology Society (EMBC) 2423-2426.
11. Cascino GD, Brinkmann BH (2021) Advances in the Surgical Management of Epilepsy: Drug-Resistant Focal Epilepsy in the Adult Patient. Neurologic clinics 39(1):181-96
12. Shum J, Friedman D (2021) Commercially available seizure detection devices: A systematic review. Journal of the Neurological Sciences 6:117611.
13. Zia S, Khan AN, Zaidi KS, Ali SE (2021) Detection of Generalized Tonic Clonic Seizures and Falls in Unconstraint Environment Using Smartphone Accelerometer. IEEE Access 9: 39432-43

14. Henze J et al (2021) Multimodal Detection of Tonic–Clonic Seizures Based on 3D Acceleration and Heart Rate Data from an In-Ear Sensor. In: International Conference on Pattern Recognition, Springer, Cham. pp. 490-502
15. Saminu S, Xu G, Shuai Z et al (2021) 11A Recent Investigation on Detection and Classification of Epileptic Seizure Techniques Using EEG Signal. Brain Sciences. (5):668.
16. Zia S, Khan AN, Zaidi KS, Ali SE (2021) Detection of Generalized Tonic Clonic Seizures and Falls in Unconstraint Environment Using Smartphone Accelerometer. IEEE Access. 4;9: 39432-43.
17. Gotman J. (1990) Automatic seizure detection: improvements and evaluation. Electroencephalography and clinical Neurophysiology 76(4) : 317-24.
18. Bong, S.Z et al (2017) Implementation of wavelet packet transform and non linear analysis for emotion classification in stroke patient using brain signals. Biomed Signal Process 36:102-112.
19. Karthick P.A, Tanaka H, Khoo HM, Gotman J (2020) Could we have missed out the seizure onset: A study based on intracranial EEG. Clin Neurophysiol 131(1):114-26.
20. Shum J, Friedman D. (2021) Commercially available seizure detection devices: A systematic review. Journal of the Neurological Sciences 6:117611.
21. Oh, S.L et al. (2018) A deep learning approach for Parkinson's disease diagnosis from EEG signals. Neural. Comput. Appl. 1-7.
22. Raghu, S et al (2020) EEG based multi-class seizure type classification using convolutional neural network and transfer learning. Neural Networks (124):202-212.
23. Rasheed et al (2020) Machine learning for predicting epileptic seizures using EEG signals: A review. IEEE Rev. Biomed. Eng 14:139-55.
24. Sinha N et al (2021) Focal to bilateral tonic–clonic seizures are associated with widespread network abnormality in temporal lobe epilepsy. Epilepsia 62(3):729-41.
25. Ochal et al (2020) The Temple University Hospital EEG Corpus : Annotation Guidelines, Institute for Signal and Information Processing Report, 1(1)
26. Shah V et al (2018) The Temple University Hospital Seizure Detection Corpus, Front. Neuroinfor vol. 12:83
27. Shah et al (2017), Optimizing channel selection for seizure detection. In: IEEE Signal Processing in Medicine and Biology Symposium (SPMB) pp. 1-5
28. Bandarabadi et al (2015) Epileptic seizure prediction using relative spectral power features. Clin Neurophysiol 126(2):237-48
29. Yuvaraj, R et al, (2014). Emotion processing in Parkinson's disease: An EEG spectral power study. Int. J. Neurosci., 124(7):491-502.
30. Bhurane et al (2019). Diagnosis of Parkinson's disease from electroencephalography signals using linear and self-similarity features. Expert Systems, p.e12472.
31. Prabhu KM. (2014) Window functions and their applications in signal processing. Taylor & Francis
32. Von Hippel PT (2005) Mean, median, and skew: Correcting a textbook rule. J. Stat. Educ. 1;13 (2)
33. Alam, MN, Ibrahimy, MI, Motakabber, SMA (2021) Feature Extraction of EEG Signal by Power Spectral Density for Motor Imagery Based BCI. In: *2021 8th* International Conference on Computer and Communication Engineering (ICCCE) pp. 234-237
34. Peeters G (2004) A Large Set of Audio Features for Sound Description (Similarity and Classification) in the CUIDADO Project. Technical Report; IRCAM: Paris, France
35. D Brinkmann, BH et al (2015) Forecasting Seizures Using Intracranial EEG Measures and SVM in Naturally Occurring Canine Epilepsy. PloS one 10(8):e0133900
36. Vance C et al (2020) Learning to detect the onset of slow activity after a generalized tonic–clonic seizure. BMC Med. Inform. Decis. Mak 20(12):1-8

Investigation of the Brain Activation Pattern of Stroke Patients and Healthy Individuals During Happiness and Sadness

Wen Yean Choong, Wan Khairunizam, M. Murugappan, Wan Azani Mustafa, Abdul Hamid Adom, Siao Zheng Bong, Ahmad Kadri Junoh, Zuradzman Mohamad Razlan, and Shahriman Abu Bakar

Abstract This study aimed to assess the emotional experiences of stroke patients and normal people using electroencephalogram (EEG) signals in happiness and sadness. The brain behaviors under both emotional states in the EEG signal were analyzed through signal processing methods. In this study, the EEG signals of normal control (NC) and stroke patients with left brain damage (LBD) and right brain damage (RBD) were analyzed through Hjorth parameters. The extracted Hjorth parameters showed significant differences between happiness and sadness in alpha, beta, and gamma frequency bands, which implied the two emotions exhibiting different brain behavior in different EEG bands. The topographic mapping of the Hjorth parameters presented different activation patterns in the subject

W. Y. Choong (✉)
Faculty of Electronic Engineering Technology, Universiti Malaysia Perlis (UniMAP), Arau, Perlis, Malaysia

W. Khairunizam · W. A. Mustafa · A. H. Adom
Faculty of Electrical Engineering Technology, Universiti Malaysia Perlis (UniMAP), Arau, Perlis, Malaysia
e-mail: wanazani@unimap.edu.my; abdhamid@unimap.edu.my

M. Murugappan
Department of Electronics and Communication Engineering, Kuwait College of Science and Technology, Kuwait City, Kuwait

S. Z. Bong
Innovation Centre, Xiamen University Malaysia, Jalan Sunsuria, Bandar Sunsuria, Sepang, Selangor, Malaysia

A. K. Junoh
Institute of Engineering Mathematics, Universiti Malaysia Perlis, Arau, Perlis, Malaysia
e-mail: kadri@unimap.edu.my

Z. M. Razlan · S. A. Bakar
Faculty of Mechanical Engineering Technology, Universiti Malaysia Perlis (UniMAP), Arau, Perlis, Malaysia
e-mail: zuradzman@unimap.edu.my; shahriman@unimap.edu.my

M. Murugappan, R. Yuvaraj (eds.), *Biomedical Signals Based Computer-Aided Diagnosis for Neurological Disorders*, https://doi.org/10.1007/978-3-030-97845-7_6

groups, and the higher frontal activation can be observed from the NC group for both emotions. Also, the Hjorth Mobility and Complexity parameters were lower in LBD and RBD in the frontal regions of the alpha band. The significant difference channels between the emotions were analyzed by statistical analysis using ANOVA. Moreover, the features of each subject group were used for emotion classification by the application of machine learning-based algorithm. The KNN classification results achieved an average accuracy of 92.35% for NC, 90.84% for LBD, and 95.59% for RBD in classifying happiness and sadness. The emotion classification showed that the emotional dominance frequency bands were the beta and gamma bands. However, the alpha band activity showed left frontal lateralization in the NC group, while right frontal lateralization in the LBD and RBD groups suggested different brain activation of the stroke groups and the controls during happiness and sadness, which reflected the emotional impairment in stroke groups.

Keywords Electroencephalography (EEG) · Stroke · Happiness · Sadness · Hjorth parameters

1 Introduction

Stroke is the second leading cause of death worldwide, approximately 11% of total death in 2019 [1]. This killer disease not only caused death cases, but its survivors often suffer from long-term disabilities that seriously affected their quality of living. Post-stroke patients have poorer functional recovery and higher rates of depression, which increased their risk of a second or a recurrent stroke [2], thus resulting in higher mortality for a depressive stroke patient. Post-stroke depression (PSD) is very common among stroke survivors, and about one-third of the stroke survivors are affected [3, 4]. Post-stroke patients often suffer from mood disorders and emotional and behavioral changes. This may be due to patients' psychological problems such as dissatisfaction with the current situation and the need to depend on others. However, the emotion deficits in stroke patients can be also caused by the damage at the brain regions responsible for emotion processing.

The damaged brain cells in stroke survivors caused the loss of some brain functions; consequently, the stroke patients lose some of their body functions after the stroke [5]. Stroke survivors or post-stroke patients with cognitive impairment often have difficulty in carrying out simple tasks, including perceiving and understanding emotional states [6–8]. Stroke patient, either with left or right brain damage, has been observed to have loss of mood control abilities to a certain extent [9]. As rated by the stroke patients themselves and their caregivers, patients with brain damage in the left hemisphere have a higher percentage of mood problems than those right hemisphere-damaged patients. However, caregivers of right hemisphere stroke patients reported the loss of emotional empathy and change in personality and behavior in patients. Moreover, the right hemisphere has been reported to have an important role in processing empathy [10].

The questionnaire-based method to evaluate emotional deficits in stroke patients has been utilized widely, for research or treatment purposes. However, there are only a few items that evaluate cognitive functions for left and right brain hemispheres [9]. Meanwhile, many of the items are used to evaluate the motor function, and few of them can be used to measure changes in left and right hemispheres' cognitive function. Therefore, the emotion deficits in stroke patients have been investigated through brain signals [11, 12]. The stroke patient, with lesion site at right posterior cerebellum, showed impaired emotional attention, suggesting the cerebellum contributes to the emotion network that relates to emotional attention [13]. The use of event-related potentials (ERPs) from the brain signals reflected the cognitive responses of the right hemisphere-affected patients when watching emotionally arousing pictures [13]. In another study, the electroencephalography (EEG) signals and the Stroke Impact Scale (SIS) were used to assess the emotions of stroke patients; the right hemisphere shows higher EEG beta power in patients that have difficulties in mood and emotional control [11]. Hence, the right hemisphere has been related to emotional deficits which can lead to depressive disorder.

Emotion is controlled and processed in the brain by the interconnection between the neural structures in different brain regions [14, 15]. The brain is divided into four brain lobes or regions, known as frontal, parietal, temporal, and occipital. These brain regions work in conjunction for most brain functions, and also indeed each lobe works a great majority for certain functions. The frontal lobe is the largest brain lobe and is responsible for cognitive functions, for instance, emotional regulation, thoughts, and voluntary actions. The parietal lobe is located behind the frontal lobe; it involves sensations such as taste and touch. The temporal lobe is important for memories and interpreting sensory perception. The occipital lobe is the visual processing center and is responsible for visual information. From previous studies, the measures of brain activity through EEG signals in stroke patients showed different characteristics with normal subjects [12, 16]. Also, the emotional states of discrete emotions can be classified by using machine learning algorithms, specifically the sadness and happiness emotions [12, 16]. In a study by Xugang Xi et al., EEG power has been investigated to be increased most in the left frontal lobe during watching happy movie clips, while it is increased most in the right frontal lobe during sad movie clips [17]. With the addition of grip strength to the subjects, the frontal lobe EEG power of different emotions showed activation in specific areas; the grip strength under neutral and happy emotions activated the EEG power of the right prefrontal lobe. Moreover, the grip strength during sadness activated the EEG power of the left frontal lobe, and to be emphasized, there was a significantly increased power in F7 channel located at the left frontal lobe [17]. From these results, happiness, as a member of positive emotion, and sadness, as a member of negative emotion, work differently and can be investigated through the activation pattern of brain activity. Hence, this current work aimed to investigate the neural correlates in happiness and sadness emotions of the stroke patients, as well as compare the activation pattern in EEG signals of stroke patients with normal subjects.

One of the challenges of EEG-based emotion assessment is the non-stationarity in EEG signals [18]. This non-stationarity of EEG signals makes it difficult for spectral analysis while increasing computational complexity using wavelet analysis. The Hjorth parameters, which were Activity (Ac), Mobility (Mo), and Complexity (Co), were proposed by Hjorth in 1970 [19]. These parameters were intended to describe the general characteristics of an EEG trace in a few quantitative terms. These parameters are entirely time-based calculations, requiring less calculation complexity as compared to frequency analysis and time-frequency analysis such as the short-time Fourier transform (STFT) [19, 20]. Furthermore, these parameters are based on the concept of variance, so that the measured values pertain also to any basic elements from which a complex curve may be composed by superposition [19].

This paper is organized as follows: Section 1 discusses the background and context of the study in emotion impairment of stroke patients and related studies of emotion processing in the brain. Section 2 focuses on the materials and methods used, and the background of the EEG database and the signal processing steps, including filtering, feature extraction, statistical analysis, and classification methods, are presented. Then, Sect. 3 presents the results and discussions. Lastly, in Sect. 4, the summary of this paper is briefly discussed.

2 Materials and Methods

2.1 Subjects' Background

The EEG dataset was taken from a previous experiment by Bong et al., collected at the Hospital Canselor Tuanku Muhriz (HCTM), Kuala Lumpur [12]. This dataset contained EEG signals from three groups of subjects, which were stroke patients with left brain damage (LBD), stroke patients with right brain damage (RBD), and normal subjects (NC). From this dataset, EEG signals of 12 subjects from each group were used for this current study. The subjects were aged between 50 and 57, suffered from stroke more than 1 year, and were not more than 3 years at the time of the experiment. As described in [12], all the subjects passed Mini-Mental State Examination (MMSE) and Beck Depression Scale, hence excluding subjects with dementia and depression. All subjects were right-handers and have normal vision or corrected to normal vision. The subjects were stimulated for six types of emotional states in the experiment; however, only happiness (H) and sadness (S) emotions were analyzed in this study as the positive and negative emotions. The audiovisual stimuli in the form of the video clip were used in the experiment, six trials were recorded with different video clips for each emotion, and a clear explanation of the experiment protocol can be referred in [12].

Fig. 1 Electrode placement of Emotiv Epoc

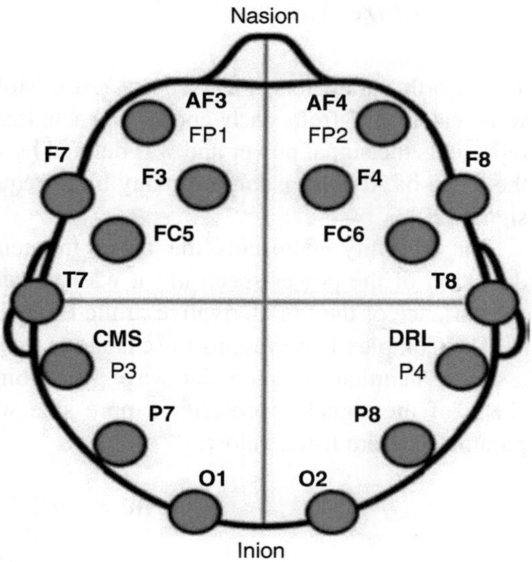

2.2 *EEG Device*

In the experiment, the Emotiv Epoc+ with 14 channels wireless headset was used to record the EEG signals during subjects' watching video clips. This wireless connection device using Bluetooth has enabled the fast and easy setup for an experiment involving stroke patients, as longer experiment time increased tiredness in them. The 14 EEG channels are placed along the entire scalp according to the international 10–20 system, and there were seven electrodes on the left and right, respectively, and with two references at P3 and P4 (labeled CMS and DRL in Fig. 1). The sampling frequency of the EEG headset was 128 Hz.

2.3 *Signal Pre-processing*

The signal processing of EEG signals was performed by using MATLAB software. The raw EEG data were pre-processed to remove artifacts and noises caused by eye blinks, body movement, and powerline interference. A sixth-order band-pass Butterworth filter was applied to filter 0.5–49 Hz from the EEG raw signals [21–23]. Then, the filtered signals were segmented to form smaller epochs for analysis. The segmentation was applied at 6 s per epoch; thus, there were 768 data in each epoch. From each epoch, the alpha (8–13) Hz, beta (13–30) Hz, and gamma (30–49) Hz frequency sub-bands were extracted for feature extraction.

2.4 Feature Analysis

The Hjorth parameters, the Activity (Ac), Mobility (Mo), and Complexity (Co), were calculated from each epoch of each frequency band. The Hjorth Activity represents the signal power and was defined by the variance of the EEG time series; the value of Activity is large if many high-frequency components exist in the EEG signal [20].

The Mobility represents the mean frequency or the proportion of standard deviation of the power spectrum; it was defined as the square root of the ratio of the variance of the first derivative of the EEG signal to its variance [20].

The Complexity represents the change in frequency; it indicates how the shape of a signal is similar to a pure sine wave. The Complexity value converges to 1 as the shape of the signal approaches a pure sine wave [20]. The equations for these parameters were listed below.

$$Ac = var\ (x(t)) \tag{1}$$

$$Mo = \sqrt{\frac{var\ (x'(t))}{var\ (x(t))}} \tag{2}$$

$$Co = \frac{Mo\ (x'(t))}{Mo\ (x(t))} \tag{3}$$

There were 12 subjects in one group, 6 trials and 6 epochs for each emotion, which was a total of 432 samples of features extracted per channel per frequency band of one emotion. To analyze the significant difference between the two emotional states for different parameters, one-way analysis of variance (ANOVA) was used. ANOVA was implemented on each subject group (LBD, RBD, and NC), and each frequency sub-band (alpha, beta, and gamma) for each Hjorth parameter. Then, the values of the features were plotted for the comparison between the two emotional states in each EEG channel; meanwhile, the significant difference between the two emotional states was presented using ANOVA with a significant value at $p \leq 0.05$.

2.5 Emotion Classification

The machine learning classification was employed to classify the happiness and sadness emotions. The KNN classifier, with the City block distance metric [24, 25], was implemented with k values of 1–10. Tenfold cross-validation was applied to split the EEG features into 90% training and 10% testing data. The KNN was implemented to classify the Hjorth parameters from alpha, beta, and gamma bands for each subject group. The KNN classification was employed for three different sets of input. Firstly, a full set of EEG channels (14) were used as the input to the classifier. Secondly, the EEG channels from frontal regions only (8), AF3, F3, F7, FC5, FC6, F8, F4, and AF4, were used as the input. And thirdly, the EEG channels that were verified by ANOVA ($p \leq 0.05$) were selected as the input to the third classification.

3 Results

3.1 Average Values of Hjorth Parameters

To quantitatively describe the emotional EEG signals of stroke patients and normal people, the Hjorth parameters were calculated; only two types of discrete emotions were analyzed, the happiness and sadness emotions, which also represented the positive emotion and negative emotion, respectively. The average of the Hjorth parameters was tabulated in Table 1 for different subject groups and frequency bands. From the results, the Ac values are higher in the beta band and lower in alpha and gamma bands for all subject groups. Both the stroke groups have relatively lower Ac than the control group in the alpha band. For average Mo, the values increased from alpha to gamma bands for all groups. The NC group has higher Mo than the stroke groups in the alpha and beta bands. Meanwhile, for Co, the highest Co is in the beta band and the gamma band has lower Co compared to other bands.

3.2 Distribution of Hjorth Parameters in Different Frequency Sub-bands

Thedistribution map of the Hjorth parameters was presented for different frequency bands in Fig. 2 for the NC group. From Fig. 2, the Ac for happiness increased in the left frontal (F7), the right prefrontal (AF4), and right temporal (T8) regions in alpha

Table 1 Average values for happiness and sadness of Hjorth parameters

		Alpha		Beta		Gamma	
		H	S	H	S	H	S
Ac	NC	43.892	53.838	58.331	71.511	19.138	22.667
	LBD	19.868	22.671	63.073	42.897	58.621	23.457
	RBD	26.080	21.881	35.383	30.111	14.928	12.803
Mo	NC	0.505	0.503	0.975	0.973	1.535	1.536
	LBD	0.503	0.502	0.974	0.967	1.542	1.543
	RBD	0.500	0.498	0.973	0.971	1.538	1.538
Co	NC	1.03141	1.03205	1.07932	1.07908	1.01079	1.01083
	LBD	1.03167	1.03091	1.07804	1.08001	1.01067	1.01082
	RBD	1.03061	1.03032	1.07883	1.07905	1.01071	1.01086

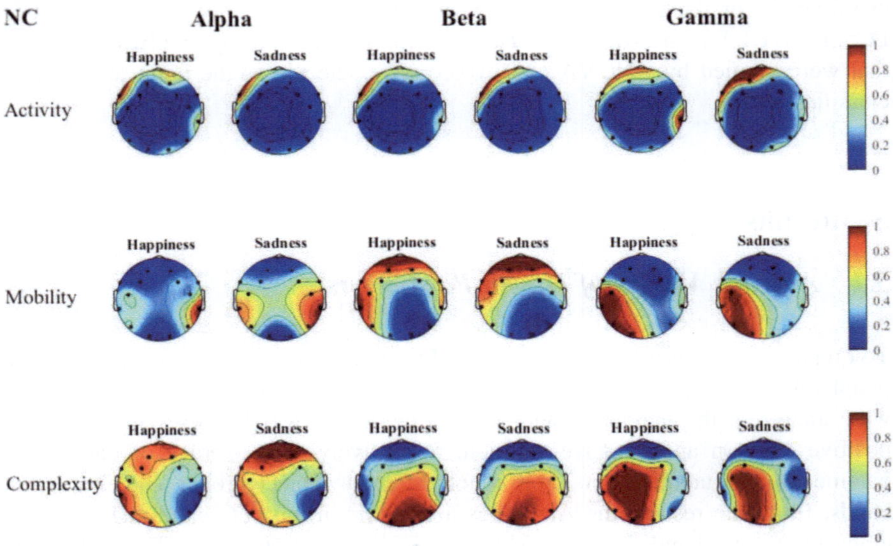

Fig. 2 The Hjorth parameters in alpha, beta, and gamma bands of the NC group

and beta bands, whereas the right temporal (T8) increased significantly in the gamma band. Meanwhile, for the sadness emotion of the NC group, the Ac increased significantly in the left frontal (F7) in all frequency bands, and the left prefrontal (AF3) has also increased in the gamma band. There was higher right Mo than left Mo in the alpha of both emotions, and Mo increased in the temporal regions (T8 and T7) in the alpha band. In the beta band, both emotions increased in the frontal regions and the left temporal and parietal, with left Mo higher than right Mo. Moreover, in the gamma band, the Mo of both emotions increased in the left hemisphere (FC5, T7, P7, and O1). Whereas for Co in the NC group, both emotions increased in most of the regions except the right parietal and temporal (P8 and T8) in the alpha band, and the beta band has a different distribution where the Co increased in the occipital and parietal regions (O1, O2, P7, and P8). The Co in gamma band was similar to Mo, where most of the channels increased in the left hemisphere (F3, FC5, T7, P7, and O1), and O2 and F4 are also higher in the right hemisphere.

Figure 3 shows the Hjorth parameter distribution for the LBD group. Unlike the NC group, the LBD group has increased Ac in the right frontal (F8) for both emotions. There was a significant increase in the left temporal (T7) in all frequency bands, especially in the gamma band. The Mo of the LBD group increased in the left temporal (T7) in the alpha band. Meanwhile, the beta band increased in the left temporal and frontal, including both left and right prefrontal (AF3 and AF4). Mo further increased in gamma band with increased value in the occipital regions. For Co in the LBD group, all the frontal regions and the left parietal and temporal have higher Co for both emotions. In the beta band, the activation of Co increased in interhemispheric of occipital and parietal and the right hemispheric has higher Co

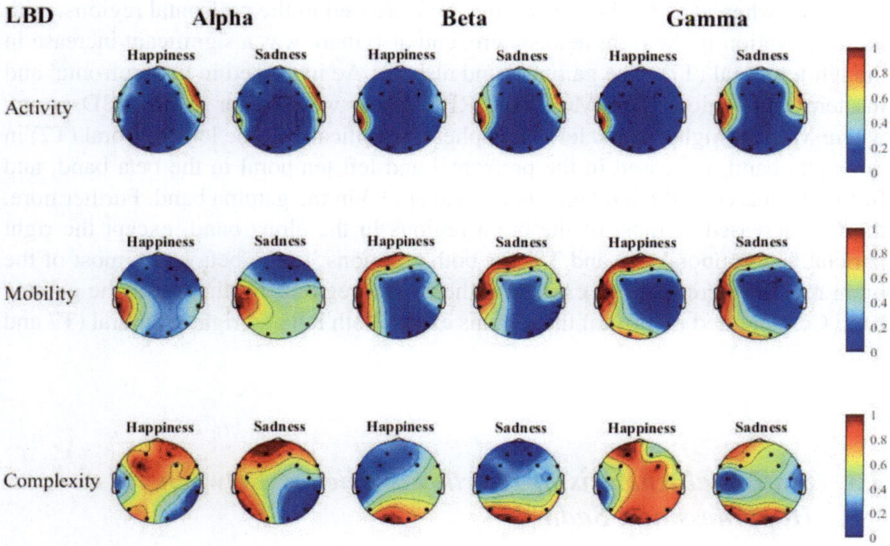

Fig. 3 The Hjorth parameters in alpha, beta, and gamma bands of LBD patients

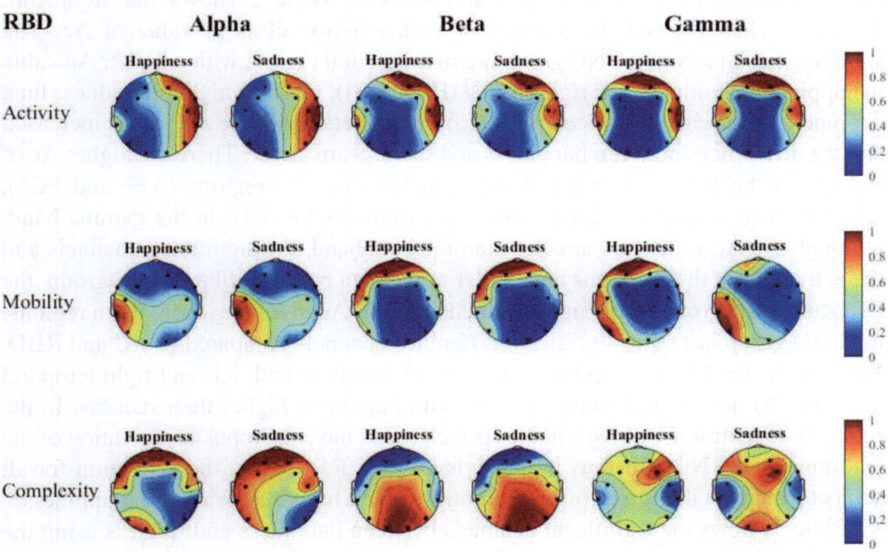

Fig. 4 The Hjorth parameters in alpha, beta, and gamma bands of RBD patients

than the left hemispheric. In the gamma band, the Co increased in most of the regions except left temporal (T7) in happiness.

Figure 4 shows the Hjorth parameter distribution of the RBD group; unlike the NC group, the RBD group has increased AC in the right hemisphere for both emotions, whereas in the beta band, the Ac increased in the prefrontal regions, with more activation in the right hemisphere, and also there was a significant increase in the left temporal (T7). The gamma band also has Ac increased in the prefrontal and the temporal regions. The Mo of the RBD group was similar to the LBD group, where Mo was higher in the left hemisphere, specifically in the left temporal (T7) in the alpha band, increased in the prefrontal and left temporal in the beta band, and further increased in the left hemisphere, except F3 in the gamma band. Furthermore, the Co increased in most of the brain regions in the alpha band, except the right parietal and temporal (P8 and T8) for both emotions. In the beta band, most of the brain regions increased Co, excluding the frontal regions. Furthermore, the gamma band Co increased in most of the regions except both left and right temporal (T7 and T8).

3.3 Statistical Analysis of Hjorth Parameters Between Happiness and Sadness

The values of the Hjorth parameters were compared using ANOVA for the significant difference between happiness and sadness. Table 2 shows the significant difference channels and the average of each emotion class in value of Ac. The significant channels of the NC group are in the frontal region, with a higher Ac value in happiness in both left and right frontal (F3 and F4), whereas higher in sadness than happiness in the right frontocentral (FC6). In the beta band, the Ac values increased and the differences between happiness and sadness are larger. There are higher Ac of sadness in the left prefrontal and the right frontocentral regions (AF3 and FC6), whereas there is higher in happiness in the right frontal (F4). In the gamma band, although the Ac values are smaller than the beta band, the significant channels and the activation of the emotions are similar to the beta band. Unlike the NC group, the LBD and RBD groups have significant channels of Ac involving more brain regions; the LBD group has more significant difference channels compared to NC and RBD. Meanwhile, the RBD group has significant channels in both left and right temporal (T7 and T8) in beta and gamma bands, with happiness higher than sadness. In the right frontocentral (FC6) region, the stroke groups have an opposite activation of the emotions to the NC, whereas in the right frontal (F4) region, the activation for all subject groups is the same for both emotions, with higher activation in happiness.

Table 3 shows the significant channels between happiness and sadness using the Mo values. In the alpha band, the NC group has higher Mo in happiness than sadness in all channels in the left frontal region (AF3, F7, F3, and FC5), both left and right occipital (O1 and O2) and right prefrontal (AF4). When in the beta band, the sadness Mo is higher than happiness in both the left and right prefrontal (AF3 and AF4). Also, there is an increase in the Mo values from alpha to gamma bands, which indicated the frequency of the emotion activation is increasing from alpha to gamma bands, slower in the alpha and faster in the gamma band. For the LBD group, the

Table 2 Significant channels and the average value for happiness and sadness in Hjorth Activity

	Alpha			Beta			Gamma		
	Channels	H	S	Channels	H	S	Channels	H	S
NC	**F3**	18.608	7.424	**AF3**	59.704	127.495	**AF3**	25.273	51.241
	FC6	20.968	30.461	**FC6**	28.113	64.435	**FC6**	10.889	19.775
	F4	15.611	12.103	**F4**	21.878	16.864	**F4**	8.106	5.805
LBD	**AF3**	10.321	12.512	**T7**	249.472	107.483	**AF3**	20.681	12.926
	O2	13.994	16.575	**P7**	32.262	15.234	**F3**	27.222	13.265
	FC6	28.924	22.447	**O1**	13.782	10.229	**T7**	328.780	55.056
				O2	32.636	22.258	**P7**	39.461	7.631
				P8	37.817	24.395	**O1**	11.163	5.884
				FC6	79.079	54.256	**O2**	27.605	11.578
				F4	29.966	22.326	**P8**	34.051	12.054
							T8	45.355	19.462
							FC6	59.920	33.755
							F4	19.818	10.265
							AF4	30.816	16.905
RBD				**T7**	47.990	30.913	**T7**	22.192	13.803
				T8	56.858	41.255	**T8**	25.279	16.177
							FC6	20.422	17.848
							F4	7.894	6.716
							F8	23.205	18.145

Table 3 Significant channels and the average value for happiness and sadness in Hjorth Mobility

	Alpha			Beta			Gamma		
	Channels	H	S	Channels	H	S	Channels	H	S
NC	**AF3**	0.503	0.499	**AF3**	0.989	0.998	**F3**	1.530	1.533
	F7	0.504	0.501	**F3**	0.974	0.981	**FC5**	1.544	1.546
	F3	0.503	0.501	**T7**	0.995	0.981	**P7**	1.544	1.547
	FC5	0.506	0.504	**P7**	0.981	0.974	**P8**	1.530	1.533
	O1	0.502	0.497	**O1**	0.962	0.954			
	O2	0.503	0.498	**T8**	0.980	0.960			
	AF4	0.503	0.499	**AF4**	0.984	0.996			
LBD	**AF3**	0.501	0.499	**F3**	0.969	0.961			
	T7	0.511	0.507	**T7**	0.993	0.984			
	P7	0.506	0.503	**P7**	0.983	0.969			
				O1	0.967	0.957			
				O2	0.959	0.945			
				P8	0.960	0.947			
				T8	0.971	0.947			
RBD	**T7**	0.508	0.505	**FC5**	0.983	0.988	**AF3**	1.542	1.540
	P7	0.504	0.502	**T7**	0.979	0.971	**F7**	1.541	1.538
				T8	0.969	0.957	**T7**	1.541	1.545
							AF4	1.538	1.536

alpha band's significant channels are in the left hemisphere; all the significant channels of the LBD group have higher Mo in happiness than sadness in alpha and beta bands. Similarly, there are more left significant channels with higher happiness than sadness in Mo for the RBD group.

As shown in Table 4, there is no significant difference channel in the alpha band using Co. In the beta band of the NC group, the Co of the frontal regions (AF3, F7, and AF4) have happiness higher than sadness, whereas other regions are opposite to them. For LBD and RBD groups, the happiness is higher than sadness in the significant channels of the alpha band; however, the LBD group has higher Co of sadness in the significant channels of beta and gamma bands.

3.4 Average Distribution over Frontal Regions of Left and Right Hemispheres

The following results presented the average distribution over the frontal regions, which were AF3, F3, F7, and FC5 for the left hemisphere and AF4, F4, F8, and FC6 for the right hemisphere, to show the involvement of frontal activity in the left and right hemisphere of both emotions. From Fig. 5a, the frontal distribution of the left hemisphere in both emotions is larger than the right hemisphere in the NC group. However, the LBD group and the RBD group have opposite results, in which the

Table 4 Significant channels and the average value for happiness and sadness in Hjorth Complexity

	Alpha			Beta			Gamma		
	Channels	H	S	Channels	H	S	Channels	H	S
NC				**AF3**	1.075	1.071	**T7**	1.0110	1.0106
				F7	1.077	1.075	**O1**	1.0111	1.0113
				T7	1.074	1.078	**P8**	1.0106	1.0108
				O1	1.085	1.087			
				T8	1.077	1.081			
				AF4	1.076	1.071			
LBD	**F3**	1.033	1.031	**F3**	1.076	1.078	**F7**	1.0106	1.0108
	O2	1.031	1.030	**T7**	1.074	1.078	**F3**	1.0107	1.0109
	P8	1.031	1.030	**P7**	1.079	1.085	**T7**	1.0105	1.0107
	T8	1.031	1.030	**O1**	1.083	1.087	**P7**	1.0106	1.0110
	FC6	1.032	1.030	**O2**	1.081	1.085	**O2**	1.0107	1.0111
	AF4	1.032	1.031	**P8**	1.080	1.083	**AF4**	1.0106	1.0108
				T8	1.078	1.083			
				F4	1.077	1.079			
RBD	**O1**	1.031	1.030	**T8**	1.077	1.081	**F7**	1.0108	1.0110
							O1	1.0108	1.0110
							T8	1.0103	1.0107
							FC6	1.0105	1.0107

right hemisphere has higher mean and larger variation compared to the left hemisphere, and with a relatively small variation compared to the distribution of the left frontal in NC.

In Fig. 5b, there is an increase in the mean and the distribution over the frontal regions for all subject groups in the beta band, especially in the left hemisphere. For the NC group, there is an increase in both emotions in the left hemisphere, while the changes of happiness are not obvious in the right hemisphere, but an increase of sadness in the right hemisphere from alpha to beta band is observable. For the LBD group, the distribution of the left frontal increased in the beta band, with a larger variation in happiness than sadness. Moreover, the changes from alpha to beta in the RBD group are the smallest among the groups and have smaller variations for both hemispheres in the beta band.

For the gamma band, as shown in Fig. 5c, all the distribution in frontal regions decreased. However, the NC group has a higher left than the right distribution for both emotions as in alpha and beta bands. Meanwhile, the frontal activity of the LBD group reacts differently in the gamma band; although there are drops in values from beta to gamma band, the distribution of the emotions, especially the happiness in the right hemisphere, is larger than the NC group. Also, the smallest variation of the frontal distribution in the gamma band still is the left hemisphere of the RBD group.

For the Hjorth Mo in the frontal region, as shown in Fig. 6a, the NC group has a higher mean of happiness than sadness in both left and right hemispheres in the alpha

(a)

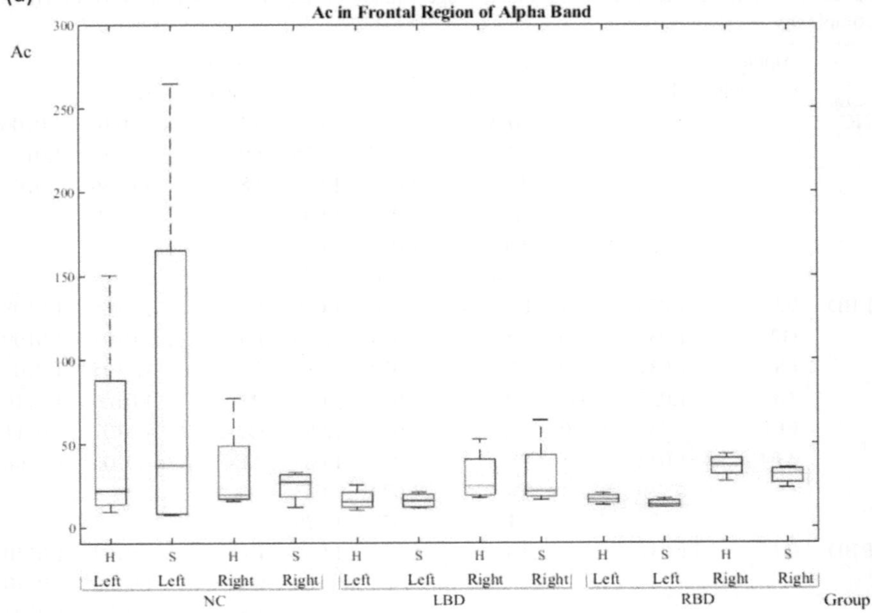

(b)

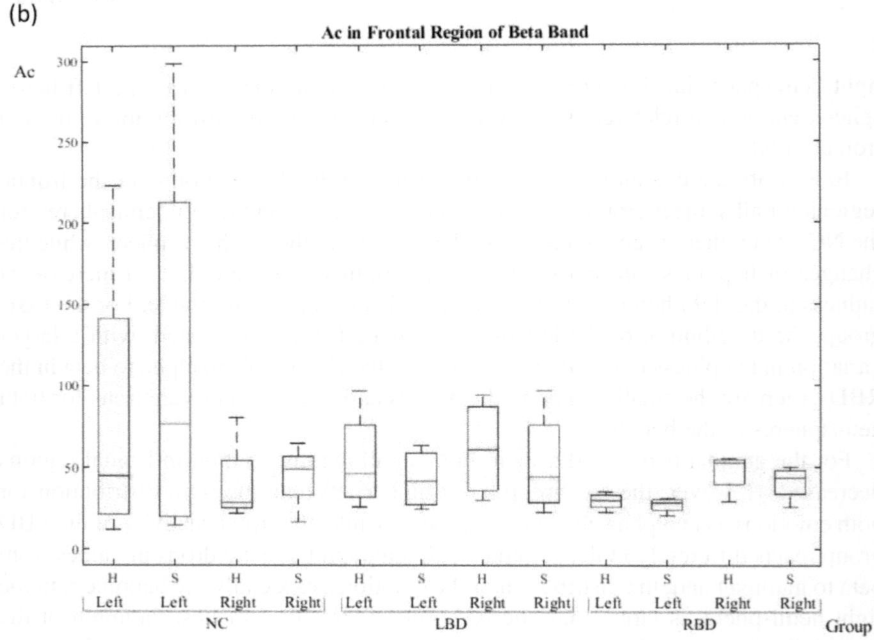

Fig. 5 Frontal distribution of Hjorth Activity of the left hemisphere (AF3, F3, F7, and FC5) and right hemisphere (AF4, F4, F8, and FC6) in happiness and sadness emotions in (**a**) alpha, (**b**) beta, and (**c**) gamma bands

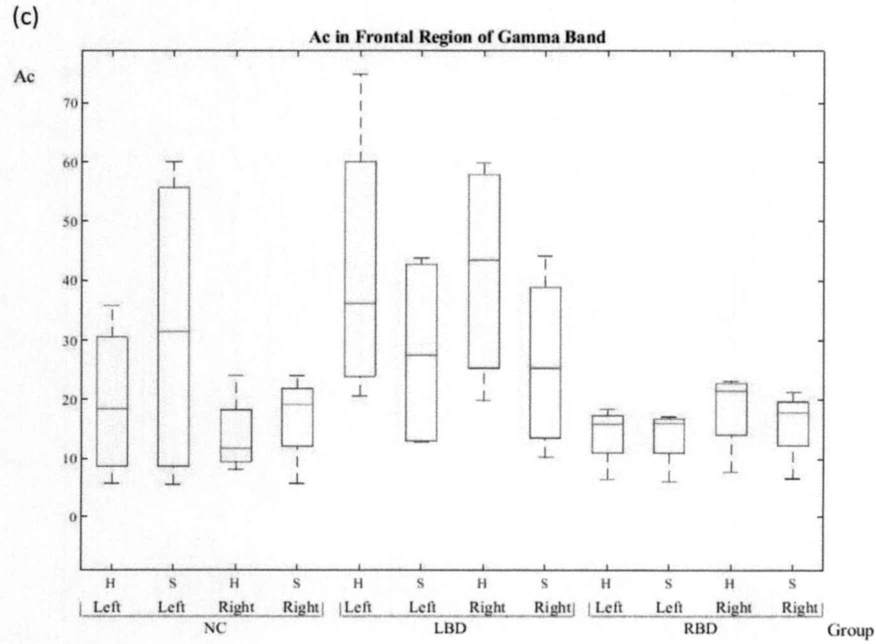

Fig. 5 (continued)

band, while the LBD group has higher left than right Mo for both emotions and the right frontal Mo of LBD are relatively lower than the NC group. However, for the RBD group, there is slightly greater left than right frontal Mo, but the Mo of RBD was the lowest among the groups.

In Fig. 6b, the NC group exhibits greater left than right beta frontal Mo, but the sadness Mo was higher than the happiness Mo. For the LBD group, right frontal Mo for both emotions was relatively lower than left frontal Mo. Meanwhile, for the RBD group, there is greater left than right frontal Mo, and the variations of Mo are larger for both emotions in the left frontal of RBD.

In Fig. 6c, the NC group has larger variations but a lower mean compared to the stroke groups. On the other hand, the LBD group has the highest mean in left frontal Mo and higher frontal Mo compared to the NC group. Similarly, the RBD group has a higher mean for both emotions in the left and right hemispheres compared to the NC group, but a lower mean than the LBD group. All subject groups have greater left than right frontal Mo in the gamma band.

The Hjorth Co in the frontal regions were plotted in Fig. 7. The NC group has higher left frontal Co in the alpha band, with sadness Co higher than happiness Co in both hemispheres in Fig. 7a. However, the LBD group has higher happiness than sadness Co in both hemispheres. The RBD group has lower frontal Co in the alpha band compared to the NC and LBD, with greater Co in happiness.

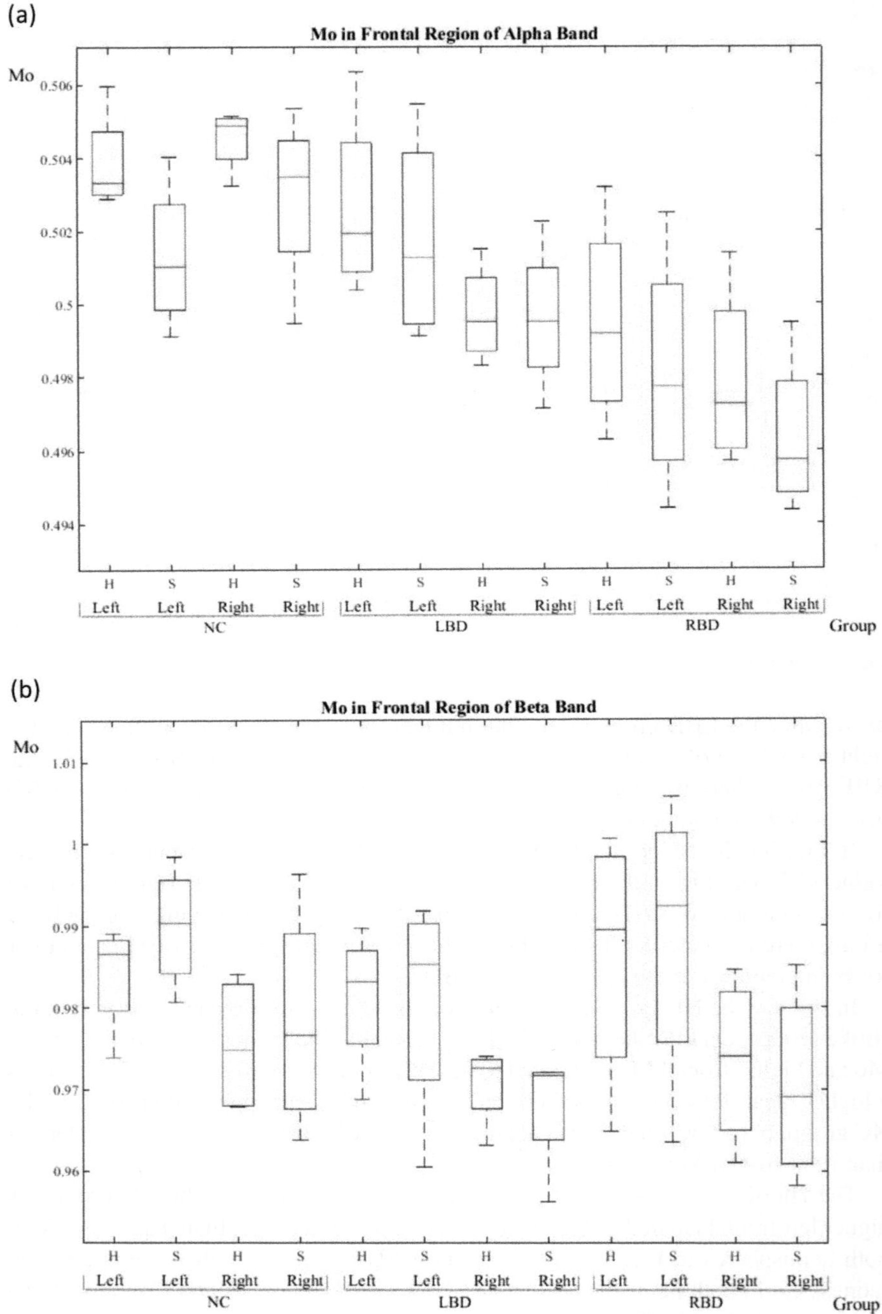

Fig. 6 Frontal distribution of Hjorth Mobility of the left hemisphere (AF3, F3, F7, and FC5) and right hemisphere (AF4, F4, F8, and FC6) in happiness and sadness emotions in (**a**) alpha, (**b**) beta, and (**c**) gamma bands

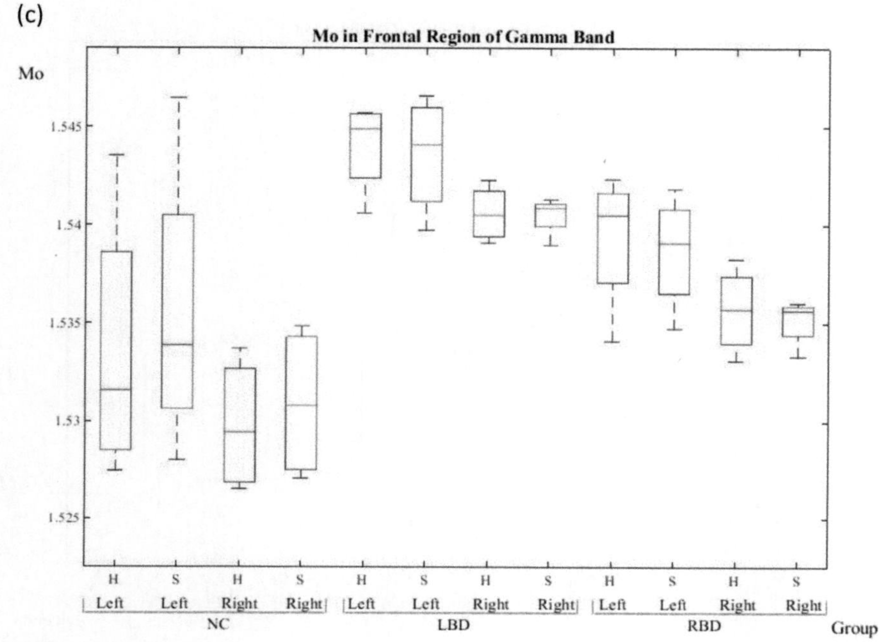

Fig. 6 (continued)

From Fig. 7b, there is a greater right than left frontal Co in the beta band of the NC group; the mean of frontal Co was higher in happiness, but larger variations of Co in sadness. Also, both LBD and RBD groups have a greater right than left frontal Co in both emotions; there was a remarkably small variation in the left frontal Co in the happiness of LBD.

In Fig. 7c, the NC group has observable greater left than right frontal Co in the gamma band. For the LBD group, the variations of frontal Co in the gamma band were smaller compared to the NC group; the sadness Co was higher than the happiness Co for both hemispheres. Meanwhile, for the RBD group, the frontal Con in the left hemisphere, particularly the happiness emotion, has a smaller variation compared to the NC group, while the right frontal Co has higher means and larger variations compared to the NC group.

3.5 *Emotion Classification Using KNN Classifier*

Tables 5, 6, and 7 show the classification accuracy of the happiness and sadness emotions in different frequency bands and Hjorth parameters using all EEG channels, frontal EEG channels, and selected EEG channels from ANOVA ($p \leq 0.05$), respectively. The Mo and Co have lower classification accuracy than Ac. In Table 5,

(a)

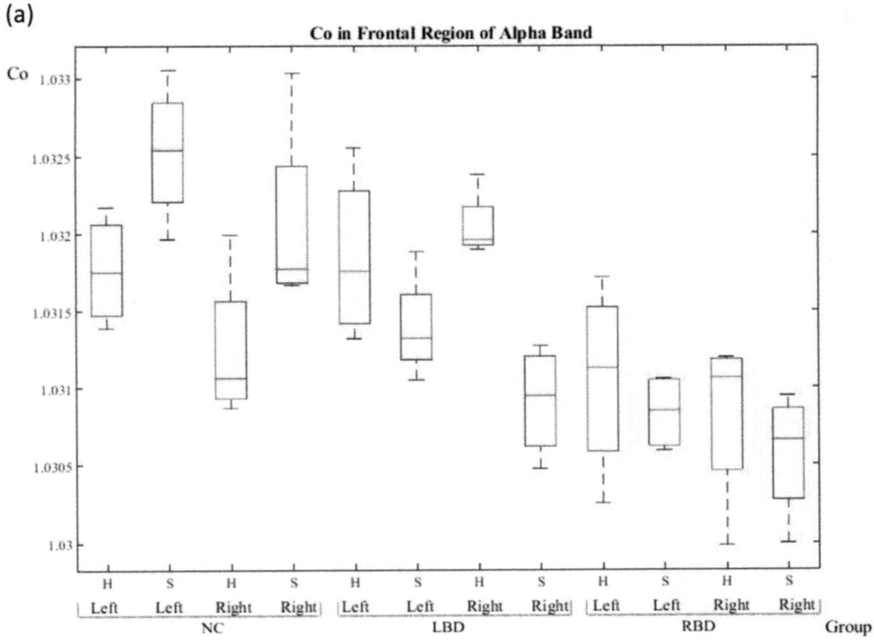

(b)

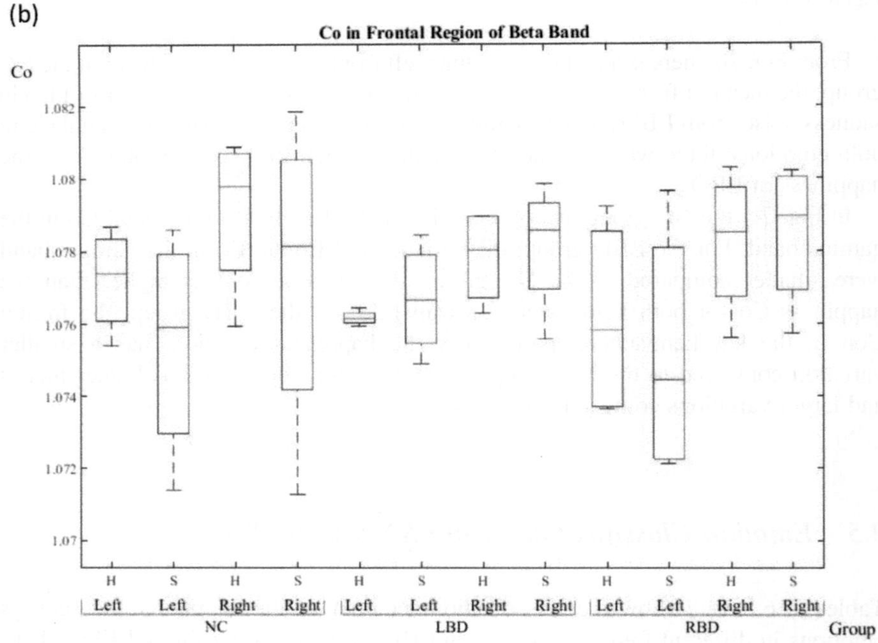

Fig. 7 Frontal distribution of Hjorth Complexity of the left hemisphere (AF3, F3, F7, and FC5) and right hemisphere (AF4, F4, F8, and FC6) in happiness and sadness emotions in (**a**) alpha, (**b**) beta, and (**c**) gamma bands

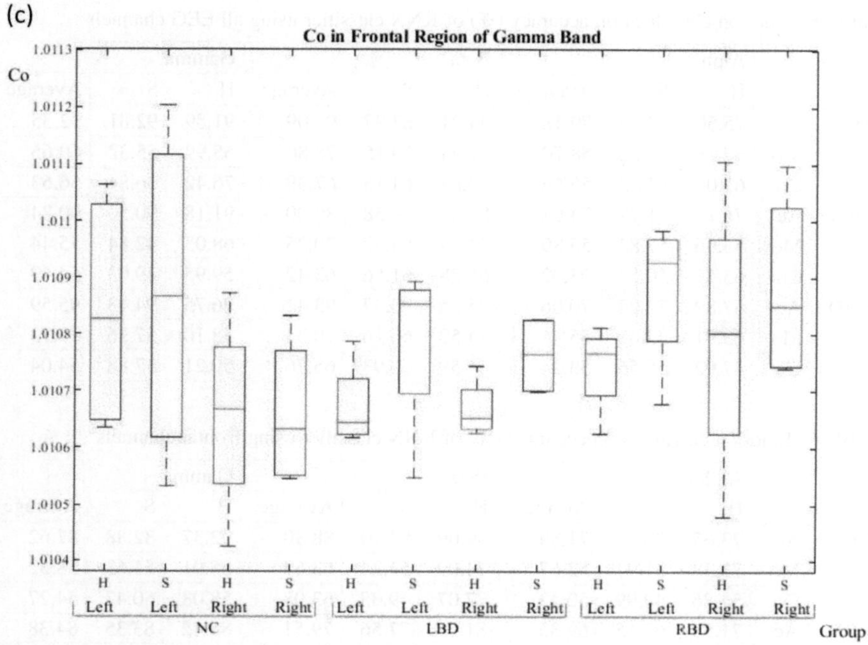

Fig. 7 (continued)

the highest happiness accuracy for the NC group is in the beta band, 94.21% ($k = 2$), whereas the highest sadness accuracy is in the gamma band, 92.81% ($k = 1$). The LBD and RBD groups have the highest happiness and sadness accuracy in the gamma band, with the RBD group having the highest classification accuracy using a k value of 6 (average accuracy = 95.59%, happiness = 96.75%, and sadness = 94.43%).

In Table 6, when the EEG channels from frontal regions only were used to classify the emotions, the performance of the classification is similar to Table 5, when all EEG channels were used. All the Ac have higher classification accuracy compared to Mo and Co. The highest accuracy for each emotion is observed in the beta and gamma band; NC achieved 92.37% ($k = 4$) happiness in the gamma band and 87.50% ($k = 8$) sadness in the beta band; LBD achieved 85.42% ($k = 3$) and 83.35% ($k = 3$) for happiness and sadness, respectively, in the gamma band; and RBD achieved 90.04% ($k = 8$) happiness in gamma band and 86.77% ($k = 3$) sadness in the beta band.

The emotion classification of selected EEG channels evaluated by ANOVA as presented in Table 7 shows similar results with Tables 5 and 6, with a slight drop in the accuracy of some features; however, the drop in the accuracy is not significant. Similarly, the Ac achieved higher classification accuracy compared to Mo and Co, and the highest classification accuracy is in the beta and gamma bands. NC achieved 85.15% ($k = 4$) happiness in the gamma band and 79.18% ($k = 3$) sadness in the beta

Table 5 Emotion classification accuracy (%) of KNN classifier using all EEG channels

		Alpha			Beta			Gamma		
		H	S	Average	H	S	Average	H	S	Average
NC	Ac	75.50	82.85	79.18	94.21	87.97	91.09	91.89	92.81	92.35
	Mo	67.18	50.22	58.70	72.44	79.15	75.80	65.99	55.32	60.65
	Co	62.00	49.32	55.66	70.63	64.15	67.39	76.42	36.84	56.63
LBD	Ac	76.18	71.77	73.98	88.42	89.58	89.00	91.18	90.50	90.84
	Mo	53.96	57.82	55.89	74.99	65.52	70.25	68.03	42.84	55.44
	Co	65.51	40.53	53.02	65.28	61.56	63.42	59.95	49.03	54.49
RBD	Ac	67.89	72.23	70.06	95.85	90.97	93.41	96.75	94.43	95.59
	Mo	63.91	47.00	55.46	74.52	66.16	70.34	58.10	57.36	57.73
	Co	47.92	58.56	53.24	64.59	66.93	65.76	50.21	57.88	54.04

Table 6 Emotion classification accuracy (%) of KNN classifier using frontal channels

		Alpha			Beta			Gamma		
		H	S	Average	H	S	Average	H	S	Average
NC	Ac	73.65	70.13	71.89	89.09	87.50	88.30	92.37	82.88	87.62
	Mo	73.38	31.97	52.67	74.09	53.24	63.67	65.01	51.63	58.32
	Co	54.86	49.99	52.43	77.07	49.08	63.08	58.08	50.47	54.27
LBD	Ac	71.54	67.15	69.35	81.46	77.56	79.51	85.42	83.35	84.38
	Mo	53.03	56.65	54.84	73.85	58.57	66.21	58.97	49.96	54.47
	Co	68.29	44.47	56.38	62.49	58.07	60.28	81.46	28.92	55.19
RBD	Ac	61.82	64.39	63.10	83.10	86.77	84.93	90.04	86.35	88.20
	Mo	52.57	54.89	53.73	57.84	67.38	62.61	70.37	43.07	56.72
	Co	58.79	46.73	52.76	75.70	49.76	62.73	63.22	37.96	50.59

Table 7 Emotion classification accuracy (%) of KNN classifier using statistically significant channels analyzed from ANOVA

		Alpha			Beta			Gamma		
		H	S	Average	H	S	Average	H	S	Average
NC	Ac	65.98	72.67	69.33	78.74	79.18	78.96	85.15	67.38	76.27
	Mo	56.51	56.52	56.52	72.46	71.52	71.99	66.91	49.33	58.12
	Co				65.52	63.44	64.48	56.47	57.63	57.05
LBD	Ac	72.66	50.70	61.68	83.12	81.48	82.30	91.20	88.41	89.81
	Mo	53.45	51.64	52.55	66.43	66.46	66.44			
	Co	54.89	54.16	54.53	52.80	62.07	57.43	54.41	50.67	52.54
RBD	Ac				71.28	69.01	70.15	85.18	90.73	87.96
	Mo	52.56	53.26	52.91	74.99	40.77	57.88	59.24	53.76	56.50
	Co	51.93	58.12	55.03	77.80	28.22	53.01	52.52	55.30	53.91

band, LBD achieved 91.20% ($k = 1$) happiness and 88.41% ($k = 1$) sadness, both in the gamma band, and RBD achieved 85.18% ($k = 5$) happiness and 90.73% ($k = 5$) sadness, also both in the gamma band.

4 Discussion

4.1 Emotion Impairment Analyzed Using Hjorth Parameters

The lower alpha activation in the stroke groups reflected the presence of cognitive impairment [26–28], whereas the beta band has been shown to achieve better performance in emotion recognition [16, 29, 30]. The lower Mo in both stroke groups compared to the control group indicates the slowing of brain activity in stroke patients. From a previous study, Mo has been related to drowsiness, where a higher Mo is related to lower drowsiness and higher cognitive function [31].

As the Co shows the comparison of EEG signal to a sin wave, with a value approaching 1 as more similar to it. In this work, the higher Co indicated the frequency was changing rapidly during the emotional states; hence, the beta band has faster changes of brain activation. From previous studies, there was higher Co in the patient with insomnia than normal people; this was due to the difficulty of maintaining sleep in the patient [32]. However, there were past studies that reported lower Co in patients with pathological disorders such as depression, schizophrenia, post-traumatic stress disorder, and epilepsy compared to normal [33].

4.2 Region of Emotional Activation in the Brain

In this work, F7 of NC has high EEG power in alpha, beta, and gamma bands, and this finding was consistent with the past studies that F7 plays a vital role in emotion recognition and the alpha power increased over left frontal for happiness [17, 34]. Also, the greater left than right frontal activation has been associated with positive emotions and facets of well-being, such as life satisfaction and independence [26, 35]. Moreover, there was activation of the frontal region during music stimulation, the alpha band increased during both stimulating and calm music, and the interhemispheric of the frontal regions (F7 and F8) increased during listening to the stimulating music [36].

F7 and F8 are both related to emotional regulation, whereas the right frontal is more related to response inhibition [37, 38]. F8 is located on the right inferior frontal gyrus (rIFG) of the brain, which has great involvement in stopping many functions in the brain and acts as the brake on responses [39]. Therefore, the high activation of F8 in the alpha band may imply poor inhibition of emotion, whereas overly high activation of F8 in the beta band may imply an overly inhibited emotional response. From the results, the right frontal, specifically F8, activated significantly in LBD, and the greater right than left hemispheric activation in the RBD group may indicate problems with emotional regulation in the stoke groups.

The temporal region also has an association with impulsivity; in the beta band, the temporal region acts as the relay station for activation in the frontal regions [40]. The left temporal activation, with lesser right temporal activation, directly influenced the activation of the left and right prefrontal regions in an impulsive individual. Both left and right temporal regions (T7 and T8) also have been reported as effective channels to classify positive and negative emotions [41]. A highly significant increase of left and especially right temporal gamma power is associated with positive emotion [34, 42]. This was reflected in the present result in the NC group, which showed significantly high activation in T8, especially in the happiness emotion of the gamma band. Hence, these results suggested that the left and right temporal regions are involved differently in the processing of happiness and sadness emotions.

4.3 The Involvement of Frontal Region in Emotion Processing

The EEG channels with statistically significant differences between happiness and sadness were in the frontal regions for NC in Table 2. As reported in the literature, the frontal regions have an important role in emotion processing [43]. This indicates that, at least for the control group, the emotion activation of the frontal regions was able to discriminate between the happiness and sadness emotions. Furthermore, from the emotion classification using the KNN classifier, the frontal channel classification shows similar results with the classification using all EEG channels, hence indicating the frontal regions can discriminate the happiness and sadness emotions.

In this work, the left frontal regions have shown high involvement in emotion processing of both happiness and sadness in all frequency bands. In this result, the left frontal is activated regardless of positive or negative emotions. The greater left than right frontal activation indicated there is an important role of the left frontal in emotion processing. However, for the RBD group, though the damaged part is at the right brain hemisphere, the left frontal of the RBD group did not show emotional activation as in the NC group. Also, the damage to the right hemisphere caused emotional impairment in the RBD group, with supportive studies suggesting the right hemisphere is more dominant for emotional activity [44, 45]. Thus, this indicated that the left frontal has importance in emotion processing; however, the role of the right frontal in emotional processing is also dominant.

Meanwhile, for the stroke groups, the LBD group reacted differently with the NC group, which has small activity in the alpha and beta band and increased activity in the gamma band of the left frontal. Furthermore, there was lower Mo and Co in the frontal regions of LBD and RBD compared to the NC group in the alpha band. The high activity in the gamma band is often related to worry, stressful, and anxiety states [46, 47], and increased activity in the right hemisphere of the gamma band has been associated with unpleasant emotion [48]. There have been reported alpha left frontal lateralization in the healthy subjects, greater right frontal activity in anxious subjects, and symmetrical frontal activity in anxiety and depressed subjects [49]. Relating these findings to the present results in the alpha band, the NC group has left frontal lateralization, whereas the LBD and RBD group have a greater right than left frontal activity in the alpha band. Hence, these results suggested the stroke groups may exhibit emotional activity that is similar to the anxious individuals.

5 Conclusion

This study presented the development of the EEG-based emotion assessment system which intends to improve the efficiency of the assessment process for stroke patients. In the present work, the Hjorth parameters were used to analyze the EEG signals of stroke patients and normal control in two emotional states, happiness and sadness. These two emotions are categorized as the positive and negative emotions, or positively valence and negatively valence (on the valence scale); people often refer to them as extreme or opposite emotions. However, by analyzing the EEG signals of these two emotional states, they are not extremely opposite to each other as investigated from the features described by the Hjorth parameters. However, these parameters were able to discriminate between the happiness and sadness emotions with an average of 92.35% using a *k* value of 1 in the gamma band of the NC group. Most of the Hjorth parameters, especially the Ac and Mo, showed higher values in the control groups compared to the stroke groups. From the Hjorth parameters, the beta and gamma bands were more dominant during both happiness and sadness emotions, and the beta band has faster changes in the emotion activation.

Furthermore, the frontal activation observed in the stroke and control groups was different. In the stroke groups, there was relatively small frontal activity compared to the control group, which indicated the emotional impairment in the stroke patients. Also, this work presented the control group with left frontal lateralization in the alpha band and right frontal lateralization in both stroke groups in the alpha band. However, both the left and right frontal regions play a vital role in emotion regulation and response inhibition; impairment to either side of the brain may cause emotion deficits. On the other hand, EEG channels from other brain regions also have involvement in emotion processing, such as the temporal region of both hemispheres, which have a direct influence on frontal emotion activation.

The results of this work are based on two types of discrete emotions, happiness and sadness, representing positive emotion and negative emotion, respectively. The major limitation of this work was comparison was only made between positive emotion and negative emotion. As a result, the positive and negative emotions showed different brain activation patterns, and the differences between positive and negative emotions are observable using the statistical method. It is possible that there is less significant difference for the comparison between emotions from the same category, such as anger and fear, where both are negative emotions, or happiness and satisfaction, where both are positive emotions. Therefore, the current results do not represent all types of emotion assessment using EEG signals. However, this work confirmed the effectiveness of Hjorth parameters and the machine learning method for positive and negative emotion classification.

Though this work provides highly accurate discrimination between happiness and sadness of stroke and control groups, further work is needed to assess and validate the results with more emotions, such as anger, disgust, fear, and surprise. Moreover, the employment of feature selection techniques can reduce the computational complexity and understand the brain's underlying mechanism of emotion processing through the selection of emotion-related features. This work focused only on a machine learning method using the KNN classifier; the use of other machine learning methods or deep learning methods can be implemented to improve the efficiency of the system in the future.

Acknowledgments The author would like to thank Medyna Rehab and Services for allowing them to conduct data collection.

Funding: The author would like to acknowledge the support from the Fundamental Research Grant Scheme (FRGS) under a grant number of FRGS/1/2019/ICT04/UNIMAP/02/1 from the Ministry of Education, Malaysia.

Conflicts of Interest: The authors declare no conflict of interest.

References

1. World Health Organization. (2020). The top 10 causes of death. Retrieved September 30, 2021, from https://www.who.int/news-room/fact-sheets/detail/the-top-10-causes-of-death
2. Kutlubaev, M. A., & Hackett, M. L. (2014). Part II: predictors of depression after stroke and impact of depression on stroke outcome: an updated systematic review of observational studies. *International journal of stroke : official journal of the International Stroke Society*, *9*(8), 1026–1036. doi: https://doi.org/10.1111/ijs.12356
3. Robinson, R. G., & Jorge, R. E. (2016). Post-stroke depression: A review. *American Journal of Psychiatry*, *173*(3), 221–231. doi: https://doi.org/10.1176/appi.ajp.2015.15030363
4. Cheah, W. K., Hor, C. P., Zariah, A. A., & Looi, I. (2016). A Review of Stroke Research in Malaysia from 2000 - 2014. *The Medical journal of Malaysia*, *71*(Suppl 1), 58–69. Retrieved from http://www.ncbi.nlm.nih.gov/pubmed/27801388
5. Mineyko, A., & Kirton, A. (2019). Long-Term Outcome After Bilateral Perinatal Arterial Ischemic Stroke. *Pediatric Neurology*, *101*, 39–42. doi: https://doi.org/10.1016/j.pediatrneurol.2019.07.013
6. Stone, J., Townend, E., Kwan, J., Haga, K., Dennis, M. S., & Sharpe, M. (2004). Personality change after stroke: some preliminary observations. *Journal of neurology, neurosurgery, and psychiatry*, *75*(12), 1708–1713. doi: https://doi.org/10.1136/jnnp.2004.037887
7. Yuvaraj, R., Murugappan, M., Norlinah, M. I., Sundaraj, K., & Khairiyah, M. (2013). Review of emotion recognition in stroke patients. *Dementia and Geriatric Cognitive Disorders*, *36*(3–4), 179–196. doi: https://doi.org/10.1159/000353440
8. Aben, H. P., Reijmer, Y. D., Visser-Meily, J. M. A., Spikman, J. M., Biessels, G. J., de Kort, P. L. M., & Study Group, P. (2017). Impaired Emotion Recognition after Left Hemispheric Stroke: A Case Report and Brief Review of the Literature. *Case Reports in Neurological Medicine*, *2017*, 1–6. doi: https://doi.org/10.1155/2017/1045039
9. Hillis, A. E., & Tippett, D. C. (2014). Stroke Recovery: Surprising Influences and Residual Consequences. *Advances in Medicine*, *2014*, 1–10. doi: https://doi.org/10.1155/2014/378263

10. Yeh, Z.-T., & Tsai, C.-F. (2014). Impairment on theory of mind and empathy in patients with stroke. *Psychiatry and Clinical Neurosciences, 68*(8), 612–620. doi: https://doi.org/10.1111/pcn.12173
11. Doruk, D., Simis, M., Imamura, M., Brunoni, A. R., Morales-Quezada, L., Anghinah, R., ... Battistella, L. R. (2016). Neurophysiologic Correlates of Post-stroke Mood and Emotional Control. *Frontiers in human neuroscience, 10*(AUG2016), 428. doi: https://doi.org/10.3389/fnhum.2016.00428
12. Bong, S. Z., Wan, K., Murugappan, M., Ibrahim, N. M., Rajamanickam, Y., Mohamad, K., ... Mohamad, K. (2017). Implementation of wavelet packet transform and non linear analysis for emotion classification in stroke patient using brain signals. *Biomedical Signal Processing and Control, 36*, 102–112. doi: https://doi.org/10.1016/j.bspc.2017.03.016
13. Adamaszek, M., Olbrich, S., Kirkby, K. C., Woldag, H., Willert, C., & Heinrich, A. (2013). Event-related potentials indicating impaired emotional attention in cerebellar stroke-A case study. *Neuroscience Letters, 548*, 206–211. doi: https://doi.org/10.1016/j.neulet.2013.04.018
14. Stam, C. J., & van Straaten, E. C. W. (2012). The organization of physiological brain networks. *Clinical Neurophysiology, 123*(6), 1067–1087. doi: https://doi.org/10.1016/j.clinph.2012.01.011
15. LeDoux, J. E., & Brown, R. (2017). A higher-order theory of emotional consciousness. *Proceedings of the National Academy of Sciences, 114*(10), E2016–E2025. doi: https://doi.org/10.1073/pnas.1619316114
16. Choong, W. Y., Khairunizam, W., Mustafa, W. A., Murugappan, M., Rajamanickam, Y., Adom, A. H., ... Bakar, S. A. (2020). An emotion assessment of stroke patients by using bispectrum features of EEG signals. *Brain Sciences, 10*(10), 1–22. doi: https://doi.org/10.3390/brainsci10100672
17. Xi, X., Tao, Q., Li, J., Kong, W., Zhao, Y. B., Wang, H., & Wang, J. (2021). Emotion-movement relationship: A study using functional brain network and cortico-muscular coupling. *Journal of Neuroscience Methods, 362*(January), 109320. doi: https://doi.org/10.1016/j.jneumeth.2021.109320
18. Shen, Y. W., & Lin, Y. P. (2019). Challenge for Affective Brain-Computer Interfaces: Non-stationary Spatio-spectral EEG Oscillations of Emotional Responses. *Frontiers in Human Neuroscience, 13*(October), 1–14. doi: https://doi.org/10.3389/fnhum.2019.00366
19. Hjorth, B. (1970). EEG analysis based on time domain properties. *Electroencephalography and clinical neurophysiology, 29*(3), 306–310. doi: https://doi.org/10.1016/0013-4694(70)90143-4
20. Oh, S.-H., Lee, Y.-R., & Kim, H.-N. (2014). A Novel EEG Feature Extraction Method Using Hjorth Parameter. *International Journal of Electronics and Electrical Engineering*, *2*(2), 106–110. https://doi.org/10.12720/ijeee.2.2.106-110
21. Bhurane, A. A., Dhok, S., Sharma, M., Yuvaraj, R., Murugappan, M., & Acharya, U. R. (2019). Diagnosis of Parkinson's disease from electroencephalography signals using linear and self-similarity features. *Expert Systems*, (October). doi: https://doi.org/10.1111/exsy.12472
22. Oh, S. L., Hagiwara, Y., Raghavendra, U., Yuvaraj, R., Arunkumar, N., Murugappan, M., & Acharya, U. R. (2020). A deep learning approach for Parkinson's disease diagnosis from EEG signals. *Neural Computing and Applications, 32*(15), 10927–10933. doi: https://doi.org/10.1007/s00521-018-3689-5
23. Murugappan, M., Alshuaib, W. B., Bourisly, A., Sruthi, S., Khairunizam, W., Shalini, B., & Yean, W. (2020). Emotion Classification in Parkinson's Disease EEG using RQA and ELM. *Proceedings - 2020 16th IEEE International Colloquium on Signal Processing and its Applications, CSPA 2020*, (Cspa), 290–295. doi: https://doi.org/10.1109/CSPA48992.2020.9068709
24. Choong, W. Y., Khairunizam, W., Omar, M. I. M. I., Murugappan, M., Bong, S. Z., Shahriman, A. B., ... Ibrahim, Z. (2018). Analysis of The Distance Metrics of KNN Classifier for EEG Signal in Stroke Patients. In *2018 International Conference on Computational Approach in Smart Systems Design and Applications (ICASSDA)* (pp. 1–4). doi: https://doi.org/10.1109/ICASSDA.2018.8477601

25. Murugappan, M., Alshuaib, W., Bourisly, A. K., Khare, S. K., Sruthi, S., & Bajaj, V. (2020). Tunable Q wavelet transform based emotion classification in Parkinson's disease using Electroencephalography. *PLoS ONE, 15*(11 November), 1–17. doi: https://doi.org/10.1371/journal.pone.0242014
26. Mathewson, K. J., Hashemi, A., Sheng, B., Sekuler, A. B., Bennett, P. J., & Schmidt, L. A. (2015). Regional electroencephalogram (EEG) alpha power and asymmetry in older adults: A study of short-term test-retest reliability. *Frontiers in Aging Neuroscience*, *7*(SEP), 1–10. doi: https://doi.org/10.3389/fnagi.2015.00177
27. Koenig, T., Prichep, L., Dierks, T., Hubl, D., Wahlund, L. O., John, E. R., & Jelic, V. (2005). Decreased EEG synchronization in Alzheimer's disease and mild cognitive impairment. *Neurobiology of Aging*, *26*(2), 165–171. doi: https://doi.org/10.1016/j.neurobiolaging.2004.03.008
28. Yuvaraj, R., Murugappan, M., Omar, M. I., Ibrahim, N. M., Sundaraj, K., Mohamad, K., & Satiyan, M. (2014). Emotion processing in Parkinson's disease: An EEG spectral power study. *International Journal of Neuroscience*, *124*(7), 491–502. doi: https://doi.org/10.3109/00207454.2013.860527
29. Yuvaraj, R., Murugappan, M., Ibrahim, N. M., Omar, M. I., Sundaraj, K., Mohamad, K., ... Satiyan, M. (2014). Emotion classification in Parkinson's disease by higher-order spectra and power spectrum features using EEG signals: A comparative study. *Journal of Integrative Neuroscience*, *13*(01), 89–120. doi: https://doi.org/10.1142/S021963521450006X
30. Li, X., Song, D., Zhang, P., Zhang, Y., Hou, Y., & Hu, B. (2018). Exploring EEG features in cross-subject emotion recognition. *Frontiers in Neuroscience*, *12*(MAR), 162. doi: https://doi.org/10.3389/fnins.2018.00162
31. Chow, J. C., Ouyang, C. Sen, Chiang, C. T., Yang, R. C., Wu, R. C., Wu, H. C., & Lin, L. C. (2019). Novel method using Hjorth mobility analysis for diagnosing attention-deficit hyperactivity disorder in girls. *Brain and Development*, *41*(4), 334–340. doi: https://doi.org/10.1016/j.braindev.2018.11.006
32. Hamida, S. T.-B. Ben, Ahmed, B., & Penzel, T. (2015). A novel insomnia identification method based on Hjorth parameters. In *2015 IEEE International Symposium on Signal Processing and Information Technology (ISSPIT)* (pp. 548–552). IEEE. doi: https://doi.org/10.1109/ISSPIT.2015.7394397
33. Portnova, G. V., & Atanov, M. S. (2018). Nonlinear EEG parameters of emotional perception in patients with moderate traumatic brain injury, coma, stroke and schizophrenia. *AIMS Neuroscience*, *5*(4), 221–235. doi: https://doi.org/10.3934/NEUROSCIENCE.2018.4.221
34. Koelstra, S., Mühl, C., Soleymani, M., Lee, J. S., Yazdani, A., Ebrahimi, T., ... Patras, I. (2012). DEAP: A database for emotion analysis; Using physiological signals. *IEEE Transactions on Affective Computing*, *3*(1), 18–31. doi: https://doi.org/10.1109/T-AFFC.2011.15
35. Urry, H. L., Nitschke, J. B., Dolski, I., Jackson, D. C., Dalton, K. M., Mueller, C. J., ... Davidson, R. J. (2004). Making a life worth living: neural correlates of well-being. *Psychological science*, *15*(6), 367–372. doi: https://doi.org/10.1111/j.0956-7976.2004.00686.x
36. Iwaki, T., Hayashi, M., & Hori, T. (1997). Changes in alpha band EEG activity in the frontal area after stimulation with music of different affective content. *Perceptual and motor skills*, *84*(2), 515–526. doi: https://doi.org/10.2466/pms.1997.84.2.515
37. Dambacher, F., Sack, A. T., Lobbestael, J., Arntz, A., Brugman, S., & Schuhmann, T. (2014). A network approach to response inhibition: dissociating functional connectivity of neural components involved in action restraint and action cancellation. *The European journal of neuroscience*, *39*(5), 821–831. doi: https://doi.org/10.1111/ejn.12425
38. Aron, A. R., Robbins, T. W., & Poldrack, R. A. (2004). Inhibition and the right inferior frontal cortex. *Trends in cognitive sciences*, *8*(4), 170–177. doi: https://doi.org/10.1016/j.tics.2004.02.010
39. Lenartowicz, A., Verbruggen, F., Logan, G. D., & Poldrack, R. A. (2011). Inhibition-related activation in the right inferior frontal gyrus in the absence of inhibitory cues. *Journal of Cognitive Neuroscience*, *23*(11), 3388–3399. doi: https://doi.org/10.1162/jocn_a_00031

40. De Pascalis, V., Cirillo, G., & Vecchio, A. (2020). Resting eeg asymmetry markers of multiple facets of the behavioral approach system: A loreta analysis. *Symmetry*, *12*(11), 1–20. doi: https://doi.org/10.3390/sym12111794
41. Jatupaiboon, N., Pan-Ngum, S., & Israsena, P. (2013). Real-time EEG-based happiness detection system. *The Scientific World Journal*, *2013*. doi: https://doi.org/10.1155/2013/618649
42. Kortelainen, J., Väyrynen, E., & Seppänen, T. (2015). High-Frequency Electroencephalographic Activity in Left Temporal Area Is Associated with Pleasant Emotion Induced by Video Clips. *Computational Intelligence and Neuroscience*, *2015*, 762769. doi: https://doi.org/10.1155/2015/762769
43. Deng, X., Yang, M., & An, S. (2021). Differences in frontal EEG asymmetry during emotion regulation between high and low mindfulness adolescents. *Biological Psychology*, *158*-(November 2020), 107990. doi: https://doi.org/10.1016/j.biopsycho.2020.107990
44. Yuvaraj, R., Murugappan, M., & Palaniappan, R. (2017). The Effect of Lateralization of Motor Onset and Emotional Recognition in PD Patients Using EEG. *Brain Topography*, *30*(3), 333–342. doi: https://doi.org/10.1007/s10548-016-0524-0
45. Gainotti, G. (2012). Unconscious processing of emotions and the right hemisphere. *Neuropsychologia*, *50*(2), 205–218. doi: https://doi.org/10.1016/j.neuropsychologia.2011.12.005
46. Dominguez-Rodriguez, A., Chavez-Valdez, S. M., Avitia, G. C., & Valencia-Espinoza, L. C. (2020). Unified protocol for anxiety disorders in two cities of Mexico measuring gamma activity: Study protocol for a randomized controlled trial. *Contemporary clinical trials communications*, *18*, 100556. doi: https://doi.org/10.1016/j.conctc.2020.100556
47. Campos, A. C., Vaz, G. N., Saito, V. M., & Teixeira, A. L. (2014). Further evidence for the role of interferon-gamma on anxiety- and depressive-like behaviors: Involvement of hippocampal neurogenesis and NGF production. *Neuroscience Letters*, *578*, 100–105. doi: https://doi.org/10.1016/j.neulet.2014.06.039
48. Keil, A., Müller, M. M., Gruber, T., Wienbruch, C., Stolarova, M., & Elbert, T. (2001). Effects of emotional arousal in the cerebral hemispheres: A study of oscillatory brain activity and event-related potentials. *Clinical Neurophysiology*, *112*(11), 2057–2068. doi: https://doi.org/10.1016/S1388-2457(01)00654-X
49. Mathersul, D., Williams, L. M., Hopkinson, P. J., & Kemp, A. H. (2008). Investigating Models of Affect: Relationships Among EEG Alpha Asymmetry, Depression, and Anxiety. *Emotion*, *8*(4), 560–572. doi: https://doi.org/10.1037/a0012811

A Novel Parametric Nonstationary Signal Model for EEG Signals and Its Application in Epileptic Seizure Detection

Gokul Krishnan and Pradip Sircar

Abstract In this article, we present a novel parametric nonstationary signal model for electroencephalogram (EEG) signals and its application in epilepsy classification. The EEG signals are considered as signatures of neural activities and represent the electrical activity inside the brain. The EEG signals are considered predominantly nonlinear and nonstationary in nature; thus, obtaining meaningful inferences from the EEG signal has been a strenuous task. We discuss a mono-component amplitude and frequency modulated (AFM) signal model for EEG signal representation. Our model decomposes the EEG signal into a low-frequency trend component and a high-frequency AFM signal component, which can be useful for classification of various neurological disorders. We consider a potential application of the proposed approach in epileptic seizure classification. The smoothing prior's approach has been used for extracting the low-frequency average component, while the three-peak cluster in the frequency domain has been used for segmenting the EEG signals in order to capture its instantaneous parameters. The study has been done on the publicly available EEG database of the University of Bonn. Our study suggests that the proposed approach helps in providing meaningful insights for EEG signal analysis using the time-varying model and aids in automated classification of various neurological disorders such as epilepsy.

Keywords EEG · Epileptic detection · Seizure · Feature extraction · Amplitude modulation · Frequency modulation

---

G. Krishnan
Electrical and Computer Engineering Department, University of Connecticut, Storrs, CT, USA
e-mail: gokul.krishnan_jr@uconn.edu

P. Sircar (✉)
Electrical Engineering Department, Indian Institute of Technology (ISM), Dhanbad, Jharkhand, India
e-mail: pradipsircar@iitism.ac.in; sircar@iitk.ac.in

M. Murugappan, R. Yuvaraj (eds.), *Biomedical Signals Based Computer-Aided Diagnosis for Neurological Disorders*, https://doi.org/10.1007/978-3-030-97845-7_7

1 Introduction

The EEG signal-based analysis techniques have attracted a wide interest among the signal processing and the neuroimaging communities due to their high temporal resolution and compatibility with various other neuroimaging techniques. The EEG is recorded via placing an array of electrodes over the scalp. It is a noninvasive technique having a wide range of applications in detection of neurological disorders, brain-computer interface (BCI), etc. The huge amount of EEG data recorded using the electrodes provides an interpretation challenge which necessitates the use of novel signal modeling and processing algorithms for automation and analysis. Traditionally, the high-frequency components (>80 Hz) of the EEG signals have not been considered till the recent studies depicted that the gamma oscillations (particularly greater than 30 Hz) have particular advantages in cognitive and various neurological studies [1].

The analysis of EEG signals using compact mathematical model has numerous advantages. The EEG signal is predominantly nonstationary due to the incessant variations of its amplitude and frequency contents. The spectral analysis (e.g., the Fourier transform-based analysis) of the nonstationary EEG signal reveals the frequency content of the signal over the window duration considered. However, analyzing the time instants where a particular band of frequency appears or disappears in the EEG signal (such as multi-resolution analysis) helps in providing a better understanding of the brain activity. When the data length is sufficiently small, the parametric time-series modeling approaches yield a better accuracy as compared to the nonparametric approaches based on the Fourier transform [2]. However, the evolutionary nature of the EEG spectrum may not be quantified properly while using a constant coefficient time-series model.

It has been previously demonstrated that by proper segmentation, the EEG signals can be modeled by a time-varying AM-FM component [3]. The EEG signals are nonstationary in nature, thereby having a time-varying dc component. The low-frequency time-varying dc component of the EEG signal possesses significance in applications such as sleep and epilepsy; however, in applications such as brain-computer interface (BCIs) for neurorehabilitation, the high-frequency components without the low-frequency trend of the signal play prime importance [4].

Hence, we postulate the EEG signal model as a combination of a time-varying dc component modulated with an amplitude and frequency modulated component. Thus, the EEG signal without its average variation can be considered as the real part of an analytical signal of the form $a(t)\ exp\ (j\phi(t))$. The instantaneous frequency of the signal can be obtained as $\omega(t) = \phi'(t)$. As compared to the previous approaches [3], the proposed modeling approach thus helps in obtaining the nonnegative amplitude function $a(t)$ and the instantaneous frequency (IF) directly from the model. Estimating the instantaneous parameters of the model helps in providing a suitable time-frequency representation of the EEG signals. The time-varying dc

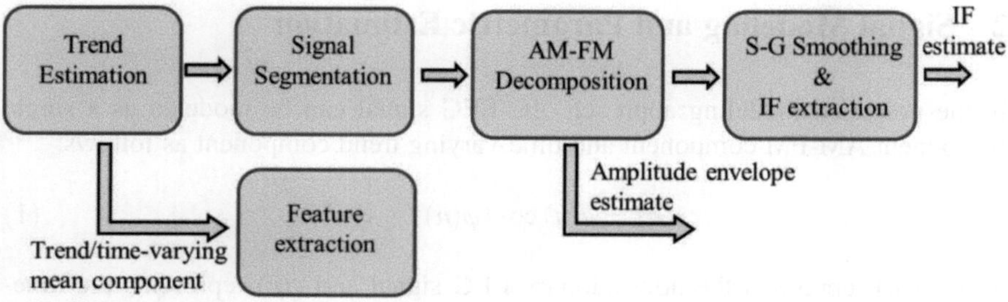

Fig. 1 Block diagram of the proposed modeling and instantaneous parameter estimation approach

component of the EEG signal can be translated as the trend of the EEG time-series data, thereby utilizing the standard trend estimation algorithms for our analysis. We have utilized smoothing priors' approach for estimating the time-varying average component, followed by segmentation and Hilbert transform analysis for extracting the instantaneous parameters of the signal. After estimating the instantaneous parameters, we have used phase unwrapping and smoothing in order to get a monotonically increasing phase. The following sections detail the estimation approach for the parameters of the model. By capturing the instantaneous parameters of the EEG signal, the proposed model is potentially useful for various applications where the insights for instantaneous parameters of the EEG signal are necessary.

In order to demonstrate the potential benefits of the proposed modeling approach, we consider epileptic signal analysis, as the epileptic seizures are characterized by instantaneous spikes. In general, epileptic seizures are caused due to the abnormality in the brain wiring or an imbalance of nerve signaling chemicals called neurotransmitters [5]. It can be categorized into two types: focal or non-focal seizures. The focal seizures originate from some circumscribed regions inside the brain, while in case of generalized seizures, most areas of the brain are involved. Continuous monitoring of the EEG dataset for epileptic seizure detection is tedious and is prone to human errors. This necessitates the use of automation algorithms for epileptic seizure detection. For signal classification, we use time domain- or frequency domain-based features. In case of nonstationary signals such as EEG, the time-frequency domain-based features have also been exploited for classification purpose. In the following subsections, we describe the estimation of the proposed model parameters and its application in epileptic seizure detection.

The block diagram of the proposed approach is as shown in Fig. 1. This article is organized into four different sections: Sect. 1 provides the introduction; Sect. 2 discusses the proposed modeling approach and its details. Section 3 deals with the simulation results; Sect. 4 discusses its application in epileptic seizure analysis. Finally, Sect. 5 provides the conclusions of the findings.

2 Signal Modeling and Parametric Estimation

In the proposed modeling approach, the EEG signal can be modeled as a single component AM-FM component and time-varying trend component as follows:

$$x(n) = a(n)\cos(\varphi(n)) + v(n) \tag{1}$$

where $x(n)$ represents the nonstationary EEG signal and $v(n)$ represents the time-varying trend component. Here, $a(n)$ and $\varphi(n)$ represent the amplitude envelope and the instantaneous phase of the signal. The separation of nonstationary EEG signal into the form represented in Eq. (1) amounts to separating the low-frequency and high-frequency modes of the EEG signal. We have used the approach similar to the one presented in [6] for estimating $v(n)$. The $v(n)$ can be modeled as

$$\mathbf{v} = \mathbf{A}\boldsymbol{\theta} + \mathbf{w} \tag{2}$$

where $\mathbf{A} \in \mathbb{R}^{\mathrm{N} \times \mathrm{m}}$ represents the observation matrix and its columns serve as the basis vectors for the fitting procedure. For simplicity, we have assumed the observation matrix $\mathbf{A} = \mathbf{I}^{\mathrm{N} \times \mathrm{N}}$. In order to estimate $\boldsymbol{\theta}$, we use the penalized log-likelihood estimation technique [7]. Using the regularized least squares method,

$$\begin{aligned} \widehat{\boldsymbol{\theta}} &= \mathbf{arg}\min_{\theta}\left(\|\mathbf{A}\boldsymbol{\theta} - \mathbf{x}\|^2 + \lambda^2\|\mathbf{L}(\mathbf{A}\boldsymbol{\theta})\|^2\right) \\ &= \left(\mathbf{A}^T\mathbf{A} + \lambda^2\mathbf{A}^{\mathrm{T}}\mathbf{L}^{\mathrm{T}}\mathbf{L}\mathbf{A}\right)^{-1}\mathbf{A}^{\mathrm{T}}\mathbf{x}. \end{aligned} \tag{3}$$

Here, the first term indicates the goodness of fit, while the second term indicates the smoothness constraint. The $\mathbf{L}$ can be chosen either as an identity matrix or discrete approximation to the kth derivative operator $D_k \epsilon (N - k) \times N$:

$$\text{where} \qquad D_k = \begin{bmatrix} C_k^0 - C_k^1 & \cdots & (-1)^k C_k^k & \ldots & 0 \\ \vdots & & \ddots & & \vdots \\ 0 & C_k^0 & \cdots & & (-1)^k C_k^k \end{bmatrix} \tag{4}$$

with the binomial coefficient C_k^i. For our analysis, we have utilized the second-order derivative matrix $D_2 \epsilon \mathbb{R}^{(\mathrm{N}-2) \times \mathrm{N}}$ as the regularizing matrix since it has been proved to be a suitable choice in time-series trend estimation [6]. The second-order derivative matrix D_2 is given by

$$D_2 = \begin{bmatrix} 1 & -2 & 1 & 0 & \cdots & & & & 0 \\ \vdots & & & & \ddots & & & & \vdots \\ 0 & & & & \cdots & 0 & 1 & -2 & 1 \end{bmatrix} \tag{5}$$

which has been obtained by the symmetric finite difference approximation to the second derivative and putting it in a matrix form. With these specific choices, we can cestimate $\widehat{\boldsymbol{\theta}}$ and $\widehat{\mathbf{v}} = \mathbf{A}\widehat{\boldsymbol{\theta}}$. With different choices of λ, the estimated trend component varies. Moreover, the estimation of λ is not trivial even by using iterative approaches as it depends on the initial value selection and the chosen λ value is subjective and application oriented. Assuming the local dc variation in the EEG signals being sufficiently smooth and continuous, we choose the λ value which will ensure the estimated trend component to be smooth and less oscillatory. This choice is made because the AM-FM features may get captured into the estimated local dc variation if we allow it to be oscillatory. Thus, we choose the λ value as the minimum value of λ which reduces the oscillatory behavior of the extracted trend component. This, in turn, can be done by plotting the number of local maxima *v*/*s* λ of $v(n)$ and choosing the λ value as the minimum value for which the plot approximately saturates. Our simulation results using synthetic data reveal that this procedure reduces the mean square error of the estimated trend component.

2.1 Segmentation and AM-FM Demodulation

Most of the naturally occurring signals such as EEG can be approximated as mono-component signals if properly segmented [3]. In order to use the demodulation algorithms such as discrete energy separation algorithm (DESA) or the Hilbert transform-based approach, the signal under consideration must be mono-component AM-FM signal. Since the mono-component AM-FM signals can be characterized by symmetric three-peak cluster in the frequency domain, we have used this characteristic as shown in Fig. 2 and the constraint on phase to be a monotonic function in time for EEG signal segmentation. After segmentation, we utilize Hilbert transform-based approach for estimating $a(n)$ and $\phi(n)$ unambiguously. Assume $x(t)$ be the continuous time signal corresponding to $x(n)$, then $x(t)$ can be expressed in its analytic form $z(t) = x(t) + j\mathcal{H}(x(t)) = a(t)e^{j\phi(t)}$, where $\mathcal{H}(x(t))$ be the Hilbert transform of $x(t)$ which can be computed as

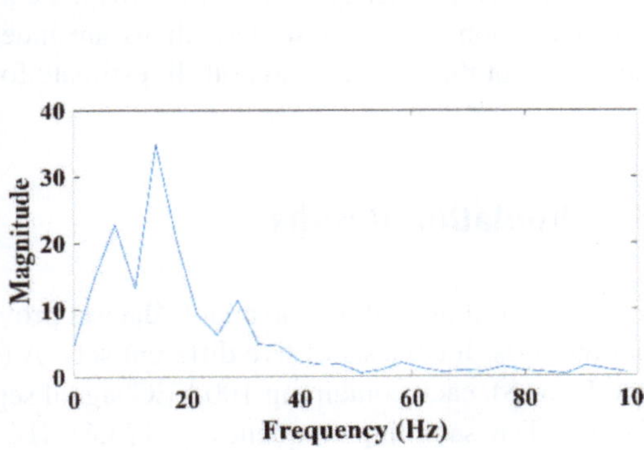

Fig. 2 FFT plot of the EEG segment (three-peak cluster)

$$\mathcal{H}(x(t)) = \frac{1}{\pi}\int_{-\infty}^{\infty} \frac{x(\tau)}{\tau - t} d\tau \tag{6}$$

The amplitude and the phase functions of the analytic form can be estimated by calculating the magnitude and argument of $z(t)$, respectively, which can be sampled to get $a(n)$ and $\phi(n)$.

2.2 Phase Smoothing and Instantaneous Frequency (IF) Estimation

Once $a(n)$ and $\phi(n)$ are estimated, in order to avoid abrupt jumps and discontinuities in $\phi(n)$, phase unwrapping and smoothing are performed. The phase unwrapping [8] is usually adopted in most of the literature for instantaneous frequency estimation. After phase unwrapping, for elimination of the discontinuities in the unwrapped phase, it is further subjected to smoothing. For this, we use Savitzky-Golay smoothing procedure due to its benefits such as preservation of high-frequency content of the signal [9]. Consider $N = 2M + 1$ samples, where N denotes the frame length and $x = [x_{-M}, \ldots, x_{-1}, x_0, x_1, \ldots, x_M]$ be the data to be fitted by a polynomial of degree m, where $N > 2m + 1$. The $m + 1$ polynomial basis vectors $s_i(n) = n^i, \ -M \leq n \leq M$. Thus, the smoothed data can be computed as

$$\widehat{x} = \sum_{i=0}^{m} c_i s_i = Sc \tag{7}$$

where $S = [s_0, s_1, \ldots, s_m]$ and $c = [c_0, c_1, \ldots, c_m]^T$. The coefficient vector c can be estimated by minimizing the cost function $J = e^T e$, where $e = x - Sc$. For Savitzky-Golay filtering, we need to choose the optimal frame length $\widehat{N}$ and polynomial order m. For a fixed polynomial order m, the optimal frame length can be obtained as the minimum frame length which ensures the estimated instantaneous frequency (IF) to be nonnegative. As described in the following section, the smoothed signal $\widehat{x}$ and the instantaneous frequency (IF) estimate which can estimated by the finite difference approximation to the phase derivatives are independent of the chosen polynomial order except that we get a smooth IF estimate for higher degree of polynomials.

3 Simulation Results

We have used the publicly available dataset provided by the University of Bonn for our analysis. It consists of five different sets: A (or Z), B (or O), C (or N), D (or F), and E (or S), each containing 100 EEG signal segments of 23.6-s duration in ASCII format. The sampling frequency is 173.61 Hz. Sets A and B are recorded from healthy volunteers, while sets C, D, and E correspond to data recorded from epileptic patients. The detailed data description can be found in [10]. Since the dataset is free of any artifacts, no prior preprocessing is required.

For analyzing the performance of the trend estimation algorithm, we have considered three different simulated signals, $v_1(t) = 2\,cos\,(10t), v_2(t) = 0.25\,cos\,(8t) + 0.5\,sin\,(5t), v3(t) = 0.5\,cos\,(8t) + 0.5\;exp\,(-t/15) + 0.75\,sin\,(5t)$ embedded on the signal $s(t) = (1 + 0.3\,cos\,(2t))\,exp\,(-t/15)\,cos\,\left(2\pi\left(2.4t + 100t^{1.2} + 0.3\,sin\,(25t)\right)\right)$. The mean square error (MSE) calculated between the estimated trend component and the original trend of the signal is shown in Fig. 3, where the red mark indicates the estimated λ value using our approach. From Fig. 3, we can see that the estimated λ value provides a trend estimate which is sufficiently close to the minimum MSE value.

The original EEG signal, its amplitude envelope and instantaneous frequency (IF) estimates, and the reconstructed signal using the proposed modeling approach are shown in Fig. 4. In order to analyze the error in the modeling approach, we have utilized various distance measures such as the signal-to-error ratio (SER) and the log-spectral distance measure between the original and reconstructed signals. The SER and the log-spectral distance measure thus obtained equal to 11.8171 dB and 0.3520, respectively, which indicates the efficiency of the proposed model.

4 Applications

As an application of the proposed modeling approach, we consider the task of epileptic seizure classification. We consider simple features extracted from the estimated parameters from the proposed model for seizure classification. Even though we consider seizure classification in our analysis, the proposed IF estimation approach will be particularly useful in cases where analysis of sudden signal transients or spikes is of prime importance. We consider the graphical-based features for the extracted trend component and simple statistical features from the AM-FM components. EEG data blocks of 4000 samples are used for our analysis, and the trend and the AM-FM components are extracted using the above-described procedure. For extracting the graphical-based features from the trend component, we use the phase space plots which have been widely used for analysis of nonlinear dynamics of various signals.

A dynamic system is characterized by a state which provides the information about a system with its dynamics which describes how the state evolves over time [11]. The phase space plots provide a visual image of the evolution of the dynamic behavior of the signal. Let the signal be $x = [x_1, x_2, \ldots, x_N]$, and then each point in the phase plot is given by $y_k = [x_k, x_{k\,+\,\tau}, \ldots, x_{k\,+\,(d\,-\,1)\tau}]$, $k = 1, 2, 3, \ldots, N - (d - 1)\tau$. In our study, we have considered $d = 2$, i.e., 2-D phase space representation (or Poincare plot) and $\tau = 1$ as considered in [12]. The standard descriptors used in case of the Poincare plot are its centroid, area, acceleration, and deceleration. The

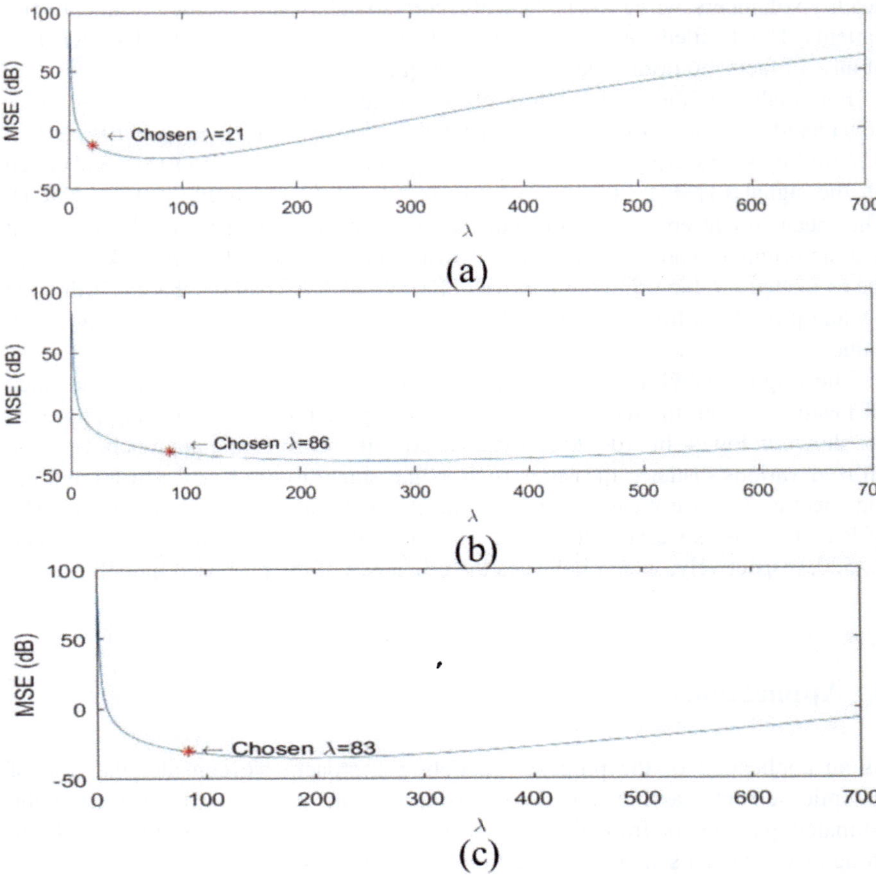

Fig. 3 Mean square error (MSE) for the trend estimate for different types of trend signals: (**a**) $v_1(t) = 2\cos(10t)$, (**b**) $v_2(t) = 0.25\cos(8t) + 0.5\sin(5t)$ and (**c**) $v3(t) = 0.5\cos(8t) + 0.5\exp(-t/15) + 0.75\sin(5t)$, embedded in a single component AM-FM signal $s(t) = (1 + 0.3\cos(2t))\exp(-t/15)\cos(2\pi(2.4t + 100t^{1.2} + 0.3\sin(25t)))$

centroid can be computed as $x_c = \bar{x}_i$ and $y_c = \bar{y}_i$, where the overbar indicates the mean. For computing other descriptors, we define the following quantities:

$$D_i^1 = \frac{|x_i - x_c - y_i - y_c|}{\sqrt{2}} \tag{8}$$

$$D_i^2 = \frac{|x_i - x_c + y_i - y_c|}{\sqrt{2}} \tag{9}$$

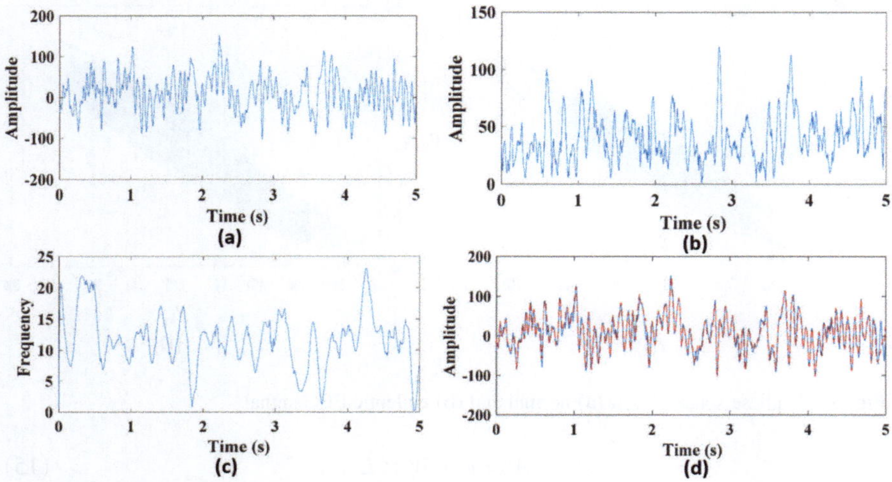

Fig. 4 (**a**) EEG signal. (**b**) Estimated amplitude envelope. (**c**) Estimated instantaneous frequency (IF). (**d**) Original and reconstructed (red) signals

In addition, for computing the measure of asymmetry of Poincare plot, about the line of identity has been defined for RR interval signals [13]. For computing such features, the following quantities are computed:

$$SD_{\text{UP}} = \frac{1}{N}\sum_{i=1}^{n_u} \left(d_U^i\right)^2 \tag{10}$$

$$SD_{\text{DOWN}} = \frac{1}{N}\sum_{i=1}^{n_D} \left(d_D^i\right)^2 \tag{11}$$

and

$$SD_1 = SD_{\text{UP}} + SD_{\text{DOWN}} \tag{12}$$

where n_U and n_D are the number of points above and below the line of identity, respectively. The d_U^i and d_D^i are the distance of the ith point above and below the line of identity. From the above quantities, we can compute the following features:

$$W = \frac{1}{N}\sum_{I=1}^{N} \left(D_i^1\right)^2 \tag{13}$$

$$L = \frac{1}{N}\sum_{i=1}^{N} \left(D_i^2\right)^2 \tag{14}$$

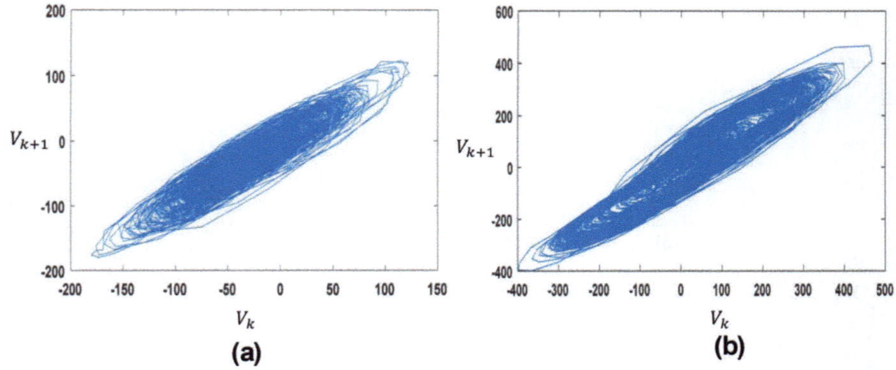

Fig. 5 2D phase space plot of (**a**) normal and (**b**) epileptic EEG signal

$$A = \pi \times W \times L \tag{15}$$

$$D_{\mathrm{UP}} = \frac{SD_{\mathrm{UP}}^2}{SD_1^2} \tag{16}$$

$$D_{\mathrm{DOWN}} = \frac{SD_{\mathrm{DOWN}}^2}{SD_1^2} \tag{17}$$

where W, L, and A represent the width, length, and the area of the Poincare plot and D_{UP} and D_{DOWN} represent the acceleration and deceleration, respectively. In addition to these features, simple features such as the mean, variance, skewness, kurtosis, etc. of the estimated instantaneous components are also computed in order to select the features with the best discrimination capability. For example, Fig. 5 shows the phase space plot of the normal and epileptic EEG signal. From Fig. 5, we can see that the features of the phase space could be potentially used for discriminating different classes of epileptic EEG signals.

In order to select the best features for classification, we utilized Kruskal-Wallis test [14], which has been widely used for feature selection [15]. Thus, we selected width, length, area of the Poincare plot of the trend component, and the mean and variance of the instantaneous amplitude envelope for classification. The variation for the extracted features (e.g., the mean and the variance) of the instantaneous parameters is shown in Fig. 6.

Once the features are selected, we have used the following classifiers for performance evaluation: decision trees, bagged trees, *k*-nearest neighbor, and support vector machines (SVM).

Decision Trees: The decision trees are a nonparametric approach where a dataset is sequentially divided into smaller subdivisions according to some decision framework defined by the tree at each node and the observations are assigned the class labels according to the root nodes which it falls [16]. In Table 1, fine tree depicts a decision tree with maximum split equals to 100, while medium and coarse tree

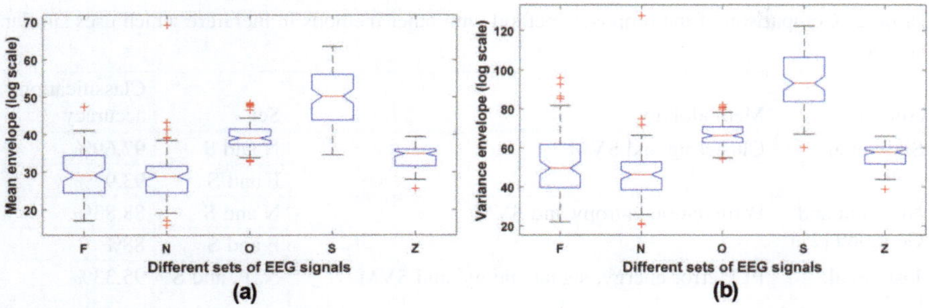

Fig. 6 Variation of the (**a**) mean and (**b**) variance of the estimated envelope of different sets of EEG signal

Table 1 The performance of different classifiers for Z, O, N, F, and S. Here k specifies the number of neighbors chosen

Classifier	Parameters/kernel used	Cross-validation	Mean accuracy	Maximum accuracy
Decision tree	Fine tree	Tenfold	95.82%	96.8%
	Medium tree		95.64%	96.8%
	Coarse tree		95.64%	96.8%
Ensemble method	Bagged trees	Tenfold	96.6%	97.2%
SVM	RBF kernel	Tenfold	95.72%	96%
k-NN	$k = 1$	Tenfold	96.88%	97.4%
	$k = 10$		96.12%	96.6%
	$k = 100$		91.02%	91.3%

denotes decision trees with maximum split of 20 and 4, respectively. For the decision trees, we have used the Gini's diversity index as the splitting criterion.

Ensemble Classifier: In this case, we have used the bagged trees. The bagged trees are an ensemble method which is implemented using Breiman's random forest algorithm [17].

k-nearest Neighbours (k-NN): The k-nearest neighbor method uses a similarity function such as a distance metric for computing the similarity between the training instances and the instances in the classification record.

Support Vector Machines (SVM): The SVM approach is a supervised learning approach which involves constructing multiple or a single hyperplane which provides the highest margin distance between the nearest points of different classes [18]. The nonlinear classification problems can be solved using kernel trick, which involves mapping the feature space into higher dimensions using appropriate kernel functions. There are different types of kernel functions used in the SVM classifiers such as linear, polynomial, Gaussian, radial basis function (RBF), etc. among which the RBF being the most popular.

Table 2 Comparison of the proposed method with other methods in literature which uses similar dataset

Authors	Methodology	Set	Classification accuracy
Siuly et al. [20]	Clustering and SVM	N and S	97.69%
		F and S	93.91%
Nicolaou and Georgiou [21]	Permutation entropy and SVM	N and S	88.83%
		F and S	88%
Joshi et al. [22]	FLP error energy, signal energy, and SVM	N, F, and S	95.33%
Zhu et al. [23]	Degree and strength of HVG and k-NN	N and S	98%
		F and S	93%
		Z, O, N, F, and S	95.4%
Swami et al. [24]	Dual tree complex wavelet transform and energy, standard deviation, Shannon entropy features using general regression neural network (GRNN)	Z, O, and S	99.18%
		N, F, and S	95.24%
		Z, O, N, F, and S	95.15%
Proposed approach	Phase space-based features of trend, mean, variance of envelope and k-NN	Z, O, and S	99.7%
		N, F, and S	95.6%
		Z, O, N, F, and S	97.4%

We have used the k-fold cross-validation (with $k = 10$) [19] for model validation. For classification, we have utilized the MATLAB Classification Learner application. From Table 1, we can see that the k-NN classifier yielded the best results.

In addition, we have compared the classification results based on the approach mentioned above with some of the existing methods in the literature which is enumerated in Table 2.

5 Conclusion and Discussion

The proposed time-varying amplitude and frequency modulated model with the associated trend component is found suitable for modeling the EEG signals. The log-spectral distance measure and the signal-to-error ratio reveal a close match, and therefore, the proposed modeling approach can be suitably used for EEG signal representation. The proposed modeling approach helps us to decompose the signal into high-frequency and low-frequency modes of the EEG signal. These different modes could be utilized for specific applications: the low-frequency model-based components may be helpful for analysis such as sleep or epilepsy analysis, while the high-frequency modes might be useful for BCI applications. Using the proposed model, we can extract the instantaneous frequency and amplitude of the EEG signal, which can be used for various neurological applications such as in Parkinson disease and emotion recognition [25, 26]. As a potential application, we consider epilepsy classification; however, this model may be suitable in a wide range of applications where the signal's accurate instantaneous frequency (IF) estimation is of prime importance. For the classification that we considered, an accuracy of about 97.4% was achieved. Recently, many deep neural network-based classification approaches have pushed the recognition accuracy to a great extent in EEG analysis [27]. However, such approaches require large amount of data, and their outputs cannot be completely comprehended even if we have access to the weights of the network, which may be crucial especially in case of certain medical applications. To handle these disadvantages, nowadays physics informed deep learning has gained increased interest among the scientific communities. Therefore, our future research aims at combining the insights gained from the proposed modeling approach to build a physics informed network which would be able to work with lesser data. Even though the Kruskal-Wallis test yields best results with these set of features, from Fig. 6, we can see that further feature reduction might be possible. The proposed model also can be potentially parameterized by expanding the time-varying amplitude and phase functions using the standard AFM model parameters or by polynomial approximation for efficient data storage.

References

1. C. Dockstader, F. Wang, E. Bouffet, and D. J. Mabbott, "Gamma deficits as a neural signature of cognitive impairment in children treated for brain tumors," J. Neurosci. **34**(26), 8813–8824 (2014).
2. S. M. Kay, *Modern Spectral Estimation : Theory and Application* (Prentice-Hall, 1988).
3. R. B. Pachori and P. Sircar, "EEG signal analysis using FB expansion and second-order linear TVAR process," Signal Processing **88**(2), 415–420 (2008).
4. G. R. Philips, M. K. Hazrati, J. J. Daly, and J. C. Principe, "Addressing low frequency movement artifacts in EEG signal recorded during center-out reaching tasks.," Annu. Int. Conf. IEEE Eng. Med. Biol. Soc. IEEE Eng. Med. Biol. Soc. Annu. Int. Conf. **2014**, 6497–6500 (2014).
5. S. Sanei and J. A. Chambers, "Seizure Signal Analysis," in *EEG Signal Processing* (John Wiley & Sons, Ltd, 2007), pp. 161–195.
6. M. P. Tarvainen, P. O. Ranta-aho, and P. A. Karjalainen, "An advanced detrending method with application to HRV analysis," IEEE Trans. Biomed. Eng. **49**(2), 172–175 (2002).
7. S. Konishi and G. Kitagawa, *Information Criteria and Statistical Modeling*, 1st ed. (Springer Publishing Company, Incorporated, 2007).
8. A. V Oppenheim and R. W. Schafer, *Discrete-Time Signal Processing*, 3rd ed. (Prentice Hall Press, 2009).
9. S. J. Orfanidis, *Introduction to Signal Processing* (Prentice-Hall, Inc., 1995).
10. R. G. Andrzejak, K. Lehnertz, F. Mormann, C. Rieke, P. David, and C. E. Elger, "Indications of nonlinear deterministic and finite-dimensional structures in time series of brain electrical activity: Dependence on recording region and brain state," Phys. Rev. E **64**(6), 61907 (2001).
11. L. D. Iasemidis, J. Chris Sackellares, H. P. Zaveri, and W. J. Williams, "Phase space topography and the Lyapunov exponent of electrocorticograms in partial seizures," Brain Topogr. **2**(3), 187–201 (1990).

12. F. Takens, "Detecting strange attractors in turbulence," in *Dynamical Systems and Turbulence, Warwick 1980,* D. Rand and L.-S. Young, eds. (Springer Berlin Heidelberg, 1981), pp. 366–381.
13. J. Piskorski and P. Guzik, "Geometry of the Poincaré plot {ofRRintervals} and its asymmetry in healthy adults," Physiol. Meas. **28**(3), 287–300 (2007).
14. Z. Wang and Z. Xie, "Infrared face recognition based on local binary patterns and Kruskal-Wallis test," in *2014 IEEE/ACIS 13th International Conference on Computer and Information Science (ICIS)* (2014), pp. 185–188.
15. R. Sharma and R. B. Pachori, "Classification of epileptic seizures in EEG signals based on phase space representation of intrinsic mode functions," Expert Syst. Appl. **42**(3), 1106–1117 (2015).
16. M. A. Friedl and C. E. Brodley, "Decision tree classification of land cover from remotely sensed data," Remote Sens. Environ. **61**(3), 399–409 (1997).
17. L. Breiman, "Random Forests," Mach. Learn. **45**(1), 5–32 (2001).
18. G. Mountrakis, J. Im, and C. Ogole, "Support vector machines in remote sensing: A review," ISPRS J. Photogramm. Remote Sens. **66**(3), 247–259 (2011).
19. R. Kohavi, "A Study of Cross-Validation and Bootstrap for Accuracy Estimation and Model Selection," in *Proceedings of the 14th International Joint Conference on Artificial Intelligence - Volume 2*, IJCAI'95 (Morgan Kaufmann Publishers Inc., 1995), pp. 1137–1143.
20. Siuly, Y. Li, and P. Wen, "Analysis and classification of EEG signals using a hybrid clustering technique," in *IEEE/ICME International Conference on Complex Medical Engineering* (2010), pp. 34–39.
21. N. Nicolaou and J. Georgiou, "Detection of epileptic electroencephalogram based on Permutation Entropy and Support Vector Machines," Expert Syst. Appl. **39**(1), 202–209 (2012).
22. V. Joshi, R. B. Pachori, and A. Vijesh, "Classification of ictal and seizure-free EEG signals using fractional linear prediction," Biomed. Signal Process. Control **9**, 1–5 (2014).
23. G. Zhu, Y. Li, and P. (Paul) Wen, "Epileptic seizure detection in EEGs signals using a fast weighted horizontal visibility algorithm," Comput. Methods Programs Biomed. **115**(2), 64–75 (2014).
24. P. Swami, T. K. Gandhi, B. K. Panigrahi, M. Tripathi, and S. Anand, "A novel robust diagnostic model to detect seizures in electroencephalography," Expert Syst. Appl. **56**, 116–130 (2016).
25. Siao Zheng Bong, Khairunizam Wan, M. Murugappan, Norlinah Mohamed Ibrahim, Yuvaraj Rajamanickam, Khairiyah Mohamad, "Implementation of wavelet packet transform and non linear analysis for emotion classification in stroke patient using brain signals", Biomedical Signal Processing and Control, Volume 36, 2017, 102-112.
26. Yuvaraj R, Murugappan M, Mohamed Ibrahim N, Iqbal M, Sundaraj K, Mohamad K, Palaniappan R, Mesquita E, Satiyan M. On the analysis of EEG power, frequency and asymmetry in Parkinson's disease during emotion processing. Behav Brain Funct. 2014 Apr 9;10:12. doi: https://doi.org/10.1186/1744-9081-10-12.
27. Yuvaraj, R., Thomas, J., Kluge, T., & Dauwels, J. (2018). A deep Learning Scheme for Automatic Seizure Detection from Long-Term Scalp EEG. Paper presented at the 52nd Asilomar Conference on Signals, Systems, and Computers, USA.

Biomedical Signal Analysis Using Entropy Measures: A Case Study of Motor Imaginary BCI in End Users with Disability

Hanif Heidari

Abstract Entropy is an important topic in biomedical signal analysis. It is widely applied to numerous practical data to determine neurological disorders. Various entropy measures have been developed from both theoretical and experimental perspectives. Approximate entropy and Shannon entropy are basic entropy measures that have been successfully applied to various biomedical data. However, these measures have some drawbacks, such as sensitivity to data length, initial parameters, and fluctuations. In an attempt to overcome these difficulties, this chapter considers NNetEn, which is based on a feedforward neural network. Unlike traditional entropy measures, the NNetEn measure does not consider a probability distribution. It applies a modified version of the LogNNET classification model to the MNIST-10 dataset to calculate the NNetEn entropy. Against this background, this chapter provides a simple description of the entropy concept, basic entropy measures, and NNetEn. Basic entropy measures are used for binary classification of 30-channel EEG of mental tasks in nine disabled users with spinal cord injury or stroke. Channel selection is another hot topic in biomedical signal analysis, which is essential for IoT devices in healthcare. The NNetEn measure is used for channel selection in the considered dataset, which shows that 22 electrode channels can be ignored without loss of accuracy. The results show that entropy-based methods are an efficient tool for detecting neurological disorders using experimental biomedical signals.

Keywords Time series · Entropy · Neural network · LogNNet · Classification · Brain-computer interface · Motor imaginary · EEG · Channel selection

---

H. Heidari (✉)
Department of Applied mathematics, Damghan University, Damghan, Iran
e-mail: heidari@du.ac.ir

M. Murugappan, R. Yuvaraj (eds.), *Biomedical Signals Based Computer-Aided Diagnosis for Neurological Disorders*, https://doi.org/10.1007/978-3-030-97845-7_8

1 Introduction

Neurological disorders (NDs) are among the greatest challenges of modern medicine. According to the WHO, any disease that affects the nervous system is considered an ND. There are more than 400 ND-related diseases. Parkinson's disease, motor neuron disease, and epilepsy are the most common NDs. NDs affect millions of people in the world every year, regardless of age, gender, education, and income. NDs are the leading cause of disability-adjusted life years (DALYs, 276 million), and they were the second leading cause of death (9.0 million, or 16.5% of global deaths) in 2019 [1]. Little wonder, then, diagnosis has become a hot topic in recent years. The analysis of the human nervous system is the starting point for diagnosis. The human nervous system can be viewed as a dynamic directed graph where neurons and synapses are nodes and vortices, respectively. This graph contains about 10^{10} nodes and 10^{12} synapses. Therefore, it is impossible to simulate and analyze it. Currently, electroencephalography (EEG), computerized tomography (CT scan or CAT scan), magnetic resonance imaging (MRI scan), electromyography (EMG), positron emission tomography (PET scan or PET imagery), arteriogram (also called angiogram), and single-photon emission computed tomography (SPECT) are used to diagnose NDs. However, the aforementioned methods are not efficient in diagnosing NDs in view of the complex behavior of the human nervous system and the many different types of NDs. Therefore, many researchers have been trying to modify the current medical methods to improve the diagnostic accuracy of NDs. For example, Oh et al. used a convolutional neural network to diagnose Parkinson's disease [2]. Murugappan et al. applied fuzzy clustering, support vector machine, and neuro-fuzzy classifier to predict sudden cardiac arrest [3]. Bhurane et al. studied early detection of Parkinson's disease using EEG and principal component analysis [4]. Shoeibi et al. investigated deep learning methods to diagnose multiple sclerosis [5]. Simons et al. compared different entropy measures for the detection of Alzheimer's disease [6]. Bong et al. applied wavelet transform on brain signals to detect emotions in stroke patients [7]. Yean et al. used bispectrum properties of EEG signals for emotion classification in stroke patients [8]. Yuvaraj et al. investigated emotion process in Parkinson's disease patient by interhemispheric EEG coherence analysis [9].

Entropy-based methods have been used to analyze time-series signals for several decades. They indicate the degree of complexity of time series and extract some hidden information from signals [10]. These methods are powerful tools for diagnosing NDs because the nervous system is complex, and the data obtained are often noisy, high-dimensional, and incomplete. For example, Yuvaraj and Murugappan studied emotional response of idiopathic Parkinson's disease patients using approximate and sample entropy [11]. Yuvaraj et al. introduced wavelet packet entropy for emotion and trajectory visualization in Parkinson's disease patients [12]. Jerritta et al. used spectral entropy for emotion detection in ECG signals [13].

Against this background, the goal of this chapter is twofold. First, it introduces the concept of entropy and various entropy measures as a powerful tool for analyzing biomedical signals. Second, it applies some entropy measures to an experimental neurological signal dataset to compare the efficiency of each entropy measure in real-world problems. The remainder of this chapter is organized as follows. Section 2 explains the concept of entropy and some different entropy measures. The efficiency of some different entropy measures in diagnosing NDs is compared using experimental datasets. Finally, Sect. 4 concludes the chapter and suggests some works for the future.

2 Entropy

In this section, the concept of entropy and some different entropy measures are introduced. Complex mathematical proofs are left out since our goal is to introduce the concept and formulas of entropy in such a way that biologists and medical doctors can understand without much knowledge of mathematics. However, appropriate references are given for the interested readers.

2.1 History

The concept of entropy was introduced in the nineteenth century by the German mathematician and physicist Clausius. The traditional definition of entropy is associated with the second law of thermodynamics and is as follows:

$$ds = \frac{dq_{rev}}{T}, \tag{1}$$

where T is the Kelvin temperature of the closed system and q_{rev} is the incremental reversible heat transfer. Boltzmann defined entropy from the microscopic physical point of view. The Boltzmann entropy formula is

$$S = K_B Ln(\theta), \tag{2}$$

where K_B is the Boltzmann constant and θ is the number of equally probable microstates that make up a given thermodynamic macrostate. The higher the Boltzmann entropy, the more information about the number of microstates is required to determine the thermodynamic properties of the system. In other words, the entropy measure is directly related to the lack of knowledge. Gibbs introduced an equivalent definition of entropy using the probability measure of the microscopic state as follows [14]:

$$S = -K_B \sum_i p_i \ln p_i, \tag{3}$$

where p_{i} is the probability of microstate i and N is the number of microscopic states whose physical properties affect the macroscopic state of the system under consideration.

2.2 Entropy and Information Theory

Shannon developed the idea of an entropy measure of random processes in 1948. Shannon introduced a measurable quantity for information in a given time series, which heralded the birth of information technology [15]. To illustrate the Shannon entropy measure, consider the following examples.

Example 1 A system has two artificial neurons, denoted as n1 and n2. Activation of a particular neuron is required to achieve the desired state, while the right neuron is unknown. Therefore, the desired state is reached with the probability of 0.5. The right neuron is determined when the answer to the following question is found.

Is the n1 the specific neuron?

We consider each question as a piece of information. Therefore, there is a unit of missing information for the right decision.

Example 2 Consider the previous example with 128 neurons. Thus, the desired state is reached with the probability of 0.0078. Suppose the desired neuron is n111. We should ask ourselves:

Is the n1 the specific neuron?

.

.

Is the n111 the specific neuron?
But there is an optimal way as follows:
Is the specific neuron between n1 and n64?
Is the specific neuron between n64 and n96?
Does the specific neuron lie between n96 and n112?
Is the specific neuron between n96 and 104?
Is the specific neuron between n104 and 108?
Is the specific neuron between n108 and 110?
Is the specific neuron n111?

The number of questions is the amount of information required to determine the actual outcome of the system. The optimal number of questions is defined as Shannon entropy (ShEn) measure. The above examples show that entropy is a lack of knowledge. Approximate entropy (ApEn) is another well-known entropy measure that examines the unpredictability of fluctuations in a time series. A time series with a similar pattern is predictable and has a low ApEn. The fluctuations and unpredictability of time series result in a high ApEn. Please consider the time series $T1=\{1,2,3,4,1,2,3,4,1,2,3,4\}$ and $T2=\{1,3,4,2,2,1,3,1,2,3,3,4,1,2\}$. The mean and variance of the time series are identical. The elements of the time series *T*1 repeat. Therefore, *T*1 is regular with small ApEn, while *T*2 is complex with high ApEn. The mathematical formulation of the Shannon entropy and approximate entropy is discussed in Sect. 2.3.

2.3 Mathematical Formulation

The complexity of a time series can be quantified with entropy algorithms. Conventional entropy algorithms divide time series into nonoverlapping segments and calculate the conditional probability that a segment in the time series repeats. This subsection presents the mathematical formulas of traditional ShEn, ApEn, and the algorithm for calculating the recently proposed entropy measure NNetEn.

2.3.1 Shannon Entropy

Consider the discrete random variable *X* whose possible outcomes $\{x_1, x_2, \ldots, x_n\}$ occur with probability $\{p_1, p_2, \ldots, p_n\}$. The Shannon entropy is calculated as follows:

$$ShEn(X) = -\sum_{i=1}^{n} p_i \log_2 p_i. \tag{4}$$

From a data compression perspective, the Shannon entropy of a string is the average of the minimum bits required to encode it [15]. For example, consider the symbol set $S=\{A,B,C,D,E\}$ with occurrence probabilities $\{0.1,0.2,0.3,0.1,0.3\}$. The Shannon entropy formula is

$$ShEn(S) = -(0.2 \log_2 0.1 + 0.2 \log_2 0.2 + 0.6 \log_2 0.3) = 2.4464,$$

which shows that a string of five symbols can be represented by three bits if it is encoded optimally.

2.3.2 Approximate Entropy

The goal of ApEn is to introduce a value for the correlation and persistence of a given time series. Two predefined constants *m* and *r* must be considered to calculate the ApEn measure. Parameter *m* is the embedding dimension, and *r* is the tolerance value. Experimental results show that $m = 1$ or 2 and *r* between 0.1 and 0.2 are suitable for slow dynamics systems such as heart rate variability (HRV) and growth hormone release [16]. There are some methods such as false nearest neighbor and standard autoregressive to determine the value of m [17]. The procedure for calculating the ApEn is explained in Algorithm 1.

Algorithm 1. Approximate entropy (ApEn)

Input: m, r, $u = \{u(1), u(2), \ldots, u(N)\}$ (the biomedical signal with the length N)

1. Define the blocks$x(i) = \{u(i), u(i+1), \ldots, u(i+m-1)\}$, $i = 1, \ldots, N-m+1$
2. Calculate the distance between $x(i)$ and $x(j)$ as follows$d[x(i), x(j)] = \max_{k=1,2,\ldots,m}(|u(i+k-1) - u(j+k-1)|)$, $i, j = 1, .., N$
3. Calculate the value $c_i^m(r) = (\text{number of } j \leq N-m+1 \text{ such that } d[x(i), x(j)] \leq r)/(N-m+1)$
4. Define $\Phi^m(r) = \frac{1}{N-m+1} \sum_{i=1}^{N-m+1} \log C_i^m(r)$
5. Calculate the ApEn as follows

$$ApEn(0, r, N) = -\Phi^1(r), m = 0$$
$$ApEn(m, r, N)(u) = \Phi^m(r) - \Phi^{m+1}(r), \quad m \geq 1$$

6. End

2.3.3 Neural Network Entropy

Recently, Velichko and Heidari proposed a new entropy measure for one-dimensional time series based on the classification procedure of the MNIST-10 dataset [18]. Their method gives similar results to ApEn but has some advantages, such as independence from the probability distribution and robustness to noise in the data. The algorithm consists of two stages, which are shown in Fig. 1. In the first stage, the input signal (the given time series) affects the MNIST-10 and generates a modified vector as the input to the second stage. Classification is performed in this stage using a feedforward neural network (FNN). The classification accuracy is directly related to the values of the input signal. Therefore, the classification accuracy is defined as the entropy of the input signal and denoted by NNetEn.

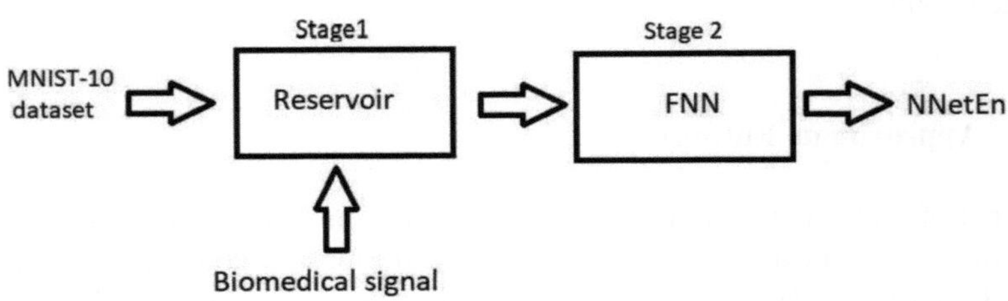

Fig. 1 The procedure of evaluating NNetEn

The procedure of calculating the NNetEn is explained in Fig. 1 and Algorithm 2.

Algorithm 2. Neural network entropy (NNetEn) algorithm

Input: MNIST-10 dataset, $u = \{u(1), u(2), \ldots, u(N)\}$ (the biomedical signal with the length N)

1. Use T-pattern 3 to convert the MNIST-10 dataset to the row vectors with 785 elements [19]
2. Make the reservoir matrix W_1 using the input signal u and the matrix filling method [18]
3. Calculate the modified input $um = W_1 Y$
4. Define um as the input of FNN
5. Train the FNN and find the accuracy of the classification
   NNetEn=The accuracy of the classification
6. End

3 Numerical Results

This section considers the EEG dataset [20] of nine disabled patients with a spinal cord injury or stroke, as shown in Table 1.

For each patient, the EEG is recorded for five mental tasks in 30 channels, as described in Tables 2 and 3, respectively.

The experiment is repeated randomly eight times for each single subject. Therefore, we have 40 trials for each mental task. A full description of the considered dataset is given in Ref. [20]. The ApEn and ShEn values of the signals in each trial are calculated to classify the mental tasks based on the EEG signals. Let us consider patient "A" and use the ApEn method to classify his mental tasks based on his EEG signals. So we have a vector with 40 elements for each class. The values for the classes "Word" and "Nav" are shown in Fig. 2, which shows a significant difference in ApEn between these classes. The values for the classes "Sub" and "Hand" are

Table 1 Patient details. The column from left to right indicates the Id, gender, age in year, duration of diagnosis in months, and the type of event for each patient

ID	Gender	Age	Months	Event
A	M	42	6	Locked-in syndrome due to brainstem stroke
C	F	31	5	Locked-in syndrome due to brainstem stroke
D	F	33	2	Spinal cord injury C5, ASIA C
E	F	40	255	Spinal cord injury C5, ASIA A
F	F	57	5	Massive hemorrhagic stroke in left hemisphere
G	F	43	27	Spinal cord injury C5, ASIA C
H	F	20	6	Hemorrhagic stroke parietotemporal, right central no cranium
J	M	36	53	Spinal cord injury C5, ASIA A
L	M	38	15	Spinal cord injury C5, ASIA A

Table 2 The name and description of each class

Class name (mental tasks)	Description
Word	Word association
Sub	Mental subtraction
Nav	Spatial navigation
Hand	Right-hand motor imaginary
Feet	Feet motor imaginary

Table 3 The 30 considered EEG channels

Ch. no.	1	2	3	4	5	6	7	8	9	10
Ch. name	AFz	F7	F3	Fz	F4	F8	FC3	FCz	FC4	T3
Ch. no.	11	12	13	14	15	16	17	18	19	20
Ch. name	C3	Cz	C4	T4	CP3	CPz	CP4	P7	P5	P3
Ch. no.	21	22	23	24	25	26	27	28	29	30
Ch. name	P1	Pz	P2	P4	P6	P8	PO3	PO4	O1	O2

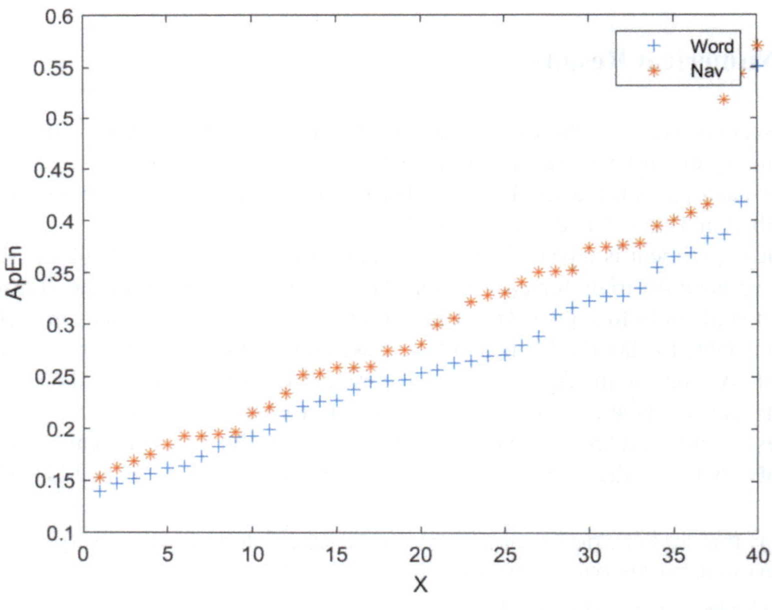

Fig. 2 The approximate entropy of the EEG signals for the classes "Word" and "Nav" in patient A

shown in Fig. 3. It shows that the similarity in ApEn between these classes is high (Fig. 3).

Therefore, ApEn is not suitable to distinguish the classes "Sub" and "Hand" in patient A. The ShEn values for these classes are shown in Fig. 4. This figure shows that there is a significant difference between values of the considered classes.

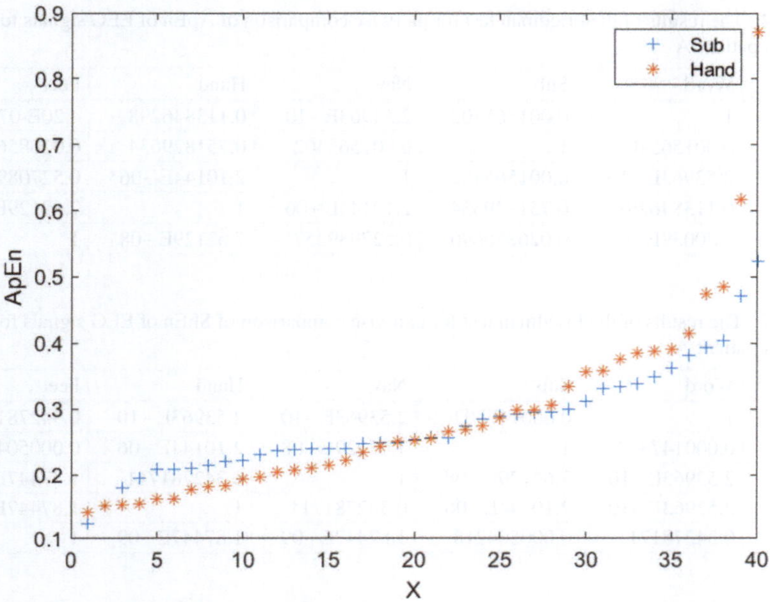

Fig. 3 The approximate entropy of the EEG signals for the classes "Sub" and "Hand" in patient A

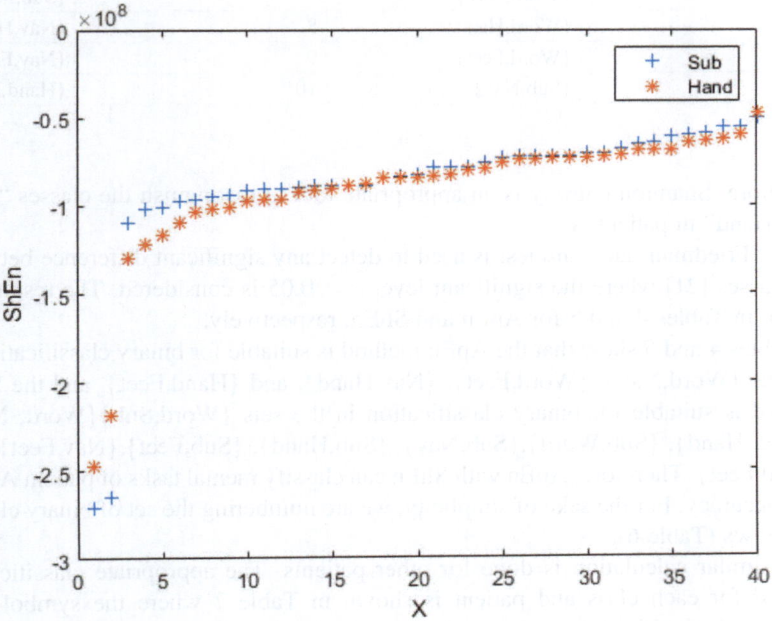

Fig. 4 The Shannon entropy of EEG signals for the classes "Sub" and "Hand" in patient A

Table 4 The results of the Friedman test for pairwise comparison of ApEn of EEG signals for each class in patient A

Class	Word	Sub	Nav	Hand	Feet
Word	1	0.001565402	2.53963E−10	0.113846298	4.20E-07
Sub	0.001565402	1	0.001565402	0.751829634	0.026856696
Nav	2.53963E−10	0.001565402	1	2.10144E−06	0.527089257
Hand	0.113846298	0.751829634	2.10144E−06	1	7.62129E−08
Feet	4.20039E−07	0.026856696	0.527089257	7.62129E−08	1

Table 5 The results of the Friedman test for pairwise comparison of ShEn of EEG signals for each class in patient A

Class	Word	Sub	Nav	Hand	Feet
Word	1	0.000147802	2.53963E−10	2.53963E−10	0.342781711
Sub	0.000147802	1	7.62129E−08	2.10144E−06	0.000504218
Nav	2.53963E−10	7.62129E−08	1	0.342781711	1.87447E−09
Hand	2.53963E−10	2.10144E−06	0.342781711	1	1.87447E−09
Feet	0.342781711	0.000504218	1.87447E−09	1.87447E−09	1

Table 6 The classification problems and its related numbers

Class number	Class set	Class number	Class set
1	{Word, Sub}	6	{Sub,Hand}
2	{Word,Nav}	7	{Sub,Feet}
3	{Word,Hand}	8	{Nav,Hand}
4	{Word,Feet}	9	{Nav,Feet}
5	{Sub,Nav}	10	{Hand,Feet}

Therefore, Shannon entropy is an appropriate tool to distinguish the classes "Sub" and "Hand" in patient A.

The Friedman rank sum test is used to detect any significant difference between the classes [21] where the significant level $\alpha = 0.05$ is considered. The results are shown in Tables 4 and 5 for ApEn and ShEn, respectively.

Tables 4 and 5 show that the ApEn method is suitable for binary classification in the sets {Word,Nav}, {Word,Feet}, {Nav,Hand}, and {Hand,Feet}, and the ShEn method is suitable for binary classification in the sets {Word,Sub}{Word, Nav}, {Word, Hand}, {Sub,Word},{Sub,Nav}, {Sub,Hand}, {Sub,Feet},{Nav,Feet}, and {Hand,Feet}. Therefore, ApEn with ShEn can classify mental tasks of patient A with high accuracy. For the sake of simplicity, we are numbering the set of binary classes as follows (Table 6):

A similar calculation is done for other patients. The appropriate classification method for each class and patient is shown in Table 7 where the symbol "×" indicates the inefficiency of the method for the considered class.

Table 7 The appropriate method for the binary classification for each patient with the significance level of 0.05

Class no.	1	2	3	4	5	6	7	8	9	10
Patient	A									
ApEN			×			×			×	
ShEn				×				×		
Patient	C									
ApEN	×						×			
ShEn	×			×				×		
Patient	D									
ApEN	×		×		×	×			×	
ShEn							×	×		
Patient	E									
ApEN			×							
ShEn					×	×				
Patient	F									
ApEN	×	×	×			×			×	
ShEn	×				×		×			
Patient	G									
ApEN			×							
ShEn			×	×						×
Patient	H									
ApEN						×	×			×
ShEn					×	×	×	×		
Patient	J									
ApEN					×					
ShEn				×	×				×	
Patient	L									
ApEN				×						
ShEn		×		×				×		

The results show that entropy-based methods are useful tools for binary classification of mental tasks in patients with neurological disabilities. To our knowledge, Ref. [20] indicates that the studied method cannot classify the movement of {Hand, Feet}, and there are no better results for this class in the dataset. The results show that the ApEn approach can classify the {Hand, Feet} with high accuracy, except for patient H, who has the hemorrhagic stroke parietotemporal problem. For this patient, the ApEn of EEG signals for the classes "Hand" and "Feet" on another day (day 2) with 35 trials are calculated and depicted in Figs. 5 and 6, respectively. The Friedman test is used to compare the efficiency of ApEn measures for the binary classification in classes {Hand, Feet} in days 1 and 2. The results show that the accuracy of the binary classification is significantly improved in day 2 (see Tables 8 and 9). This confirms the obtained results and shows that the therapy is successful and the patient returns to normal.

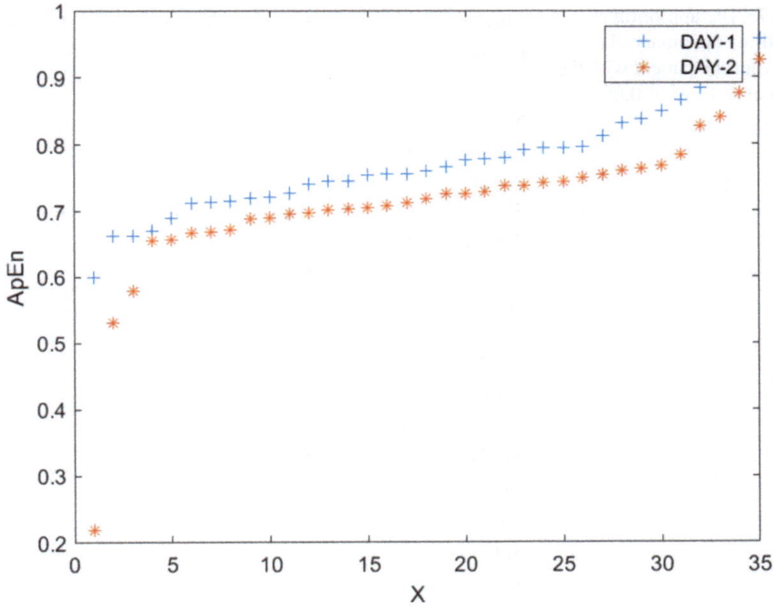

Fig. 5 The values of ApEn for the class "Hand" in patient H at two different days

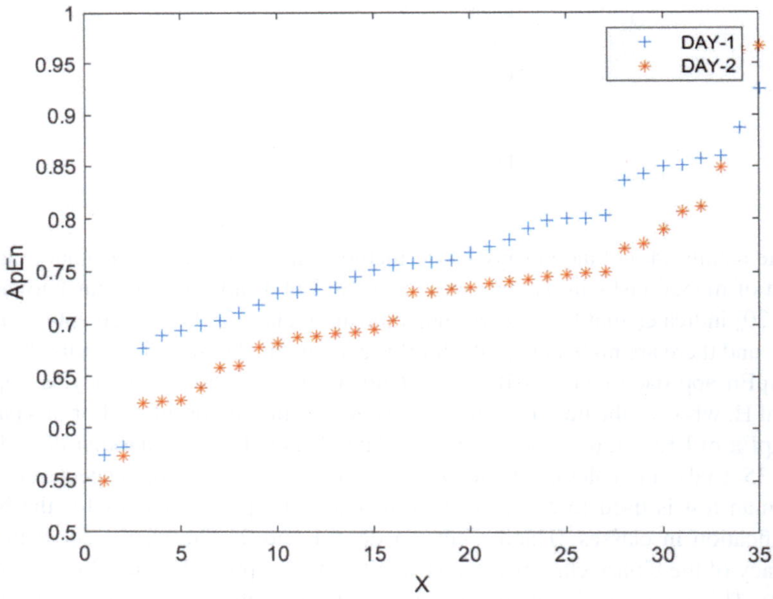

Fig. 6 The values of ApEn for the class "Feet" in patient H at two different days

Table 8 The results of the Friedman test for pairwise comparison of ApEn of EEG signals for each class in patient H in day 1

Class	Word	Sub	Nav	Hand	Feet
Word	1	0.011412036	2.10144E−06	4.20039E−07	2.10144E−06
Sub	0.011412036	1	0.000504218	0.527089257	0.342781711
Nav	2.10144E−06	0.000504218	1	0.001565402	0.001565402
Hand	4.20039E−07	0.527089257	0.001565402	1	1
Feet	2.10144E−06	0.342781711	0.001565402	1	1

Table 9 The results of the Friedman test for pairwise comparison of ApEn of EEG signals for each class in patient H in day 2

Class	Word	Sub	Nav	Hand	Feet
Word	1	0.004059196	0.865772375	0.865772375	0.128190174
Sub	0.004059196	1	5.0228E−06	2.38124E−05	0.027991815
Nav	0.865772375	5.0228E−06	1	0.011229887	0.865772375
Hand	0.865772375	2.38124E−05	0.011229887	1	0.39802472
Feet	0.128190174	0.027991815	0.865772375	0.39802472	1

3.1 Channel Selection

The human brain consists of more than 100 channels. Considering all channels in analyzing EEG signals is computationally inefficient and not suitable for practical applications. Moreover, selecting as few channels as possible is an important issue in developing an EEG headset [22, 23]. Liu et al. used Fisher's score and fractal dimension of motor imaginary EEG signals for channel selection in patients with amyotrophic lateral sclerosis [24]. Baig et al. investigated some filtering algorithms, wrapper algorithms, and embedded algorithms in channel selection of EEG signals [25]. Soka classified the channel selection methods in three categories as statistical, amplitude-based, and entropy-based methods [26]. Ghaemi et al. used binary gravitation algorithm for the channel selection in EEG signals of left- or hand-right movements in brain-computer interfaces [27]. Moctezuma and Molinas applied non-dominated genetic algorithm (NSGA) for the channel selection in epileptic-seizure patients [22]. In this paper, the considered dataset consists of 30 channels (Table 3) [20]. Scherer et al. ignored some channels for classifying the mental tasks in each patient (see Table 10) [20]. Here we use the NNetEn measure and Friedman test to select the optimal number of channels. As mentioned above, the test is performed eight times for each class and patient. One EEG of each class is randomly selected, and the NNetEn of its columns is calculated. Then, the values of the NNetEn for column i and class s are stored in a vector called NNetEni,s. The Friedman test is used to compare the NNetEni,s and the NNetEnj,s for each column *i* and *j*. There is no significant difference between columns *i* and *j* when the value of the Friedman test is equal to 1. Thus, it is possible to ignore columns *i* or *j* without compromising the accuracy of the classification. The redundant channels for each class are ignored,

Table 10 The ignored EEG channels for each patient in Ref. [20]

Patient	A	C	D	E	F	G	H	J	L
Ignored channels	1 2 10 18	1, 2 14, 27, 29	1 2 28	1 2 10 14 18 27 28	2 10	2 13 25	1 2 10 18	1 2 5 7 9 14 18 27	1 2 10 14

Table 11 The selected channels in each patient using NNetEn

Class	Ignored channels	Selected channels
Words	4,5,6,8,9,10,11,12,13,14,15,16,17,18,19,20,21,22,23,24,25, 26,27,28,29,30	1,2,3,7
Subs	2,5,6,7,8,9,10,11,12,13,14,15,16,17,18,19,20,21,22,23,24,25, 26,27,28,29,30	1,3,4
Nav	3,6,7,8,9,10,11,12,13,15,16,17,18,19,20,21,22,23,24,25, 26,27,28,29,30	1,2,4,5,14
Hand	3,4,5,6,7,8,9,10,11,12,13,14,15,16,17,18,19,20,21,22,23,24,25, 26,27,28,29,30	1,2
Feet	2,3,5,6,8,9,10,11,12,13,14,16,17,18,19,20,21,22,23,24,25, 26,27,28,29,30	1,4,7,15

and the required ones are displayed in Table 11. In addition, the NNetEn value of the EEG signal channels for each class in patients A and H is shown in Figs. 7, 8, 9, 10, 11. A comparison between these figures shows that the NNetEn value of the EEG signals in patient A is lower than the corresponding values in patient H for the "Hand" class. This suggests that the inefficiency of the binary classification in class 10 in patient H is related to the hand's imaginary motor function, and the therapist should attend to this issue. A comparison between NNetEn of the EEG signal channels in each class for patients A and H is shown in Figs. 12 and 13, respectively. Figure 13 shows a significant difference between the NNetEn of the class "Hand" and other classes in patient H. This confirms our previous finding.

4 Conclusion and Further Work

This chapter gives a brief and simple description of the concept of entropy in different sciences. Three important methods for evaluating entropy are discussed. To show the effectiveness of entropy-based methods in diagnosing NDs, the EEG of nine disabled patients with a spinal cord injury or stroke is considered. The ApEn and ShEn measures are used together with the Friedman test to classify patients' mental tasks. The results show that these methods perform better than [20]. In

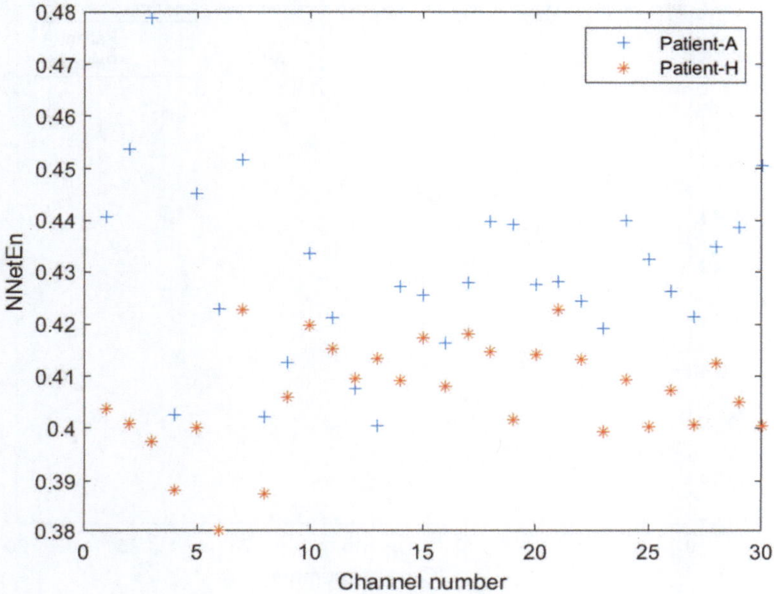

Fig. 7 The NNetEn values of the EEG signal channels for the class "Word" in patients A and H

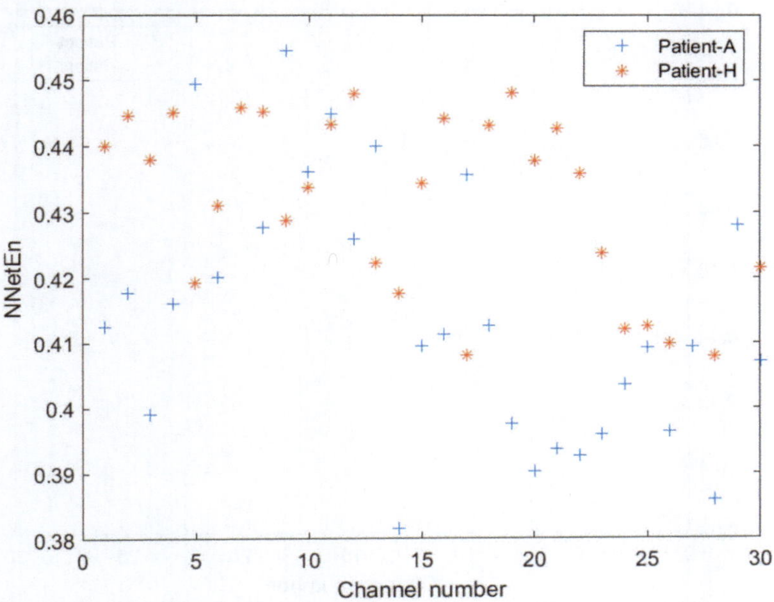

Fig. 8 The NNetEn values of the EEG signal channels for the class "Sub" in patients A and H

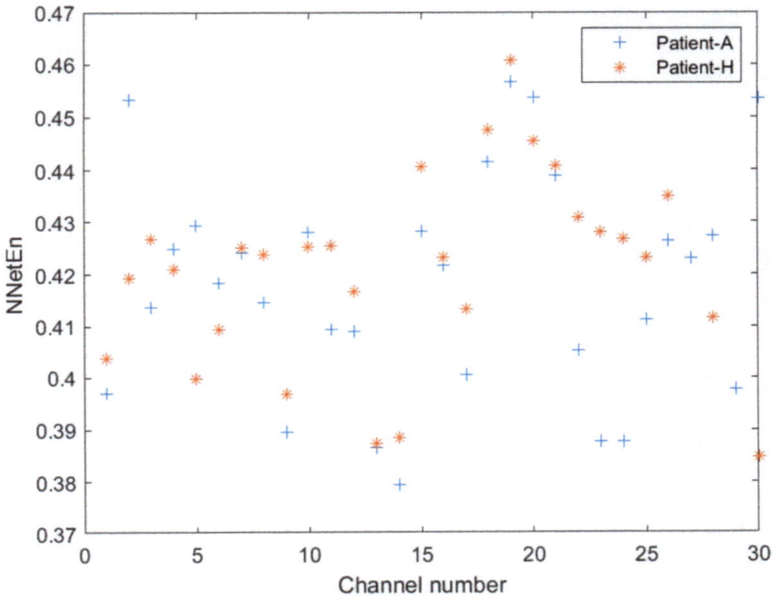

Fig. 9 The NNetEn values of the EEG signal channels for the class "Nav" in patients A and H

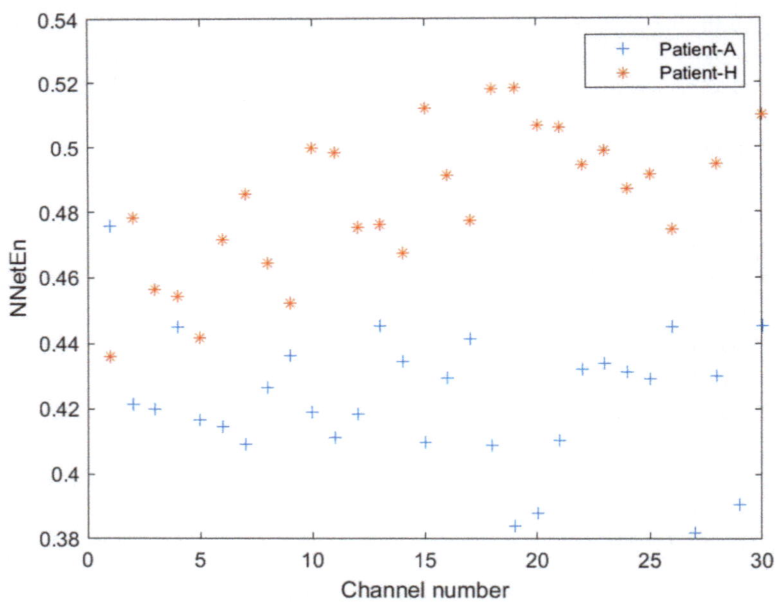

Fig. 10 The NNetEn values of the EEG signal channels for the class "Hand" in patients A and H

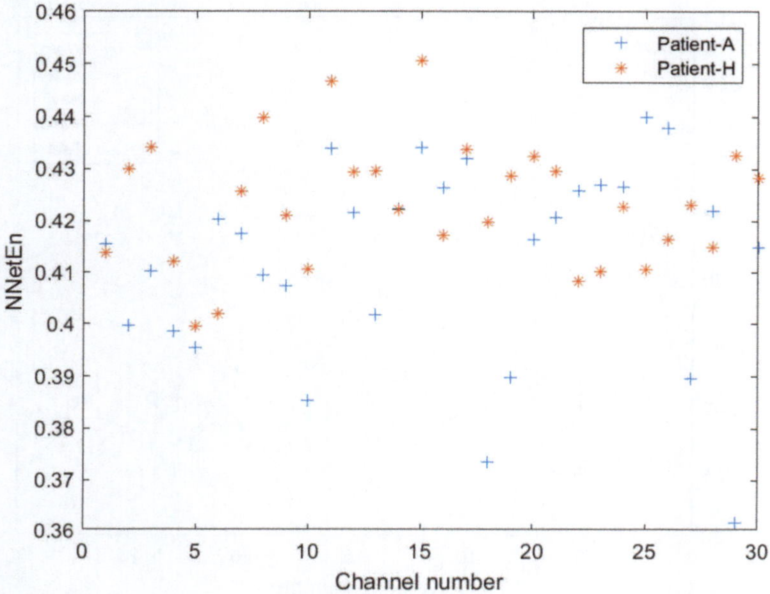

Fig. 11 The NNetEn values of the EEG signal channels for the class "Feet" in patients A and H

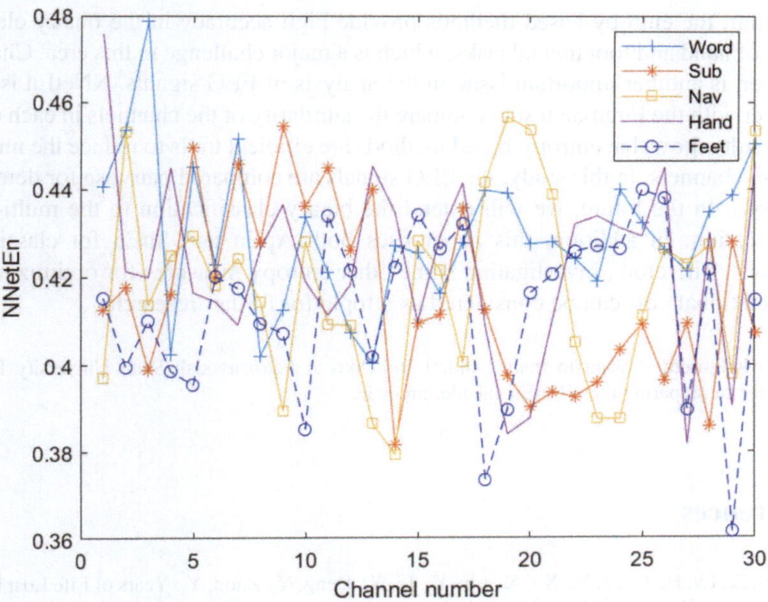

Fig. 12 The NNetEn values of the EEG signal channels for each class in patient A

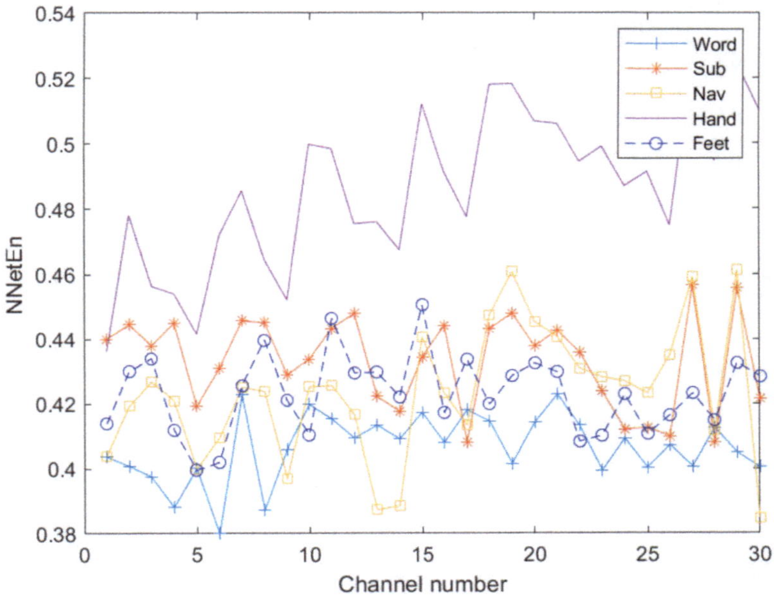

Fig. 13 The NNetEn values of the EEG signal channels for each class in patient H

particular, the entropy-based methods provide high accuracy in the binary classification of hand and foot mental tasks, which is a major challenge in this area. Channel selection is another important issue in the analysis of EEG signals. NNetEn is used together with the Fireman test to compare the similarity of the channels in each class. The results show that entropy-based methods are efficient tools to reduce the number of EEG channels. In this study, the EEG signals are compared pairwise for detecting emotions. In the future, we will extend the binary classification to the multi-class classification. In addition, this paper uses both ApEn and ShEn for classifying emotions. Therefore, investigating some other entropy measures for optimizing the computational cost can be considered as a topic for further research.

Acknowledgment I want to thank Andrei Velichko at Petrozavodsk State University for his assistance in preparing the NNetEn calculation code.

References

1. Luo, Z., Lv, H., Chen, Y., Xu, X., Liu, K., Li, X., Deng, Y., Zhou, Y.: Years of Life Lost Due to Premature Death and Their Trends in People With Selected Neurological Disorders in Shanghai, China, 1995–2018: A Population-Based Study. Front. Neurol. 12, 1–10 (2021). doi: https://doi.org/10.3389/fneur.2021.625042.
2. Oh, S.L., Hagiwara, Y., Raghavendra, U., Yuvaraj, R., Arunkumar, N., Murugappan, M., Acharya, U.R.: A deep learning approach for Parkinson's disease diagnosis from EEG signals.

Neural Comput. Appl. 32, 10927–10933 (2020). doi: https://doi.org/10.1007/s00521-018-3689-5.
3. Murugappan, M., Murugesan, L., Jerritta, S., Adeli, H.: Sudden Cardiac Arrest (SCA) Prediction Using ECG Morphological Features. Arab. J. Sci. Eng. 46, 947–961 (2021). doi: https://doi.org/10.1007/s13369-020-04765-3.
4. Bhurane, A.A., Dhok, S., Sharma, M., Yuvaraj, R., Murugappan, M., Acharya, U.R.: Diagnosis of Parkinson's disease from electroencephalography signals using linear and self-similarity features. Expert Syst. 1–12 (2019). doi: https://doi.org/10.1111/exsy.12472.
5. Shoeibi, A., Khodatars, M., Jafari, M., Moridian, P., Rezaei, M., Alizadehsani, R., Khozeimeh, F., Gorriz, J.M., Heras, J., Panahiazar, M., Nahavandi, S., Acharya, U.R.: Applications of deep learning techniques for automated multiple sclerosis detection using magnetic resonance imaging: A review. Comput. Biol. Med. 136, (2021). doi: https://doi.org/10.1016/j.compbiomed.2021.104697.
6. Simons, S., Espino, P., Abásolo, D.: Fuzzy Entropy analysis of the electroencephalogram in patients with Alzheimer's disease: Is the method superior to Sample Entropy? Entropy. 20, 1–13 (2018). doi: https://doi.org/10.3390/e20010021.
7. Bong, S.Z., Wan, K., Murugappan, M., Ibrahim, N.M., Rajamanickam, Y., Mohamad, K.: Implementation of wavelet packet transform and non linear analysis for emotion classification in stroke patient using brain signals. Biomed. Signal Process. Control. 36, 102–112 (2017). doi: https://doi.org/10.1016/j.bspc.2017.03.016.
8. Yean, C.W., Ahmad, W.K.W., Mustafa, W.A., Murugappan, M., Rajamanickam, Y., Adom, A. H., Omar, M.I., Zheng, B.S., Junoh, A.K., Razlan, Z.M., Bakar, S.A.: An emotion assessment of stroke patients by using bispectrum features of EEG signals. Brain Sci. 10, 1–22 (2020). doi: https://doi.org/10.3390/brainsci10100672.
9. Yuvaraj, R., Murugappan, M., Ibrahim, N.M., Sundaraj, K., Omar, M.I., Mohamad, K., Palaniappan, R., Satiyan, M.: Inter-hemispheric EEG coherence analysis in Parkinson's disease: Assessing brain activity during emotion processing. J. Neural Transm. 122, 237–252 (2015). doi: https://doi.org/10.1007/s00702-014-1249-4.
10. Selvaraj, J., Murugappan, M., Wan, K., Yaacob, S.: Classification of emotional states from electrocardiogram signals: A non-linear approach based on hurst. Biomed. Eng. Online. 12, 1–18 (2013). doi: https://doi.org/10.1186/1475-925X-12-44.
11. Yuvaraj, R., Murugappan, M.: Hemispheric asymmetry non-linear analysis of EEG during emotional responses from idiopathic Parkinson's disease patients. Cogn. Neurodyn. 10, 225–234 (2016). doi: https://doi.org/10.1007/s11571-016-9375-3.
12. Yuvaraj, R., Murugappan, M., Ibrahim, N.M., Sundaraj, K., Omar, M.I., Mohamad, K., Palaniappan, R.: Optimal set of EEG features for emotional state classification and trajectory visualization in Parkinson's disease. Int. J. Psychophysiol. 94, 482–495 (2014). doi: https://doi.org/10.1016/j.ijpsycho.2014.07.014.
13. Jerritta, S., Murugappan, M., Wan, K., Yaacob, S.: Electrocardiogram-based emotion recognition system using empirical mode decomposition and discrete Fourier transform. Expert Syst. 31, 110–120 (2014). doi: https://doi.org/10.1111/exsy.12014.
14. Grandy, Jr., W.T.: Entropy and the Time Evolution of Macroscopic Systems (International Series of Monographs on Physics). Oxford University Press, New York (2008).
15. Goyena, R., Fallis, A..: Handbook of Data Compression 5th. Springer-Verlag, London (2010).
16. Chon, K., Scully, C., Lu, S.: Approximate entropy for all signals. IEEE Eng. Med. Biol. Mag. 28, 18–23 (2009). doi: https://doi.org/10.1109/MEMB.2009.934629.
17. Chen, X., Solomon, I.C., Chon, K.H.: Comparison of the use of approximate entropy and sample entropy: Applications to neural respiratory signal. Annu. Int. Conf. IEEE Eng. Med. Biol. - Proc. 7 VOLS, 4212–4215 (2005). doi: https://doi.org/10.1109/iembs.2005.1615393.
18. Velichko, A., Heidari, H.: A method for estimating the entropy of time series using artificial neural network. Entropy 23, 1432 (2021). doi: https://doi.org/10.3390/e23111432
19. Velichko, A.: Neural Network for Low-Memory IoT Devices and MNIST Image Recognition Using Kernels Based on Logistic Map. Electronics. 9, 1432 (2020).

20. Scherer, R., Faller, J., Friedrich, E.V.C., Opisso, E., Costa, U., Kübler, A., Müller-Putz, G.R.: Individually adapted imagery improves brain-computer interface performance in end-users with disability. PLoS One. 10, 1–14 (2015). doi: https://doi.org/10.1371/journal.pone.0123727.
21. Corder, G.W., Foreman, D.I.: Nonparametric statistics for non-statistician. John Wiley & Sons, New Jersey (2009).
22. Moctezuma, L.A., Molinas, M.: EEG Channel-Selection Method for Epileptic-Seizure Classification Based on Multi-Objective Optimization. Front. Neurosci. 14, (2020). doi: https://doi.org/10.3389/fnins.2020.00593.
23. Majumdar, K., Jayachandran, S.: A geometric analysis of time series leading to information encoding and a new entropy measure. J. Comput. Appl. Math. 328, 469–484 (2018). doi: https://doi.org/10.1016/j.cam.2017.07.006.
24. Liu, Y.H., Huang, S., Huang, Y. De: Motor imagery EEG classification for patients with amyotrophic lateral sclerosis using fractal dimension and Fisher's criterion-based channel selection. Sensors (Switzerland). 17, (2017). doi: https://doi.org/10.3390/s17071557.
25. Baig, M.Z., Aslam, N., Shum, H.P.H.: Filtering techniques for channel selection in motor imagery EEG applications: a survey. Artif. Intell. Rev. 53, 1207–1232 (2020). doi: https://doi.org/10.1007/s10462-019-09694-8.
26. Ein Shoka, A.A., Alkinani, M.H., El-Sherbeny, A.S., El-Sayed, A., Dessouky, M.M.: Automated seizure diagnosis system based on feature extraction and channel selection using EEG signals. Brain Informatics. 8, 1–16 (2021). doi: https://doi.org/10.1186/s40708-021-00123-7.
27. Ghaemi, A., Rashedi, E., Pourrahimi, A.M., Kamandar, M., Rahdari, F.: Automatic channel selection in EEG signals for classification of left or right hand movement in Brain Computer Interfaces using improved binary gravitation search algorithm. Biomed. Signal Process. Control. 33, 109–118 (2017). doi: https://doi.org/10.1016/j.bspc.2016.11.018.

Automatic Detection of Epilepsy Using CNN-GRU Hybrid Model

S. Pravin Kumar, Janaki Ramachandran, Nanthini Narayanan, and Bhavadharani Thangavel

Abstract Epileptic seizures manifest as abnormal activity in the brain. The examination of long-term EEG signals is a time-consuming and arduous task that can be made more efficient by using deep learning which can automatically extract effective features. In this work, a 2D convolutional neural network-gated recurrent unit (CNN-GRU) hybrid model is proposed to perform binary, ternary, and quinary classification on the University of Bonn epilepsy database which consists of three different classes—normal (set A and B), interictal (set C and D), and ictal (set E). Initially, a convolutional neural network model is trained using scalogram images of the EEG signals, and 99.4% and 99.6% accuracy are obtained for two cases of binary classification (AB-CDE and ABCD-E). An accuracy of 99.4% and 99.2% is obtained for ternary (AB-CD-E) and quinary (A-B-C-D-E) classification, respectively. The proposed 2D CNN-GRU hybrid model further improves the accuracy to 100% for binary and ternary classifications and to 99.4% in the case of the quinary classification. It is ascertained that the hybrid model outperforms the CNN model in all cases.

Keywords Epilepsy detection · Epilepsy classification · 2D CNN · 2D CNN-GRU · Deep learning · Scalogram

1 Introduction

Epilepsy is regarded as unusual neural activity causing seizures. The three main phases of seizures are preictal, ictal and postictal. Interictal is the period between seizures. An electroencephalograph (EEG) records the electrical activity of the brain. It is an essential clinical tool for classifying neurological disorders, including diagnosis and classification of epilepsy [1, 2].

S. P. Kumar (✉) · J. Ramachandran · N. Narayanan · B. Thangavel
Department of Biomedical Engineering, Centre for Healthcare Technology, Sri Sivasubramaniya Nadar (SSN) College of Engineering, Kalavakkam, Tamil Nadu, India
e-mail: pravinkumars@ssn.edu.in

M. Murugappan, R. Yuvaraj (eds.), *Biomedical Signals Based Computer-Aided Diagnosis for Neurological Disorders*, https://doi.org/10.1007/978-3-030-97845-7_9

The University of Bonn dataset [3] is often utilized for epilepsy classification studies [4]. Binary classification of the EEG recordings into healthy and epileptic has been frequently attempted. The morphological EEG features during normal and epileptic intervals are used for this analysis. Often, epilepsy patients may show normal states called the interictal period, in between the seizure events. Thus, differentiating between normal EEG from a healthy individual and interictal EEG signals from an epilepsy patient can be used to clinically detect and diagnose epilepsy, while accurately detecting the ictal EEG for treatment [5]. Classification of EEG signals for epilepsy thus has great clinical significance. Going further, quinary classification of the University of Bonn dataset attempts to additionally differentiate between interictal data—set C and set D. This is more complicated than binary and ternary classification due to the similar EEG patterns. Nevertheless, it can be used to localize the focal point of the seizure inside the brain.

However, the classification problems mentioned above require screening of the wave morphology to detect different states of epileptic seizure from the EEG signal by trained experts, which is laborious and time-consuming. Hence, an automated approach that can make this process more efficient, accurate, and less demanding on the clinician is required. Different approaches have been undertaken for classifying EEG for epilepsy detection and are summarized below.

Deep learning models such as RNN, LSTMs, and CNNs have been implemented for binary and ternary classification of EEG data. Ullah et al. [6] detected epilepsy using pyramidal 1D CNN. The University of Bonn dataset was used, and 99.1% accuracy was achieved for the ternary classification (AB-CD-E). Liang et al. [7] achieved recognition and localization of epileptogenic zones by employing a long-term recurrent convolutional network (LRCN) and achieved 99% accuracy. Abdelhameed et al. [8] implemented a 1D deep convolutional autoencoder for extracting features and reducing dimensionality. Classification of normal and epileptic data achieved 100% results. An average of 99.33% was the result for ternary classification. Acharya et al. [9] used higher-order statistical (HOS) features from normal, interictal, and epileptic signals for classification with developed classifiers and attained 98.5% overall accuracy with support vector machine (SVM). Akut [10] implemented five-level wavelet decomposition on the Bonn dataset, and the coefficients were given to the classifier. They reached 100% accuracy for binary classification and 99.4% accuracy for ternary classification. Doi et al. [11] used temporal CNNs (TCN) and observed 100% accuracy for two-class (interictal vs ictal) epilepsy detection. Lin et al. [12] implemented a novel framework of stacked sparse autoencoders (SSAE) with a softmax classifier. They obtained 100% accuracy for A–E classification. Bhattacherjee [13] utilized a CNN classifier by applying the preprocessing steps on the dataset of the EEG signals. A multicolumn CNN was created by concatenating the average of the CNN columns with the four-level discrete wavelet transform operated on the preprocessed EEG signal with power spectrum calculation value for each frequency band by using machine learning techniques, and this model achieved 99.99% recognition rate. Abdelhameed and Bayoumi [14] implemented a new approach on the CHB-MIT dataset to classify ictal and interictal data. Highest accuracy of 98.79% was obtained by the supervised deep convolutional autoencoder (SDCAE) model with a Bi-LSTM-based classifier. Nogay and Adeli [15] converted the University of Bonn EEG data into spectrogram images and used a pre-trained 2D CNN model (AlexNet) and transfer learning to obtain 100% accuracy for binary and ternary classification. Quinary classification was not implemented.

Apart from homogenous neural networks models, hybrid models have also been implemented for epilepsy classification. Chen et al. [16] used a double deep NN with 1D CNN, LSTM and GRU, respectively. This method was evaluated using the University of Bonn dataset for three-class classification by segmenting each record into two second long segments. Accuracy of 97.28%, 96.82%, and 96.87% were obtained, respectively. Affes et al. [17] proposed a convolutional-gated recurrent neural network (CGRNN) using EEG features. The CHB-MIT dataset was used for evaluation and a mean accuracy of 75.6% was obtained. The ChronoNet is a recurrent neural network (RNN) architecture proposed by Roy et al. [18]. The TUH abnormal EEG corpus dataset was used to evaluate this model, and an accuracy of 86.57% was attained. Choi et al. [19] implemented a DNN model for seizure detection by mapping transformed features into images. This model was evaluated using two databases and acquired an accuracy of 99.4% and 99.2% in each case.

Quinary classification of EEG signals using the University of Bonn dataset has been attempted by very few researchers. Lyapunov exponents and wavelet coefficients were used by Guler and Ubeyli [20] with an SVM and a probabilistic neural network (PNN) achieving notable classification accuracies of 99.28% and 98.05%, respectively. In 2009, Übeyli [21] used statistical features and maintained an accuracy of 99.2%. Using wavelet approximate entropy with SVM, Shen et al. [22] obtained 99.97% accuracy. By integrating novel statistical features and multi-class SVM, Siuly and Li [23] reached 99.99%. In the method proposed by Hussein et al. [24], EEG data from the Bonn dataset was segmented and was used with a LSTM network. An accuracy of 100% was observed. Additionally, spectral rhythms of EEG were used by Muthanantha Murugavel et al. [25] as delegate features along with MSVM, for EEG classification, and this attained a classification accuracy of 96.00%. Übeyli [26] implemented an eigenvector-based method for feature extraction and using SVM multiclass with ECOC obtained 99.30% classification accuracy. Türk and Özerdem [27] proposed a convolutional neural network that used 2D scalograms by applying continuous wavelet transform, and an accuracy of 93.6% was seen for quinary classification after cross-validation.

Convolutional neural networks are effective in various fields of healthcare, especially medical analysis and automatic clinical diagnosis [28–31]. In this study, scalogram-based 2D CNN and hybrid 2D CNN-GRU models were implemented for detection and classification of epileptic states from EEG. The 2D CNN implemented in this work was adapted from the Türk and Özerdem [27] model to improve the accuracy in classifying scalograms of the EEG. The hybrid 2D CNN-GRU model was implemented to improve the accuracy acquired by the homogenous model. Our approach overcomes the limitation of machine learning [9, 22, 23, 25, 26] and doesn't require additional preprocessing of the data [13, 14, 19] prior to scalogram conversion. Thus, it can be used to aid neurologists in the diagnosis of epilepsy and can make the process more efficient. Three cases of EEG classification were done on the University of Bonn dataset: (i) binary, healthy and epileptic (AB-CDE) and normal and active seizure (ABCD-E); (ii) ternary, healthy, non-seizure epileptic and seizure epileptic (AB-CD-E); and (iii) quinary, healthy eyes open, healthy eyes closed, interictal from different epileptogenic zones and ictal (A-B-C-D-E).

In the section that follows, the methodology of this work is elaborated including details about choice of dataset, architecture, and training of the proposed networks. The result section discusses the outcomes of the models in classifying the test sets in each classification problem. Performance of the two proposed networks is compared. Inferences are made and the factors that could have contributed to the obtained results are discussed. A comparison of the obtained results with those achieved by related research is also discussed. Lastly, the paper concludes by outlining the scope of this work in clinical settings. In addition, limitations of this work and suggestions to overcome them in the future are also discussed.

2 Methodology

2.1 Dataset

The objective of this work is to develop a deep learning-based assistive tool that can aid clinicians in effective and accurate diagnosis of epilepsy. Therefore, we focused on choosing a reliable open-source dataset that is frequently used by researchers and contains artefact-free, properly labelled adult EEG recordings. The dataset must contain recordings belonging to different classes such that it provides scope to train the network in not just distinguishing epileptic EEG from normal but also distinguish similar EEG data recorded from different zones. In commonly used open datasets like the Bern-Barcelona EEG database [32] (focal and non-focal EEG) and Temple University EEG corpus [33] (normal and abnormal EEG), the recordings are only composed of data belonging to two classes. On the other hand, even though the Neurology and Sleep Centre, New Delhi, EEG dataset [34] has recordings of ictal, pre-ictal, and interictal stages, it does not provide healthy recordings. The single electrode EEG data of healthy and epileptic patients [35] was recorded from 10 healthy and 10 epilepsy patients, on similar acquisition settings as that of University of Bonn dataset. However, the dataset is available as 20 files each for subjects as such, and active seizure and seizure-free events are not marked explicitly. Taking all the above-discussed factors into consideration, the University of Bonn dataset was regarded fit for this work.

The University of Bonn dataset by Andrzejak [3] was collected from five patients and consists of three different sets, namely, normal (sets A and B), interictal (sets C and D), and ictal (set E). The sets contain 100 single-channel EEG, where each segment is of 23.6 s duration. Digitization of the recordings was done at a sampling rate of 173.61 Hz with a resolution of 12 bits. It was filtered using a 0.53–40 Hz bandpass filter. The normal data, taken from five healthy subjects, was obtained

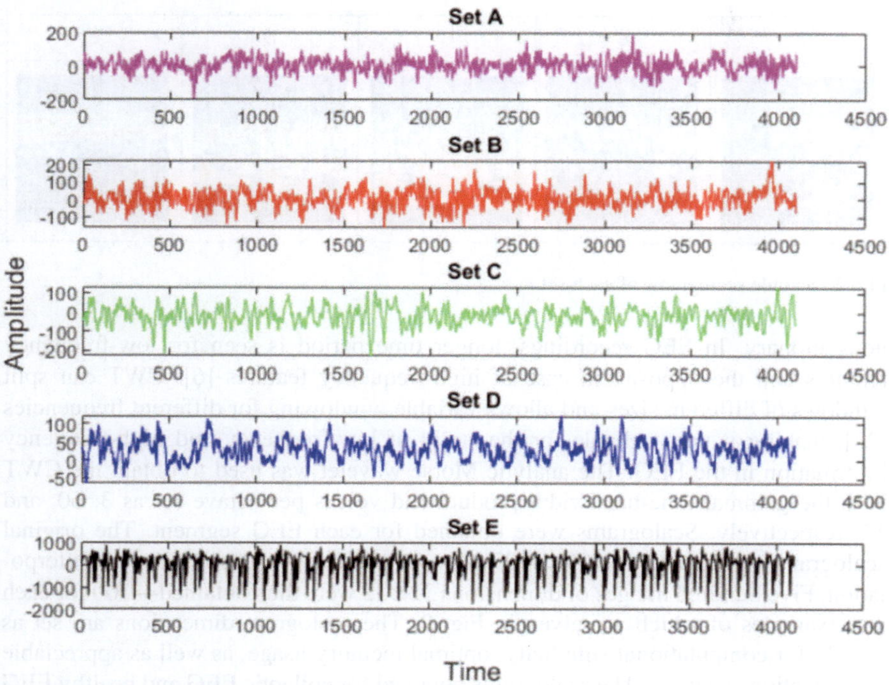

Fig. 1 Sample raw EEG of each set

extracranially using standard surface electrodes. The interictal data were taken from five patients, with no seizure incidences during acquisition, from the epileptogenic zone and hippocampal formation of the opposite hemispheres of the brain using depth electrodes. The ictal data, taken from five epileptic patients, was obtained from basal and lateral regions in the neocortex using strip electrodes along with depth electrodes mentioned above. No additional preprocessing methods were applied. Sample raw EEG signals are given in Fig. 1 for each set in the dataset.

The scalograms of the data were generated in MATLAB, and implementation of the proposed models was carried out with keras using a GPU. Tenfold cross-validation was performed for both models, and their accuracy, specificity, sensitivity, and f-scores were evaluated.

2.2 Scalogram

A scalogram is the absolute value of the coefficients obtained from continuous wavelet transform (CWT), plotted as a function of frequency and time. CWT is used to decompose signals into wavelets and provides time and frequency-localization information. Time-frequency transforms are ideal for signals that are

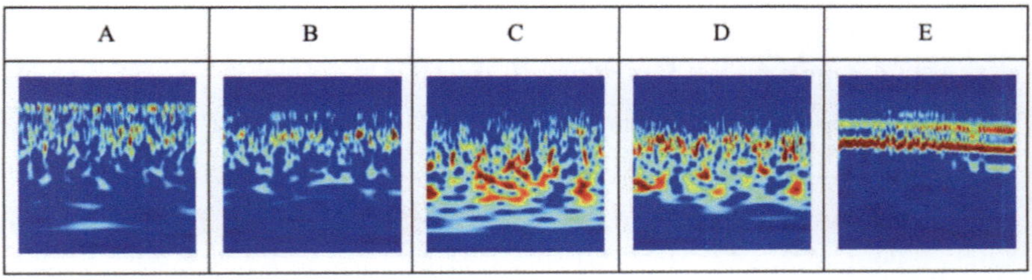

Fig. 2 Sample scalograms of each set

nonstationary. In EEG recordings, longer time period is seen for low-frequency features and the opposite in case of high-frequency features [6]. CWT can split windows of different sizes and allows variable windowing for different frequencies [27], making it very suitable for the study of low-frequency and high-frequency information in the EEG. The analytic Morse wavelet was used to obtain the CWT with the gamma, time-bandwidth product and voices per octave set as 3, 60, and 12, respectively. Scalograms were obtained for each EEG segment. The original scalograms of dimensions 109×4097 were resized to 32×32 using bicubic interpolation. Five hundred images of dimensions 32×32 were thus obtained—100 for each set, examples of which are given in Fig. 2. The scalogram dimensions are set as 32×32 for computational simplicity, optimal memory usage, as well as appreciable classification accuracy. The scalograms obtained for epileptic EEG and healthy EEG can be differentiated based on the coloration which correlates to the magnitude of the CWT coefficient.

2.3 Architecture

Two models were implemented: 2D CNN and 2D CNN-GRU. The hyperparameters like learn rate, epochs, stride, dropout, and number of convolutional filters of the 2D CNN model were modified from the Türk and Özerdem [27] model. A batch normalization layer was also added.

The 2D CNN architecture (Figs. 3 and 4, and Table 1) comprises two blocks. Each has a 2D convolution layer, a nonlinear ReLU layer, and a max pooling layer.

The input image is convolved with a kernel (filter) by the convolution layer in order to extract its features which results in a map of activations (feature map). Every point $p(i,j)$ in the input image is convolved with the convolutional kernel $w(i,j)$ to obtain the output at the location (i,j) as $z(i,j)$ (Eq. (1)):

$$z(i,j) = p(i,j) * w(i,j) = \sum_{s=-a}^{a} \sum_{t=-b}^{b} p(s,t) \cdot w(i-s, j-t) \tag{1}$$

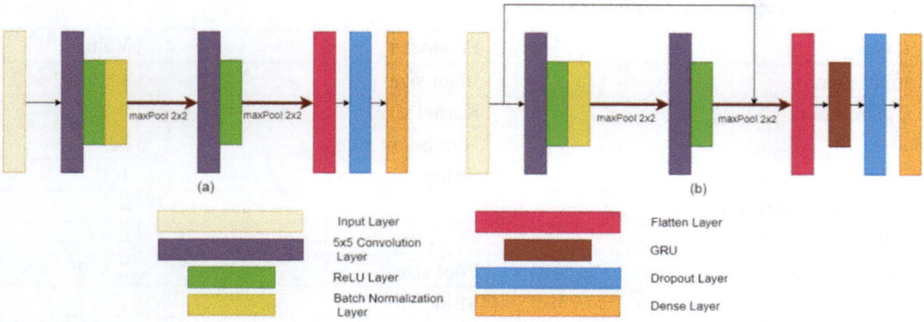

Fig. 3 Architecture of the proposed (**a**) 2D CNN and (**b**) 2D CNN-GRU

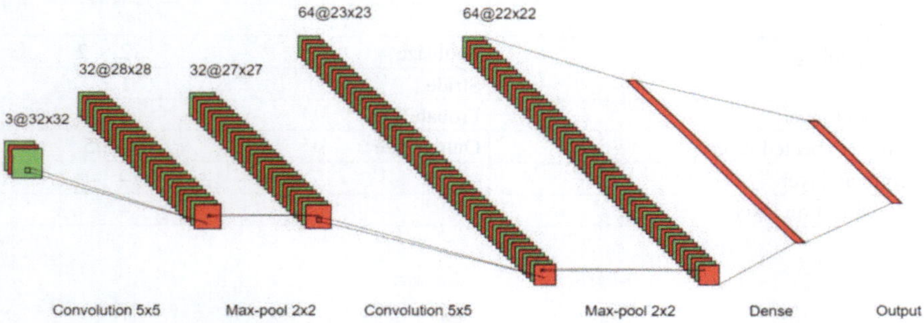

Fig. 4 Feature maps of proposed 2D CNN at each layer

The first convolution layer has 5 × 5 kernel size and performs convolutions with the image input layer resulting in 32 feature maps. A batch normalization layer is present between the first convolution layer and ReLU layer which standardizes the input to a layer for each mini-batch. The ReLU function is used for activation. The batch normalized features learnt by the convolution layer after passing through the ReLU activation layer can be expressed as in Eq. (2):

$$z_i^k = \sigma\left(BN\left(z_i^{k-1}\right)\right) \tag{2}$$

where z_i^k is the i-th output feature at the k-th layer, z_i^{k-1} is the i-th output feature from the $(k - 1)$-th layer, and $\sigma()$ and BN() represent the ReLU function and batch normalization function, respectively. Pooling operations summarize features in the subregions covered by the sliding filter by using functions such as average or maximum value. The number of parameters and computation is reduced by down-sampling. Max pooling with pool size 2 × 2 is applied to each feature map. The second convolution layer that has a kernel size of 5 × 5 and 64 feature maps is followed by another max pooling layer. After a dropout layer of rate 0.15, the output is then flattened and given as the input to a fully connected layer. It receives all

Table 1 Parameters of proposed 2D CNN

Layer	Parameter	Value
Input layer	Input size	32 × 32 × 3
Convolution	Kernel size	5 × 5
	Number of filters	32
	Stride	1
Batch normalization layer	–	–
ReLU	–	–
Max pooling	Pool size	2 × 2
	Stride	1
Convolution	Kernel size	5 × 5
	Number of filters	64
	Stride	1
ReLU	–	-
Max pooling	Pool size	2 × 2
	Stride	1
Dropout layer	Probability	0.15
Fully connected layer	Output size	2/3/5
Softmax layer	–	–
Classification layer	–	–

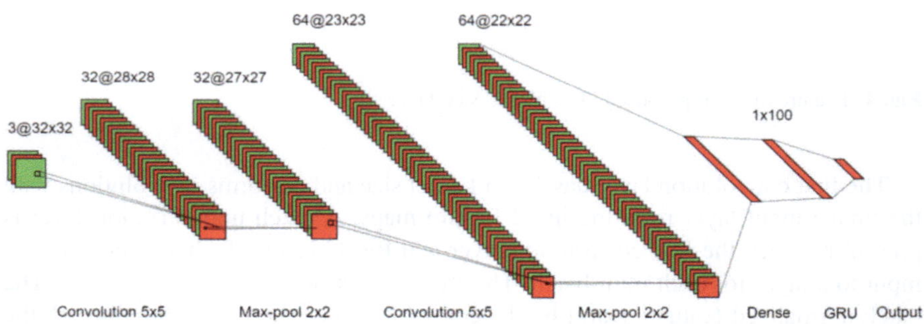

Fig. 5 Feature maps of proposed 2D CNN GRU at each layer

neurons from the previous layer and then predicts the probability for each label. The fully connected layer has 2, 3, and 5 neurons in the case of binary, ternary, and quinary classification, respectively. Lastly, a softmax classifier with neurons equal to the number of classes required is present.

GRU is a gating mechanism and is a variation of RNN. It is similar to LSTM but only has two gates (rest and update gates) which decide the information to be passed on to the output. GRU was used in this work instead of LSTM as it gives comparable results and computes faster. In the 2D CNN-GRU hybrid model (Figs. 3, 5, and Table 2), additionally a GRU with 100 hidden units is present after the flatten layer. The dropout rate is set to 0.3.

Table 2 Parameters of proposed 2D CNN-GRU

Layer	Parameter	Value
Input layer	Input size	32 × 32 × 3
Convolution	Kernel size	5 × 5
	Number of filters	32
	Stride	1
Batch normalization layer	–	-
ReLU	–	-
Max pooling	Pool size	2 × 2
	Stride	1
Convolution	Kernel size	5 × 5
	Number of filters	64
	Stride	1
ReLU	–	–
Max pooling	Pool size	2 × 2
	Stride	1
Gated recurrent unit	Hidden units	100
ReLU	–	–
Dropout layer	Probability	0.3
Fully connected layer	Output size	2/3/5
Softmax layer	–	–
Classification layer	–	–

Table 3 Distribution of data from each class for testing and training

Classification type	Binary AB-CDE		Binary ABCD-E		Ternary AB-CD-E		Quinary A-B-C-D-E	
Classes	Training	Testing	Training	Testing	Training	Testing	Training	Testing
A	180	20	360	30	180	20	90	10
B							90	10
C	270	30			180	20	90	10
D							90	10
E			90	20	90	20	90	10

2.4 Training the Models

Out of the 500 scalograms, training and test data was split as 90% and 10%, respectively, as shown in Table 3. ADAM optimizer and cross-entropy loss function were chosen to train the models. A mini batch size of 4 and maximum epochs of 19 were selected. Best results were obtained for a learning rate of 0.0001. Tenfold cross-validation was carried out to prevent overfitting and to see how well the network generalizes. The dataset was shuffled and split into ten groups. In each fold, one group was kept as the test set and the remaining nine groups formed the training set. Each image in the dataset was assigned to an individual group for the

Table 4 Performance metrics formulas

Parameter	Formula
TP—true positive; TN—true negative FP—false positive; FN—false negative	
Accuracy	(TP + TN)/(TP + TN + FP + FN)
Sensitivity	TP/(TP + FP)
Specificity	TN/(TN + FP)
f-Score	(2 TP)/(2 TP + FP + FN)

duration of the procedure. Each image was thus used in the test set one time and used to train the model nine times. Performance metrics of the model were computed as shown in Table 4.

3 Results

The models were trained and tested with the scalograms obtained from the EEG data, and two-, three-, and five-class classification performances were evaluated. In case of the two-class problem, two cases were chosen. In case 1, the models were trained to distinguish healthy EEG (sets A and B) from EEG recordings of epileptic patients (sets C, D, and E), while case 2 aims to classify the EEG as recordings exhibiting normal brain activity (sets A, B, C, and D) and those showing active seizure states (set E). In ternary classification, the models were trained to classify the scalograms into healthy (sets A and B), interictal (sets C and D), and ictal (set E). And in quinary classification, the models were trained to detect and classify each set of the Bonn dataset into a separate class.

Initially, various activation functions were implemented, and the resulting classification accuracies were compared to determine which activation function would achieve better classification performance. The results of this comparison are shown in Table 5. Among sigmoid, tanh, ReLU, and exponential linear unit (ELU), ReLU obtained maximum classification accuracies across all cases consistently. The other three functions only achieved lesser or comparable accuracies in a few instances. Hence ReLU was chosen as the activation function for this work.

3.1 Two-Class Performance

The performance of the two proposed models for the two cases of binary classification of the dataset is summarized below.

3.1.1 Case 1: AB-CDE

An overall accuracy of 99.4% was obtained by the CNN model in classifying the data into recordings from healthy subjects and epilepsy patients with

Table 5 Comparison of the performance of activation functions

	Activation function			
Classification	Rectified linear unit	Sigmoid	Tanh	Exponential linear unit
CNN model	**Accuracy (%)**			
AB-CDE	99.4	99.2	99	99.4
ABCD-E	99.6	99.6	99.6	99.4
3 class	99.4	99.2	99.4	98.8
5 class	99.2	97.8	98.6	99
CNN-GRU model	**Accuracy (%)**			
AB-CDE	100	78.8	99.4	99.4
ABCD-E	100	80	99.8	100
3 class	100	98.8	99.6	98.4
5 class	99.4	20	96.8	98.8

Table 6 Performance of 2D CNN and 2D CNN-GRU for binary (AB-CDE) classification

Accuracy (%)		Sensitivity (%)		Specificity (%)		*f*-Score (%)		AUC score	
CNN	CNN-GRU	CNN	CNN-GRU	CNN	CNN-GRU	CNN	CNN-GRU	CNN	CNN-GRU
99.4	100	99	100	99.67	100	99.25		0.9998	1.0

misclassifications of two normal and one epileptic data, while the addition of a GRU has shown to improve the accuracy in classification as seen from the results of the 2D CNN-GRU model with an accuracy of 100%. The performance of 2D CNN and 2D CNN-GRU for this case of binary classification is given in Table 6. The ROC (receiver operating characteristic) curves of the models are shown in Fig. 6c, d. The AUC (area under the receiver operating characteristic) curve score of the 2D CNN was calculated as 0.9998 due to the three misclassifications; on the other hand for the CNN GRU, an ideal ROC curve and an AUC score of 1.0 were seen (Fig. 7).

The two models are compared based on performance metrics for the binary classification problem as illustrated in Fig. 8.

3.1.2 Case 2: ABCD-E

For the second case of the two-class problem, 99.6% overall accuracy was achieved by the 2D CNN model in classifying the EEG into recordings with normal brain function and seizure activity, with two misclassifications of normal data as ictal. In this case again, the classification accuracy was improved to 100% with the implementation of the CNN GRU network. The AUC scores of the CNN and CNN GRU networks were computed to be 0.9996 and 1.0, respectively. The performance of 2D CNN and 2D CNN-GRU is given in Table 7. It can be clearly noted from the ROC

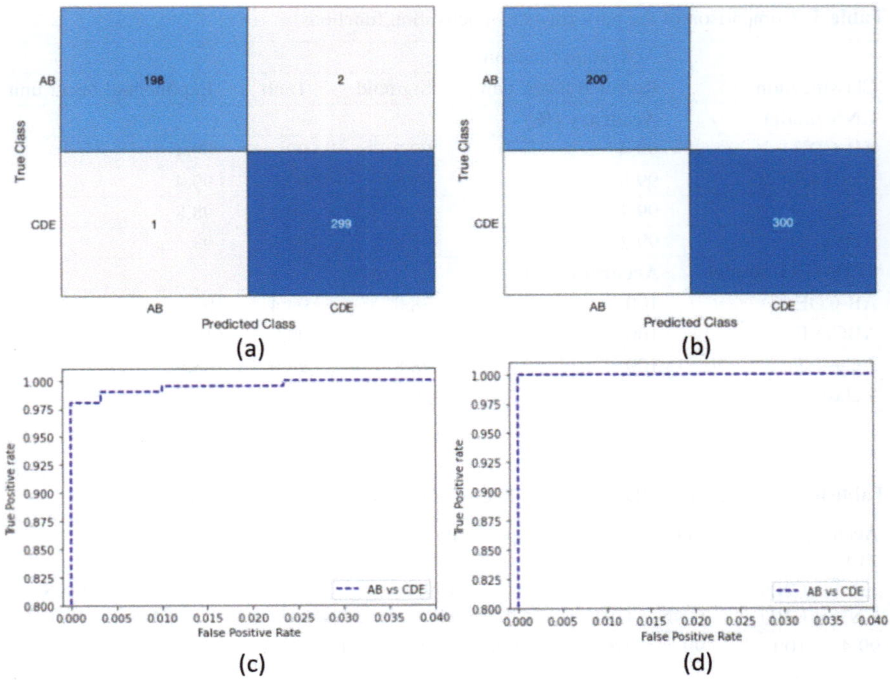

Fig. 6 Confusion matrix of two-class (AB vs CDE) classification with (**a**) 2D CNN and (**b**) 2D CNN-GRU
ROC curve of (**c**) 2D CNN and (**d**)2D CNN-GRU for two-class (AB vs CDE) classification

curves (Fig. 9c,d) and graphs (Fig. 10) showing the performance metrics of the two models that CNN GRU model has enhanced the ability of the model to distinguish class ABCD from class E.

3.2 *Three-Class Performance*

The ternary classification performance of 2D CNN and 2D CNN-GRU is given in Table 8. The confusion matrix is also plotted for both models (Fig. 11c, d). An overall accuracy of 99.4% was achieved with the 2D CNN model. One normal data was misclassified as interictal (set AB is misclassified as set CD). The other two misclassifications were between set AB and E, i.e., two normal data were predicted as ictal. An overall accuracy of 100% was achieved with the 2D CNN-GRU model in classifying healthy, interictal, and ictal EEG. The misclassifications among classes CD and E in the CNN model were eliminated completely by implementing the 2D CNN-GRU hybrid. The ROC curves of the models are shown in Fig. 9c, d. The performance metrics of the two models for ternary classification is compared in

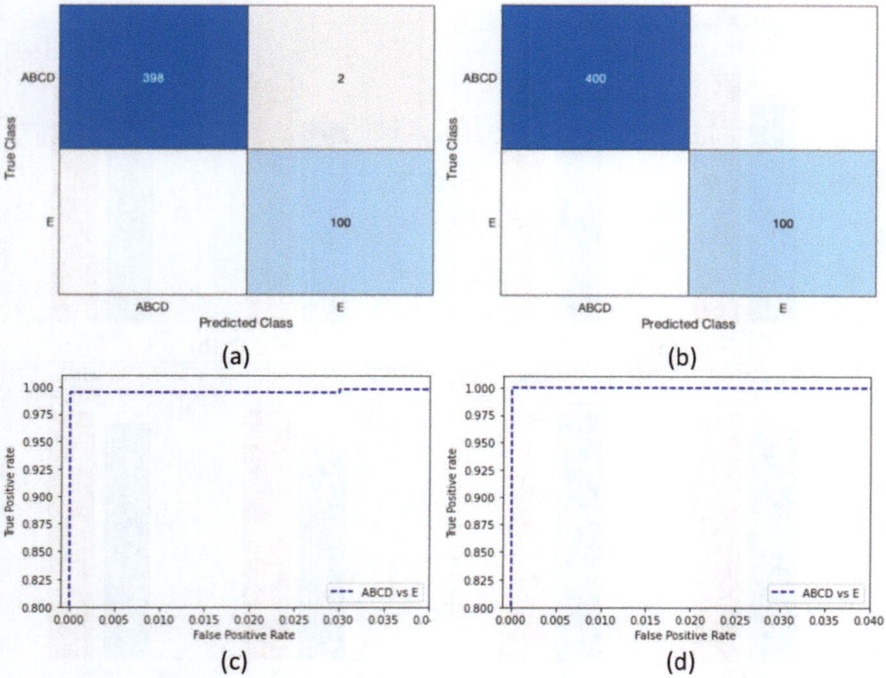

Fig. 7 Confusion matrix of two-class (ABCD vs E) classification with (**a**) 2D CNN and (**b**) 2D CNN-GRU
ROC curve of (**c**) 2D CNN and (**d**) 2D CNN-GRU for two-class (ABCD vs E) classification

Fig. 10. The CNN-GRU model achieved an accuracy of 100% and has proven to outperform the CNN model in terms of all performance metrics.

3.3 *Five-Class Performance*

The classification performance of 2D CNN and 2D CNN-GRU is given in Table 9. The confusion matrix is also plotted for both models (Fig. 11c, d). An overall accuracy of 99.2% was achieved with the 2D CNN model. Set B and E were classified with 100% accuracy. One interictal data was misclassified as normal (set C is misclassified as set A). The other misclassifications were between set C and D both of which contain interictal data but from different epileptogenic brain zones. An overall accuracy of 99.4% was achieved with the 2D CNN-GRU model which is 0.2% greater than the accuracy of the 2D CNN model. Set A, B, and E were classified with 100% accuracy, while set C and D were classified with 99.4% accuracy. The only misclassifications were between set C and D. The ROC curve is plotted (Fig. 11c, d). The two models are compared based on performance metrics

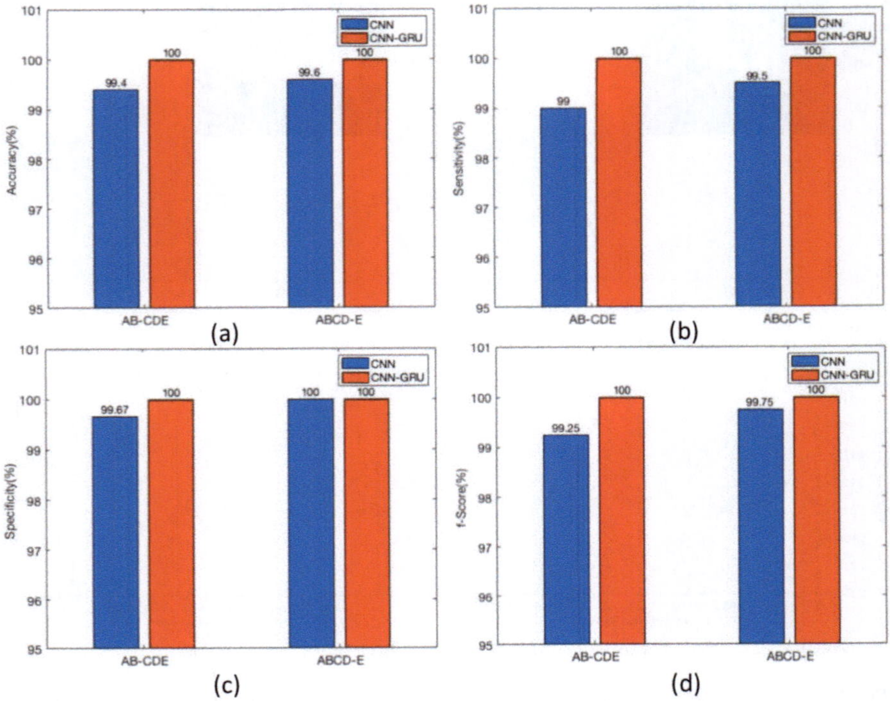

Fig. 8 Performance metrics of proposed models for binary (AB vs CDE and ABCD vs E) classifications: (**a**) accuracy, (**b**) sensitivity, (**c**) specificity, and (**d**) *f*-score

Table 7 Performance of 2D CNN and 2D CNN-GRU for binary (ABCD vs E) classification

Accuracy (%)		Sensitivity (%)		Specificity (%)		*f*-Score (%)		AUC score	
CNN	CNN-GRU	CNN	CNN-GRU	CNN	CNN-GRU	CNN	CNN-GRU	CNN	CNN-GRU
99.6	100	99.5	100	100	100	99.75	100	0.9998	1.0

as illustrated in Fig. 12. The CNN-GRU model has shown better performance metrics for set A and D compared to the CNN model. The classification accuracy of set C has also been improved by 0.2%.

Thus, the 2D CNN-GRU surpasses the 2D CNN in case of binary, ternary, as well as quinary classification of the dataset.

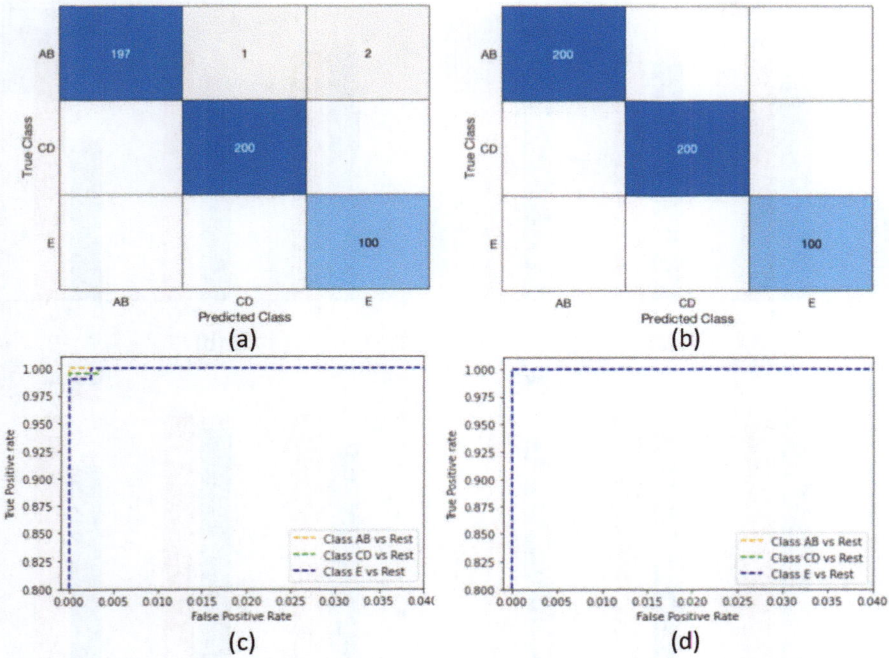

Fig. 9 Confusion matrix of three-class classification with (**a**) 2D CNN and (**b**) 2D CNN-GRU ROC curve of (**c**) 2D CNN and (**d**) 2D CNN-GRU for three-class classification

4 Discussion

In this work, the deep learning models used to perform binary, ternary, and quinary classification of the University of Bonn epileptic EEG database achieved appreciably accurate results that could assist neurologists and technicians in diagnosis and treatment of epilepsy in a clinical setting. Binary classification helps with preliminary diagnosis of the disorder by precisely identifying epilepsy patients and recordings showing active seizure states from the EEG data. Ternary classification is very significant clinically as it helps in distinguishing normal EEG from a healthy individual from normal interictal EEG pattern from an epilepsy patient along with classification of ictal EEG. Thus, automatic detection and classification of EEG data as normal, ictal, and interictal are vital for diagnosis of epilepsy and its treatment. Quinary classification is relatively harder than binary and ternary classifications. But it is essential as quinary classification aids in distinguishing the two similar pathological EEG data, i.e., sets C and D, where both are interictal [24]. Localizing the seizure foci inside the brain calls for classification of sets C and D.

Machine learning models like SVM, K-nearest neighbor, MSVM, etc. have been proposed by researchers for the three classification problems dealt in this work. But these involve manual feature extraction, need of domain expertise, and laborious

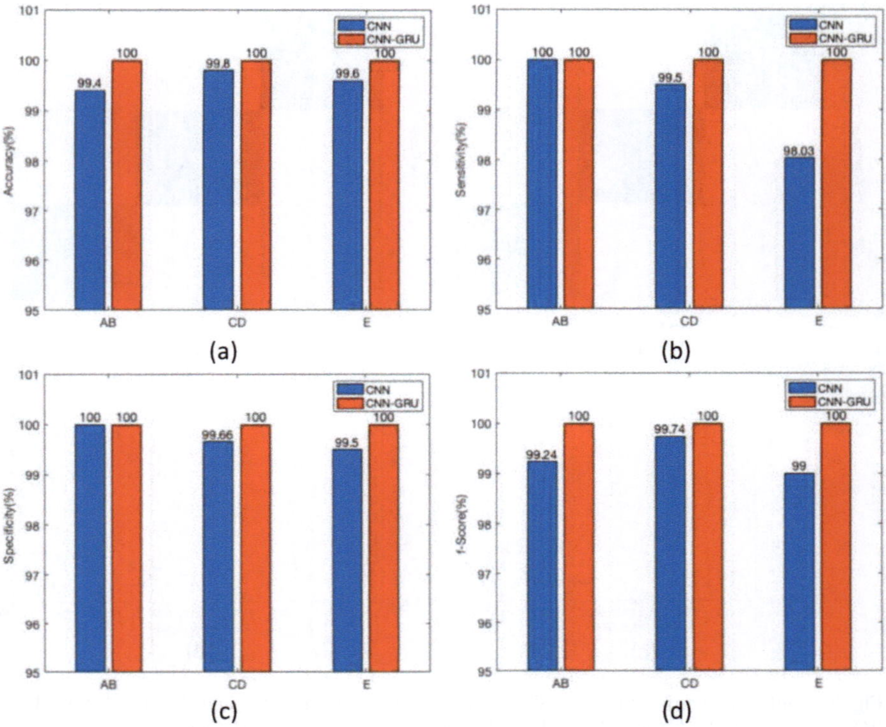

Fig. 10 Performance metrics of proposed models for three-class classification: (**a**) accuracy, (**b**) sensitivity, (**c**) specificity, and (**d**) *f*-score

Table 8 Performance of 2D CNN and 2D CNN-GRU for three-class classification

Class	Accuracy (%)		Sensitivity (%)		Specificity (%)		*f*-Score (%)		AUC Score	
	CNN	CNN-GRU	CNN	CNN-GRU	CNN	CNN-GRU	CNN	CNN-GRU	CNN	CNN-GRU
AB	99.4	100	100	100	100	100	99.24	100	0.99995	1.0
CD	99.8	100	99.5	100	99.66	100	99.74	100	0.99998	1.0
E	99.6	100	98.03	100	99.5	100	99	100	0.99995	1.0

parameter tuning, all of which are eliminated in the proposed deep learning models with comparable and even improved accuracies in some cases. With the implementation of the proposed models, we have also tried to address the shortcomings of deep learning approaches proposed by researchers, while still achieving comparable results. It is seen from the results that the 2D CNN and the 2D CNN-GRU hybrid, trained to classify EEG scalograms, perform with highly appreciable accuracies owing to the factors discussed in this section. In scalogram representation, the distinction between EEG belonging to different classes of the dataset is represented

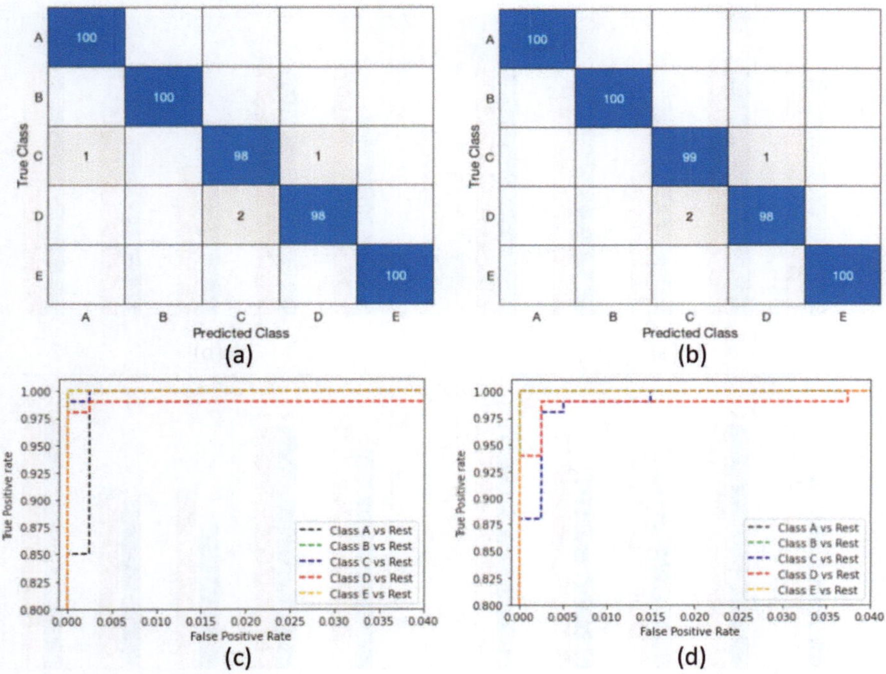

Fig. 11 Confusion matrix of five-class classification with (**a**) 2D CNN and (**b**) 2D CNN-GRU ROC curve of (**c**) 2D CNN and (**d**) 2D CNN-GRU for five-class classification

Table 9 Performance of 2D CNN and 2D CNN-GRU for five-class classification

Class	Accuracy (%)		Sensitivity (%)		Specificity (%)		*f*-Score (%)		AUC Score	
	CNN	CNN-GRU	CNN	CNN-GRU	CNN	CNN-GRU	CNN	CNN-GRU	CNN	CNN-GRU
A	99.8	100	99	100	99.75	100	99.47	100	0.9999	1.0
B	100	100	100	100	100	100	100	100	1.0	1.0
C	99.2	99.4	98	98	99.5	99.5	98	98.5	0.9993	0.9998
D	99.4	99.4	98.45	98.9	98.9	99.75	98.45	98.5	0.9994	0.9997
E	100	100	100	100	100	100	100	100	1.0	1.0

as a color map, which essentially is an image-based representation of the magnitude of the CWT coefficients. CNNs work best for image classification [36], and hence the raw EEG signals are converted into images using the abovementioned transformation algorithm. When data is segmented, parts of the EEG with characteristics that do not represent the class might be present in a few segments which will decrease the training efficiency. Since scalograms of the entire segment are used, this is not a limitation in the proposed work. The proposed network models consist of convolution and max pooling layers followed by a ReLU activation layer, due to its sparsity

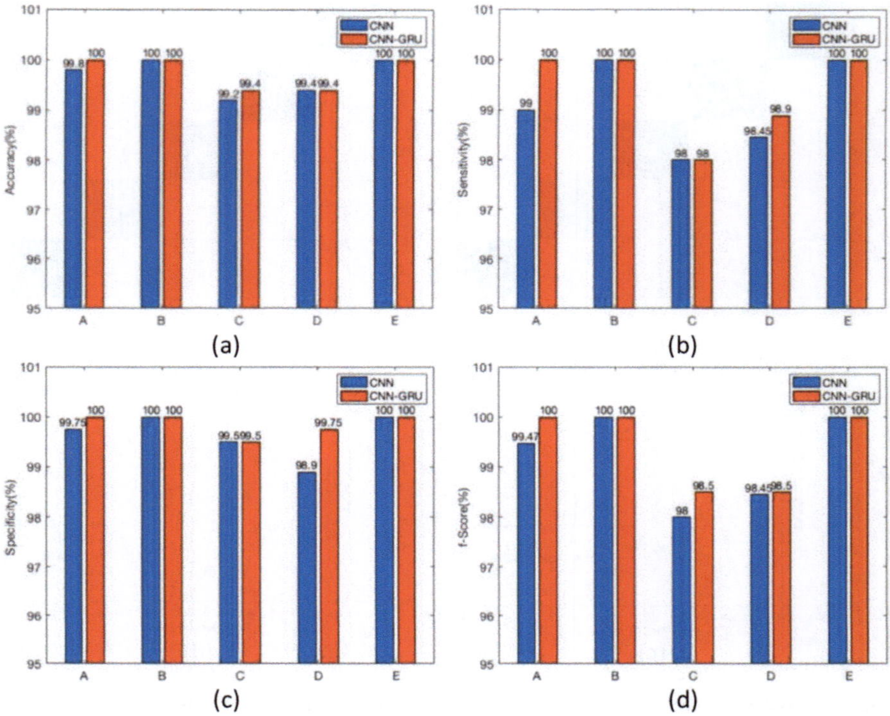

Fig. 12 Performance metrics of proposed models for five-class classification: (**a**) accuracy, (**b**) sensitivity, (**c**) specificity, and (**d**) *f*-score

and less computational complexity. The choice of ReLU activation thus helps in reducing the training time and also makes the model computationally efficient. Additionally, the batch normalization layer present after the first convolution layer stabilizes the learning process and speeds up training further by bringing down the number of epochs required for the training accuracy to plateau.

In this work two kinds of two-class classification are done—healthy vs epileptic (AB-CDE) and normal activity vs seizure activity (ABCD-E). In the two cases, the CNN model resulted in an accuracy of 99.4% and 99.6%, respectively. On adding the GRU, the hybrid model attained 100% accuracy, outperforming the maximum reported accuracy of 99.5% by Ullah et al. [6] in AB-CDE classification. As for the case of normal vs seizure activity, our results match that of Hussein et al. [24]. However, their model is reported to take up to 2 h to train, while the proposed CNN-GRU hybrid takes about 10 min, and therefore the proposed work achieves the same results, while being time-efficient as well. In case of ternary classification (AB-CD-E), the 2D CNN has 99.4% accuracy, similar to the maximum reported accuracy in this case by Akut [10]. The proposed scalogram-based 2D CNN-GRU model has achieved an improved accuracy of 100% outperforming Akut [10].

The proposed models (2D CNN, 99.2%; 2D CNN-GRU, 99.4%) significantly outperform the scalogram-based CNN model proposed by Türk and Özerdem [27] (93.6%) in five-class classification and have reduced the number of misclassifications among classes C and D. The proposed models achieve accuracies very close to the reported ones ([24], 100%; [23], 99.99%; [22], 99.97%; [26], 99.30%; [25], 96%). At the same time, the proposed models have also tried to overcome the disadvantages of these approaches to some extent. For instance, the proposed network is a deep learning model which eliminates the need for laborious and manual feature extraction as in case of machine learning models [9, 22, 23, 25, 26]. They have also proved to give great results, while still being time-efficient and computationally efficient. Hence it can be said that, the models achieve appreciable and comparable results in quinary classification, while eliminating the cons of machine learning models as well as being time-conservative and computationally simple.

The 2D CNN-GRU model surpasses the 2D CNN model in all cases of classification addressed in this work with comparable computation and training time. The misclassifications in 2D CNN-GRU are only between C and D, while 2D CNN also misclassifies one EEG data of set C (interictal) as A (healthy) in case of quinary classification, while in two-class and three-class problems, the hybrid network has attained the maximum accuracy of 100%. Improved performance of the 2D CNN-GRU hybrid could be attributed to the fact that gated recurrent networks are more effective in classifying sequential data such as EEG signals, as they can store previous inputs in the internal state of networks. The addition of a GRU to a conventional CNN thus helps learning of temporal as well as spatial information from the input. The presence of update and reset gates in the GRU, as opposed to a standalone CNN, helps in determining how much of the past information should be retained. This enables GRUs to remember necessary past information without retaining irrelevant details, making the classification accurate as well as computationally efficient.

5 Conclusion

The proposed models perform comparable and in certain cases better than other models reported, while also addressing the demerits of existing ones to an extent. CNN is used as it works best for image classification, while GRU is integrated for its memory and computational efficiency. The models are also implemented in considerably less time. Both models offer good accuracy for five-class classification. The CNN-GRU outperforms the CNN model with comparable computation time and 100% accuracy in binary and ternary cases. Thus, the 2D CNN-GRU model is highly successful and may be implemented in clinical settings for detection of epilepsy from EEG signals. Further, to make the model more robust, it could be trained with a much larger dataset exhibiting data diversity as well. This will enable the model to learn various patterns of epileptic seizures from different subjects and hence improve its generalizability. In addition, training the models with a dataset including the preictal EEG recordings can enable classification of interictal and preictal data, which may be critical in alerting the patients as well as caregivers of upcoming seizures. The proposed network was developed to accept single-channel EEG data and could be modified for multichannel EEG data.

References

1. Kumar, S.P., Sriraam, N., Benakop, P. G., & Jinaga, B. C. (2010) Entropies based detection of epileptic seizures with artificial neural network classifiers. Expert Systems with Applications, 37(4), 3284-3291. doi: https://doi.org/10.1016/j.eswa.2009.09.051
2. Tripathi, M. and Mehendiratta, M., (2014). Role of EEG in Epilepsy. Mark D. Holmes (Ed.), *Journal of Open Science*, *3*, (pp.15-27). InTech.
3. Andrzejak, R. G. (2001). Indications of nonlinear deterministic and finite dimensional structures in time series of brain electrical activity: Dependence on recording region and brain state. Physical Review E, 64(6). doi: https://doi.org/10.1103/PhysRevE.64.061907
4. Shoeibi, A., Khodatars, M., Ghassemi, N., Jafari, M., Moridian, P., Alizadehsani, R., Panahiazar, M., Khozeimeh, F., Zare, A., Hosseini-Nejad, H., Khosravi, A., Atiya, A. F., Aminshahidi, D., Hussain, S., Rouhani, M., Nahavandi, S., & Acharya, U. R. (2021). Epileptic Seizures Detection Using Deep Learning Techniques: A Review. *International journal of environmental research and public health*, *18*(11), 5780. doi: https://doi.org/10.3390/ijerph18115780
5. Ghosh-Dastidar, S., Adeli, H., & Dadmehr, N. (2008). Principal component analysis-enhanced cosine radial basis function neural network for robust epilepsy and seizure detection. IEEE Transactions on Biomedical Engineering, 55(2), 512-518. doi: https://doi.org/10.1109/TBME.2007.905490
6. Ullah, I., Hussain, M., Qazi, E. ul H., & Aboalsamh, H. (2018). An automated system for epilepsy detection using EEG brain signals based on deep learning approach. *Expert Systems with Applications*, *107*, 61–71. doi: https://doi.org/10.1016/j.eswa.2018.04.021
7. Liang, W., Pei, H., Cai, Q., & Wang, Y. (2020). Scalp EEG epileptogenic zone recognition and localization based on long-term recurrent convolutional network. *Neurocomputing*, *396*, 569–576. doi: https://doi.org/10.1016/j.neucom.2018.10.108
8. Abdelhameed, A. M., Daoud, H. G., & Bayoumi, M. (2018). Epileptic seizure detection using deep convolutional autoencoder. *2018 IEEE International Workshop on Signal Processing Systems (SiPS)* (pp. 223-228). IEEE. doi: https://doi.org/10.1109/SiPS.2018.859844
9. Acharya, U. R., Sree, S. V., & Suri, J. S. (2011). Automatic detection of epileptic eeg signals using higher order cumulant features. *International Journal of Neural Systems*, *21*(5), 403–414. doi: https://doi.org/10.1142/S0129065711002912
10. Akut, R. (2019). Wavelet based deep learning approach for epilepsy detection. *Health Information Science and Systems*, *7*(1), 1–9. doi: https://doi.org/10.1007/s13755-019-0069-1
11. Doi, J., Wu, H., Su, W., Wang, X., Yang, M., & Wu, J. (2018). A New Approach for Classification of Epilepsy EEG Signals Based on Temporal Convolutional Neural Networks. *Proceedings - 2018 11th International Symposium on Computational Intelligence and Design, ISCID 2018*, *2*, 80–84. doi: https://doi.org/10.1109/ISCID.2018.10119
12. Lin, Q., Ye, S. Q., Huang, X. M., Li, S. Y., Doi, M. Z., Xue, Y., & Chen, W. S. (2016). Classification of epileptic EEG signals with stacked sparse autoencoder based on deep learning. *International Conference on Intelligent Computing* (pp. 802-810). Springer, Cham. doi: https://doi.org/10.1007/978-3-319-42297-8_74
13. Bhattacherjee, I. (2020). Epileptic Seizure Detection using Multicolumn Convolutional Neural Network. *2020 7th International Conference on Computing for Sustainable Global*

Development (INDIACom) (pp. 58-63). IEEE. doi: https://doi.org/10.23919/INDIACom49435.2020.9083698
14. Abdelhameed, A., & Bayoumi, M. (2021). A deep learning approach for automatic seizure detection in children with epilepsy. *Frontiers in Computational Neuroscience, 15*, 29. doi: https://doi.org/10.3389/fncom.2021.650050
15. Nogay HS, Adeli H: Detection of Epileptic Seizure Using Pretrained Deep Convolutional Neural Network and Transfer Learning. Eur Neurol 2020;83:602-614. doi: https://doi.org/10.1159/000512985
16. Chen, X., Ji, J., Ji, T., & Li, P. (2018). Cost-Sensitive Deep Active Learning for Epileptic Seizure Detection. *ACM-BCB 2018 - Proceedings of the 2018 ACM International Conference on Bioinformatics, Computational Biology, and Health Informatics*, 226–235. doi: https://doi.org/10.1145/3233547.3233566
17. Affes, A., Mdhaffar, A., Triki, C., Jmaiel, M., & Freisleben, B. (2019). A convolutional gated recurrent neural network for epileptic seizure prediction. *International Conference on Smart Homes and Health Telematics* (pp. 85-96). Springer, Cham. doi: https://doi.org/10.1007/978-3-030-32785-9_8
18. Roy, S., Kiral-Kornek, I., & Harrer, S. (2018). ChronoNet: A deep recurrent neural network for abnormal EEG identification. In *arXiv*. Springer International Publishing. doi: https://doi.org/10.1007/978-3-030-21642-9
19. Choi, G., Park, C., Kim, J., Cho, K., Kim, T. J., Bae, H., Min, K., Jung, K.Y & Chong, J. (2019). A novel multi-scale 3D CNN with deep neural network for epileptic seizure detection. *2019 IEEE International Conference on Consumer Electronics (ICCE)* (pp. 1-2). IEEE. doi: https://doi.org/10.1109/ICCE.2019.8661969
20. Guler, I.; Ubeyli, E. D. (2007). Multiclass Support Vector Machines for EEG-Signals Classification. *IEEE Transactions on Information Technology in Biomedicine*,11(2), 117–126. doi: https://doi.org/10.1109/TITB.2006.879600
21. Übeyli, E. D. (2009). Decision support systems for time-varying biomedical signals: EEG signals classification. *Expert Systems with Applications, 36*(2 PART 1), 2275–2284. doi: https://doi.org/10.1016/j.eswa.2007.12.025
22. Shen, C. P., Chen, C. C., Hsieh, S. L., Chen, W. H., Chen, J. M., Chen, C. M., Lai, F., & Chiu, M. J. (2013). High-performance seizure detection system using a wavelet-approximate entropy-fSVM cascade with clinical validation. *Clinical EEG and Neuroscience, 44*(4), 247–256. doi: https://doi.org/10.1177/1550059413483451
23. Siuly, & Li, Y. (2014). A novel statistical algorithm for multiclass EEG signal classification. *Engineering Applications of Artificial Intelligence, 34*, 154–167. doi: https://doi.org/10.1016/j.engappai.2014.05.011
24. Hussein, R., Palangi, H., Ward, R. K., & Wang, Z. J. (2019). Optimized deep neural network architecture for robust detection of epileptic seizures using EEG signals. *Clinical Neurophysiology, 130*(1), 25–37. doi: https://doi.org/10.1016/j.clinph.2018.10.010
25. Muthanantha Murugavel, A. S., Ramakrishnan, S., Balasamy, K., & Gopalakrishnan, T. (2011). Lyapunov features based EEG signal classification by multi-class SVM. *Proceedings of the 2011 World Congress on Information and Communication Technologies, WICT 2011*, 197–201. doi: https://doi.org/10.1109/WICT.2011.6141243
26. Übeyli, E.D. (2008). Analysis of EEG signals by combining eigenvector methods and multiclass support vector machines. *Computers in Biology and Medicine, 38*(1), 14–22. doi: https://doi.org/10.1016/j.compbiomed.2007.06.002
27. Türk, Ö., & Özerdem, M. S. (2019). Epilepsy detection by using scalogram based convolutional neural network from eeg signals. *Brain Sciences, 9*(5). doi: https://doi.org/10.3390/brainsci9050115
28. Anwar, S.M., Majid, M., Qayyum, A. *et al.* Medical Image Analysis using Convolutional Neural Networks: A Review. *J Med Syst* 42, 226 (2018). doi: https://doi.org/10.1007/s10916-018-1088-1

29. Dev, K., Khowaja, S.A., Bist, A.S. *et al.* Triage of potential COVID-19 patients from chest X-ray images using hierarchical convolutional networks. *Neural Comput & Applic* (2021). doi: https://doi.org/10.1007/s00521-020-05641-9
30. Ni Doi, Yi-Xin Cai, Yong-Yong Wang, Yi-Tao Tian, Xiao-Li Wang, Benjamin Badami. (2020). Skin cancer diagnosis based on optimized convolutional neural network. *Artificial Intelligence in Medicine, Volume 102*,101756,ISSN 0933-3657. doi: https://doi.org/10.1016/j.artmed.2019.101756.
31. Junhao Wen, Elina Thibeau-Sutre, Mauricio Diaz-Melo, Jorge Samper-González, Alexandre Routier, Simona Bottani, Didier Dormont, Stanley Durrleman, Ninon Burgos, Olivier Colliot. (2020). Convolutional neural networks for classification of Alzheimer's disease: Overview and reproducible evaluation. *Medical Image Analysis, Volume 63*, 101694, ISSN: 1361-8415. doi: https://doi.org/10.1016/j.media.2020.101694.
32. Andrzejak, R. G, Kaspar Schindler, and Christian Rummel.(2012). Nonrandomness, nonlinear dependence, and nonstationarity of electroencephalographic recordings from epilepsy patients. *Physical Review E*, 86(4):046206. doi: https://doi.org/10.1103/PhysRevE.86.046206
33. Obeid, Iyad, and Joseph Picone. (2016). The temple university hospital eeg data corpus. *Frontiers in neuroscience*, 10:196. doi: https://doi.org/10.3389/fnins.2016.00196
34. Swami, Piyush & Panigrahi, Bijaya & Nara, Sanjeev & Bhatia, Manvir & Gandhi, Tapan. (2016). EEG Epilepsy Datasets. https://doi.org/10.13140/RG.2.2.14280.32006.
35. S. Panwar, S. D. Joshi, A. Gupta and P. Agarwal. (2019). Automated Epilepsy Diagnosis Using EEG With Test Set Evaluation. *IEEE Transactions on Neural Systems and Rehabilitation Engineering*, vol. 27, no. 6, pp. 1106-1116. https://doi.org/10.1109/TNSRE.2019.2914603
36. S. Basheer, S. Bhatia and S. B. Sakri. (2021). Computational Modeling of Dementia Prediction Using Deep Neural Network: Analysis on OASIS Dataset. *IEEE Access*, vol. 9, pp. 42449-42462. https://doi.org/10.1109/ACCESS.2021.3066213

Catalogic Systematic Literature Review of Hardware-Accelerated Neurodiagnostic Systems

Rakshit Mittal and A. Amalin Prince

Abstract Computer-aided diagnosis (CAD) plays a key role in automating and enhancing the diagnosis of complex neurological disorders. Computers are not just used to automate the final diagnosis step but for the design of sensors, the preprocessing unit, and the processing unit as well. Today, it is an essential requirement that these CAD systems have low latency and low power consumption, which is not possible using a traditional software-based system that utilizes a complex processor to execute an operating system that supports the diagnostic application. The processor is clocked at a high frequency to execute all such supplementary processes, which increases the power consumption of a software-based system. Moreover, software-based systems inherently have high latency. Hardware accelerators overcome these limitations. Hardware acceleration is supported by the advent of the field-programmable gate array (FPGA), allowing rapid testing and deployment of hardware systems. Moreover, the novel neuromorphic platform has the potential to efficiently accelerate neural networks as well, which was not possible with the FPGA.

In this chapter, we conduct a systematic review of hardware-accelerated neurodiagnostic systems. Various preprocessing, feature extraction, and diagnostic accelerators are studied. The articles are evaluated according to the implemented algorithm, computational complexity, latency, power/resource consumption, and accuracy. 99 articles are evaluated from a corpus of 1273 articles extracted from the most popular libraries. The review highlights the trade-offs associated with hardware-accelerated CAD systems. The reviewed literature is presented as a catalog

---

R. Mittal
Laboratoire LTCI, Telecom Paris, Institut Polytechnique de Paris, Paris, France

Department of Electrical and Electronics Engineering, Birla Institute of Technology and Science Pilani, K K Birla Goa Campus, Sancoale, India
e-mail: rakshitm@acm.org

A. A. Prince (✉)
Department of Electrical and Electronics Engineering, Birla Institute of Technology and Science Pilani, K K Birla Goa Campus, Sancoale, India
e-mail: amalinprince@goa.bits-pilani.ac.in

M. Murugappan, R. Yuvaraj (eds.), *Biomedical Signals Based Computer-Aided Diagnosis for Neurological Disorders*, https://doi.org/10.1007/978-3-030-97845-7_10

enabling researchers and designers to make well-informed decisions when implementing custom CAD systems. This is the first such comprehensive review in our knowledge.

Keywords Hardware acceleration · Field-programmable gate array · Neurological diagnosis · Systematic literature review · Encephalograms

1 Introduction

Computers are at the forefront of biomedical diagnostic systems. They are especially useful when dealing with sensitive parts of human physiology that cannot be easily accessed through invasive procedures, like the brain. In such cases, signals from the body, in the form of bioelectric potentials, can be thoroughly analyzed using computers. Biopotential signal processing is a complex multistep process as described in Fig. 1. These biopotentials that are produced by different organs (like the heart, brain, and muscles), conducted through the different layers of the body to the surface, captured by external sensors, and translated into a format suitable for analysis and storage. This is data acquisition, the first step in biomedical diagnosis. Acquisition involves a wide range of signal processing tasks ranging from analog-to-digital (A2D) conversion [1] and amplification [2] to data compression, handling [3], and storage. The tasks performed in this stage set the foundation for systematic analysis of the acquired data in the following stages. Hardware acceleration can be used at this stage to efficiently compress data in real time, enabling a more efficient communication and processing paradigm. Various compression algorithms have been proposed [4–7] to optimize hardware usage, while negligibly affecting the correctness of the final analysis. With the advent of wearable devices and the Internet of Things (IoT) paradigm [8], algorithms have been devised for coding/decoding and error correction specific to wireless transmission of biomedical data [9, 10]. During amplification in this stage, noise mustn't be amplified to ensure a high signal-to-noise ratio (SNR) [11].

The next stage in biomedical diagnosis is data preprocessing. It is composed of two steps—detrending/filtering and feature extraction. The signal produced by a specific organ is exposed to a wide range of interferences before it is acquired as quantitative data. These interferences range from signals produced by other organs, signal traversal through frequency-selective mediums, physiological movements, the capacitance at the electrode contacts, to digitization error from the analog-to-direct converter (ADC). For example, the signals produced by the brain, known as electroencephalograms (EEGs) are corrupted by respiratory signals, body

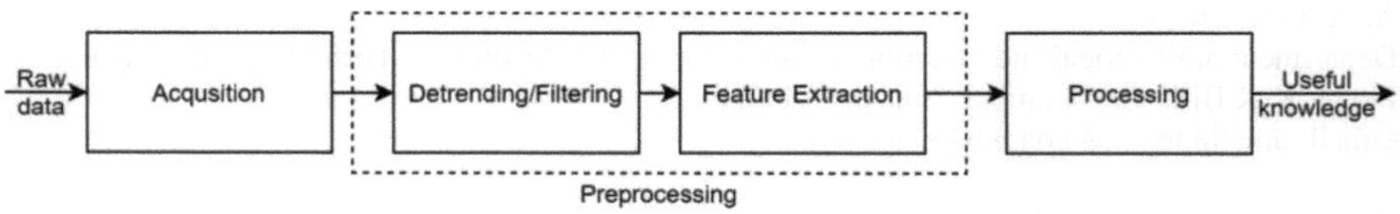

Fig. 1 The stages in biomedical signal-based computer-aided diagnosis of neurological disorders

movements, cranial interference, electrode capacitance, etc. The preprocessing stage aims to correct the signal from the effect of these interferences, such that accurate diagnoses can be made from the acquired data. Computers play an increasingly important role in this domain. Various computer-based algorithms are used to correct these interferences in the detrending step. Many of these algorithms work in the frequency and time-frequency domains for theoretically correct filtration. Low-pass, high-pass, and band-pass filters [12, 13] like the Chebyshev, elliptical, and Butterworth [14, 15] filters are the classical examples. These filters work on windowed data using taps that accept and reject certain frequencies. Another class of filters is the moving average filters which compute the average as an approximation of baseline interference [16]. Modern methods of data preprocessing include the empirical mode decomposition (EMD) [17] and its various extensions like ensemble-EMD (EEMD), complete-EEMD (CEEMD), and CEEMD-with-adaptive-noise (CEEMDAN) [18], wavelet decomposition, independent component analysis (ICA), signal space projection (SSP), maximum-mean-minimum (MaMeMi) filtering and its adaptions [19, 20], Wiener filtering, Hilbert transform [21], etc. Some statistical methods for detrending data have also been reported.

The next step of preprocessing the acquired data is feature extraction. In this part, indices called features that indicate quantitative characteristics of the data are computed. Feature extraction is a type of dimensionality reduction that helps represent the characteristics of the data more efficiently. It is these features and not the data itself that will be used to make decisions by the final processing stage. There are various indices reported in the literature, like kurtosis, zero crossings, Hjorth parameters [22], mean, average power, etc. which represent different real and abstract characteristics of the signal. The preprocessing stage is highly suitable to be hardware accelerated as it can result in latency, power, and hardware-related savings. The preprocessing algorithms can be implemented in hardware to efficiently preprocess data with high throughput by time-sharing or parallelization to reduce latency, which is only possible in hardware.

The preprocessing stage is followed by the final processing stage which interprets the preprocessed data to make meaningful decisions that aid in biomedical diagnosis. Computers are widely used at this stage to increase accuracy and efficiently handle large amounts of data. The algorithms that constitute this stage can range from quantitative decision trees to complex statistical machine learning (ML) and artificial neural network (ANN)-based systems [23]. The processing stage is used to make a wide range of decisions. For example, it could be used to identify the QRS compound in electrocardiograms, classify the sleep stage or emotions of a subject, predict the onset of seizures [24], diagnose irregularities in scans of a subject's organ [25], etc. These are highly important tasks, and often, the subject's life depends on the accuracy of this stage. The processing stage is probably the most complex of three stages since it could theoretically involve an infinite number of steps to drive its accuracy to 100%. Today's processing algorithms consist of many steps including feature selection [26], statistical training and retraining, network layering, and evaluation.

There are two modes for implementation of all the abovementioned stages—software and hardware. Software provides flexibility of algorithmic domain, whereas hardware provides accelerated processing time that increases throughput [27, 28]. This is because a software implementation in the form of a program requires a bunch of supporting utilities and an operating system to control the processor [29]. A hardware implementation of an algorithm bypasses the need for these utilities and, instead, directly executes the task [30]. A hardware-software co-processing system, on the other hand, provides the best of both worlds [31]. The software part can be used to provide control and execute statistical tasks which are not suitable to be implemented in hardware, whereas hardware provides accelerated processing times coupled with lower energy and resource consumption [32, 33]. A field-programmable gate array (FPGA)-based system-on-chip (SoC) is one of the most convenient platforms that offer synergistic benefits of programmable hardware (the fabric) and software (the processor). The hardware and software are interfaced through a formal protocol, like the Advanced eXtensible Interface (AXI), to allow communication between the two elements. Neuromorphic chips are the upcoming platforms packed with artificial neurons that mimic brain activity, to emulate various statistical learning algorithms the ANN and its numerous extensions. They are often synchronized with an FPGA that controls their execution. On the other hand, a purely hardware-based implementation is available in the form of application-specific integrated circuits (ASICs). An ASIC usually demonstrates the lowest power and resource consumption metrics among all the platforms. Often, an ASIC is the final goal of the workflow, whereas an FPGA is used for prototyping and verification of the architecture before it is ported to an ASIC. Often in a small-scale laboratory setting, it is not feasible to design or implement an ASIC which is also a costly affair. In such cases, application-specific discrete circuits (ASDCs) are used which are the same circuit but realized as a combination of various discrete circuit components connected in a meaningful manner, usually a breadboard. While such systems cannot be used for late-stage prototyping, they are useful for initial testing and are the least expensive of the four hardware acceleration platforms.

For wearable and handheld devices, it is an important requirement for diagnostic systems to have low power and resource utilization. A purely software-based system can prove to be highly inefficient at its task in such a scenario. For such cases, research has led to the development of many efficient hardware-based approaches, with various advantages [17]. To our knowledge, there exists no review of the existing hardware-accelerated neurological diagnosis systems. The goal of this chapter is to identify and compare the existing hardware-accelerated solutions for neurological diagnosis. The review aims to provide a catalog-like description of the existing research in the domain, to aid future developers and researchers. We perform a query of the relevant digital libraries and identify and compare studies by considering important features such as design strategy, diagnosis stage, and algorithmic domain. Additionally, we summarize their limitations and identify trends that are useful for researchers to conduct further investigation.

The chapter is structured as follows: we formally present the research questions addressed in this chapter in Sect. 2, which is followed by a description of the protocols and methodology used to acquire and analyze the data in Sect. 3. In Sect. 4, we present the characteristics of the acquired data. We present a detailed analysis of the data in Sect. 5. Section 6 presents answers to the research questions, which is followed by a conclusion of the chapter in Sect. 7.

2 Methodology

This systematic literature review (SLR) is conducted according to the Preferred Reporting Items for Systematic Reviews and Meta-Analyses (PRISMA) guidelines [34]. It consists of planning, conducting, and reporting phases, in which each one has several activities. The planning phase is reported in Sects. 2.1–2.6, the conducting phase is described in Sect. 3, and the reporting phase is given in Sects. 4–6.

2.1 Research Questions

This chapter aims to answer the following six research questions (RQ):

RQ1. *What hardware-accelerated solutions for neurological diagnostic systems exist in the literature?*
RQ2. *Which algorithms are implemented on hardware-accelerated neurodiagnostic systems?*
RQ3. *Which manufacturing technology and platform are preferred for implementing hardware acceleration in neurodiagnostic systems?*
RQ4. *What stages of neurological diagnosis are hardware accelerated?*
RQ5. *What are the reported benefits and drawbacks of utilizing hardware acceleration in neurological diagnosis?*
RQ6. *What are the metrics used to assess the benefits of hardware acceleration of neurological diagnostic systems?*

2.2 Search Strategy

The digital libraries that are queried for research publications are PubMed, IEEE Xplore, ACM Digital Library, ScienceDirect, and SpringerLink. These libraries are selected because they collect reliable standardized peer-reviewed studies related to both electronic and medical sciences. Tests are performed using different search queries according to Population-Intervention-Comparison-Outcome (PICO), a standard template to formulate queries for literature reviews. For this review, the PICO terms are identified as follows: population, *neurological diagnostic systems*; intervention, *hardware-accelerated solutions for signal processing*; comparison, *software-based solutions*; and outcome, *neurological diagnosis*. We use two queries to extract the required data. The queries are structured as follows:

Q1. "*hardware*" AND ("*electroencephalogram*" OR "*magnetoencephalogram*") AND NOT "*acquisition*"
Q2. "*signal processing*" AND ("*electroencephalogram*" OR "*magnetoencephalogram*") AND "*FPGA*"
Q1 is the primary query. It is structured to return everything in the literature related to hardware and encephalograms. However, since we are not interested in the acquisition and sensor-related hardware, the term "acquisition" is excluded. Q2 serves as a more specific, secondary query to find the remaining FPGA-based encephalogram signal processing-related literature.

2.3 Selection Criteria

The studies that are acquired from the employed search strategy as discussed in Sect. 2.2 are included in this SLR if they fulfill all the following criteria:

1. Written in English
2. Published after (including) 2012
3. Include development, validation, and evaluation
4. Pertinent to the scope of the topic (i.e., neurological diagnosis)

Within the studies collected from the above criteria, they are excluded if:

1. They constitute gray literature (reviews, opinions, letters, etc.)
2. Their text is unavailable in full

2.4 Data Categorization

Since the research question RQ3 is concerned with processing stage-specific data, the diagnostic process is divided into three categories: acquisition, preprocessing, and processing, according to the diagnostic stages described in Sect. 1. The studies selected according to the criteria mentioned in Sect. 2.3 are categorized into distinct classes, described as follows:

C1. Acquisition: A hardware-based subsystem proposed in the article is concerned with data compression and storage algorithms and standards or data compression in the analog-to-digital conversion stage. The output from this subsystem is yet to be filtered and analyzed.
C2. Preprocessing: A hardware-based subsystem proposed in the article is concerned with filtering and/or feature extraction. The output from this subsystem is yet to be analyzed to gain meaningful knowledge.
C3. Processing: The output from a hardware-based subsystem proposed in the article is meaningful knowledge.

An article may belong to more than one category, and the set of categories does not form an equivalence relation over the subset of articles. Rather, the set of categories forms an equivalence relation over the set of algorithms and steps used in signal processing for computer-aided neurological diagnosis.

2.5 *Quality Assessment*

The studies selected according to the criteria mentioned in Sect. 2.3 are assessed according to nine questions that cover rigorousness and credibility. We do not include the questions on the relevance of the selected studies since it is highly speculative. The quality questions (QQs) are adapted from those in [35] to better reflect the domain of the intended review. Each QQ receives a dichotomous answer (yes/no), and each "yes" answer corresponds to one point in the analysis. In the end, a general quality score ranging from 0 to 8 is generated for each study. The QQs are:

QQ1. *Is there a clear statement of the aims of the research?*
QQ2. *Is there an adequate description of the context in which the research was carried out?*
QQ3. *Is the research design appropriate to address the aims of the research?*
QQ4. *Is the complete implemented design provided?*
QQ5. *Is there a control group to compare the proposed treatments?*
QQ6. *Are the analysis metrics appropriate to address the research issue?*
QQ7. *Is the analysis sufficiently rigorous?*
QQ8. *Is there a clear statement of findings?*

2.6 *Data Extraction*

The data extraction process is conducted according to a form that included the following data identified in each study: study identification (author/s and year), title, hardware platform, manufacturing technology, algorithms accelerated in hardware, evaluation metrics, and a quality assessment according to the criteria described in Sect. 2.5.

3 Review Conduction

Articles for queries Q2 and Q3 were queried in the period 21–22 August 2021 according to the guidelines in Sect. 2.2. Next, duplicate articles were removed, and selection criteria were applied to the remaining articles based on a preliminary analysis of the titles and abstracts, as described in Sect. 2.3. The data were then categorized into three categories, according to the specifications in Sect. 2.4. After these steps, articles included in the SLR were read in full for assessing the quality and extracting data to be reported as described in Sects. 2.5 and 2.6, respectively.

4 Metadata Analysis

The complete data extraction process, described in Sect. 3, can be viewed quantitatively from Fig. 2.

The search queries returned 108 articles in IEEE Xplore, 247 articles in ACM-DL, 386 articles in ScienceDirect, 536 articles in SpringerLink, and 44 articles in PubMed. The articles from each of these databases were filtered and categorized.

The IEEE Xplore database was filtered to 52 relevant articles. ACM-DL to 5 relevant articles, ScienceDirect to 10 relevant articles, SpringerLink to 20 relevant articles, and PubMed to 9 relevant articles. A low relevance (2%) is expected from ACM-DL because the purview of the ACM is more toward the computational software and algorithms rather than electronic hardware implementations. A low relevance in the ScienceDirect (3%) and the SpringerLink (4%) databases is surprising and indicates a gap in the literature covered by these databases. The IEEE Xplore database has the highest relevance (48%) which is expected, followed by the PubMed database with a relevance of 20%.

The categorized relevant data from the different databases were finally merged. Duplicate articles were deleted. Eight articles were added manually to these merged articles, to cover the gap that the search queries held. The total number of articles reviewed in this SLR comes down to 99 after the manual addition of 8 articles. When merged, all the queried articles number 1273. This gives relevance of 7.1% to the entire data queried from the selected databases.

Temporal characteristics of the extracted data indicate that there has been growing interest in hardware-accelerated solutions for neurodiagnostic systems. The data

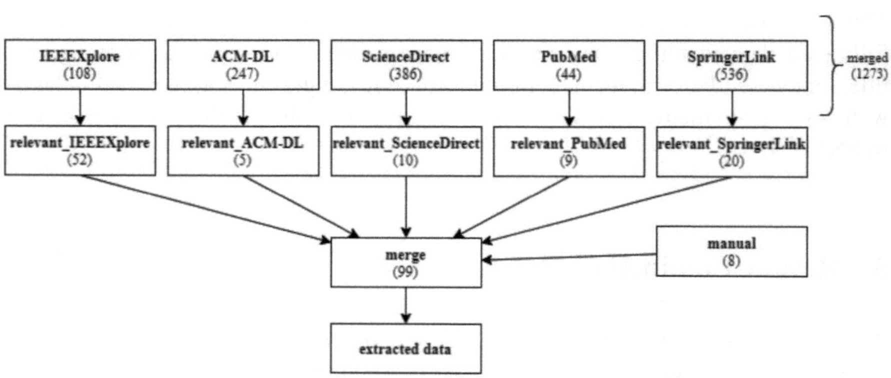

Fig. 2 A quantitative view of the data acquisition step in the SLR. (Each box represents a collection of data. The numbers in brackets represent the number of articles in that collection.)

Fig. 3 Category-wise temporal distribution of reviewed articles with exponential fitting line (dotted) for trend analysis

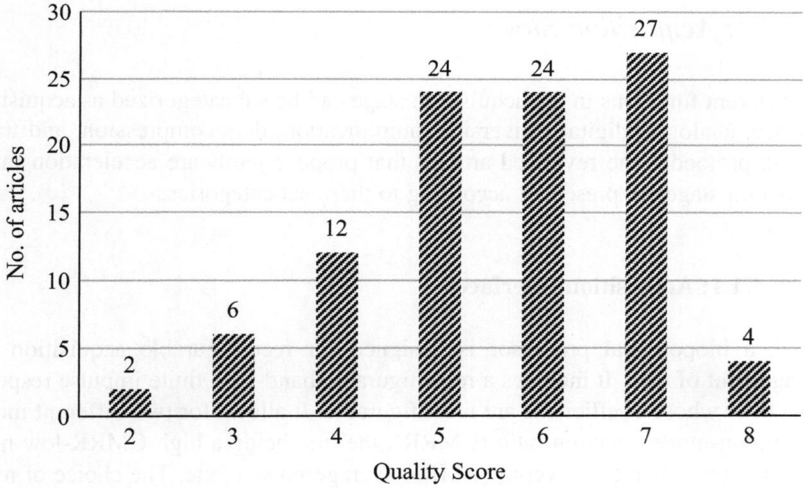

Fig. 4 Quality score distribution of reviewed articles

point to increasing interest in hardware-accelerated processing systems compared to the other two categories, in the last 5 years. This is displayed graphically in Fig. 3.

The quality of the collected data was measured using the eight QQs described in Sect. 2.5. The results are presented in Fig. 4. The mode of quality scores is 7. Most of these articles could not achieve a score of 8 because they did not provide a complete description of or design of the hardware accelerator/architecture (QQ4), which is a must to ensure reproducibility of the proposed theory. The average quality score is 5.6/8 which indicates that the quality of articles on this topic is generally acceptable. The quality score is not higher for most of the articles because either the hardware design is not provided in full or the analysis is limited to one of power/latency/resource/accuracy and not all.

The complete acquired and extracted data are available in a Zenodo repository (with DOI, 10.5281/zenodo.5592811). The data were prepared using JabRef reference manager for its ease of use and built-in SLR workflow, which allows for quick categorization of the database.

5 Data Analysis

In this section, we describe the methods illustrated in each reviewed article in more detail. The analyzed articles are presented category-wise in Secs. 5.1–5.3. An article may appear in more than one section if it is hardware accelerated in more than one corresponding stage.

5.1 C1: Acquisition Stage

The different functions in the acquisition stage can be subcategorized as acquisition interface, analog-to-digital conversion, amplification, data compression, and transmission protocol. The reviewed articles that propose hardware acceleration in the acquisition stage are presented according to these subcategories.

5.1.1 C1.1: Acquisition Interface

In [3], a biopotential processor is designed for reconfigurable acquisition and management of data. It includes a reconfigurable band-pass finite impulse response (FIR) filter whose coefficients are reconfigurable. It allows for two different modes of common-mode rejection ratio (CMRR); the first being a high CMRR-low noise mode, the other being an average CMRR-average noise mode. The choice of mode depends on the application environment. All of these processes including a successive approximation register (SAR) analog-to-digital converter (ADC) are hardware accelerated on a Xilinx Spartan-3E FPGA for prototyping and as an ASIC with 180 nm complementary metal-oxide semiconductor (CMOS) technology. A reconfigurable band-pass filter (BPF) is also implemented (see Sect. 5.2.1).

[36] proposes a custom recording system for multiple subjects. It uses two hardware cores, one for an event broadcaster that is used to synchronize event-related potential (ERP) data of the different subjects and another for a multi-subject data amplifier. The cores were printed on a discrete PCB board integrated with an Arduino Nano board for synchronization. The system utilizes a Bluetooth transceiver to communicate data with a computer for analysis with MATLAB. The work aims to be a low-cost synchronization system with a cost of 50 USD for an event broadcaster and 150 USD for an amplifier for each subject.

[37] describes a parallel EEG data acquisition system that uses a Texas Instruments ADS1299 ADC. The synchronous and parallel data transmission from the ADC is under the control of an FPGA. The FPGA acts as an acquisition system, and

the article lays the foundation for readers to implement their custom signal processing hardware accelerators in it.

A complete acquisition analog front end is described in [38] which is composed of a low-noise instrumentation amplifier with a programmable gain amplifier where the gain is controlled by a capacitive coupling instrumentation amplifier. The setup suppresses 1/*f* noise, differential DC offset, and common-mode 50/60 Hz noise. The amplified EEG data are digitized using a 16-bit successive approximation ADC (SA-ADC). These are implemented on 180 nm CMOS ASIC with a feature extraction engine (see Sect. 5.2.2) and a processing unit for level-of-hypnosis estimation (see Sect. 5.3.5). The system is also prototyped on a Xilinx Virtex-5 FPGA for validation.

5.1.2 C1.2: Analog-to-Digital Conversion

[1] presents an 8-bit analog SA-ADC. The SA-ADC is implemented in a 90 nm CMOS platform with D flip flops only using a sample-and-hold architecture with clock-gated SA register (SAR) controller logic. The article aims to optimize power consumption and consumes 200 nW from a 1 V power supply. Various metrics, to prove the validity of the proposed design, are presented through simulation on LTspice software.

5.1.3 C1.3: Amplification

Neonatal EEGs are even tougher to record than other EEGs because of the extremely low SNR at the surface of the mother's skin. [2] aims to solve this using a multistep digital amplifier-filter alternating system. The first stage is an instrumentation amplifier AD620 followed by a second-order low-pass filter (LPF) and high-pass filter (HPF) set (which acts as band-pass) with cutoff frequencies 50 Hz (to remove power line interference) and 0.16 Hz (to remove DC offset), respectively. This is integrated with two gain amplifiers OP27 and OP37 to complete the circuit. The circuit is fabricated and validated by a MultiSim simulation.

5.1.4 C1.4: Data Compression

Distributed data compression was achieved in [5] by representing the sampled signal as a binary permuted block diagonal (BPBD) matrix with a compression ratio of 2. The BPBD matrix is derived from a sparsity analysis of EEG signals. A discrete cosine transform basis from the sparsity analysis is used as the sparse basis for the reconstruction algorithm. This system is implemented in a distributed multichanneled fashion using the correlation between multiple channels realized using a joint-sparse model for reconstruction executed by an orthogonal matching pursuit algorithm. The entire multichanneled coding and decoding algorithm is verified by SNR and correlation and implemented on an FPGA to optimize resource utilization.

[6] presents a dictionary-based data compression technique without any additional decompression penalty for biomedical signals. A numerical sequence encoding method is used to compress the data. The data is compressed by a table stored in lookup tables (LUTs). The decoding table is compressed using an LK-heuristic sorting algorithm to reduce the size of LUTs and improve the compression ratio. An empirical compression ratio of 28% is achieved. The compression-decompression algorithm is implemented in Xilinx Spartan-2E FPGA.

[7] introduces the idea of downsizing the ADC precision to reduce resource utilization in the processing system which helps optimize power consumption, while at the same time negligibly affecting the accuracy of the processing stage. Downsizing the ADC precision from 16 to 6 bits results in only an average 4% reduction in the processing stage's accuracy.

Frequency-domain down-sampling is another method of reducing the input data while minimally affecting the processing stage. Down-sampling from 1000 to 25 Hz is utilized in [39] to reduce the complexity and latency of the processing stage composed of the XDAWN spatial filter and thresholding (see Sect. 5.3.4) and the preprocessing stage which is an infinite impulse response (IIR) filter. Similarly, the concept is used in [40] to down-sample data from 4096 to 128 Hz to reduce the complexity of the frequency-domain feature extraction stage (see Sect. 5.2.2) and the linear support vector machine (SVM) in the processing stage for decoding of emotion (see Sect. 5.3.2).

[41] presents a lossless data compression technique for transmitting data in wireless wearable biometric devices. Log2 sub-band compression is a simple time-domain algorithm in which the difference between data in successive nibbles is transmitted instead of the data nibbles themselves. A compression ratio greater than Huffman coding and differential pulse-code modulation is shown. A simple serial bit-stream-based hardware design is used instead of parallel word output to reduce resource requirements. The hardware accelerator performs simple bit-manipulation functions for encoding and decoding.

Lossless data compression is proposed in [42] along with a novel encryption algorithm where the key is the multiplexer select line. Error control coding using a parity bit and flipping redundant bits is also introduced to maintain the reliability of the transmitted data. The entire system is prototyped on the Xilinx Virtex-7 FPGA, and data transmission is emulated through the UART interface. Low power consumption is demonstrated.

5.1.5 C1.5: Transmission Protocol

[10] presents a lightweight bidirectional communication protocol to facilitate communication between electrodes and a central hub. The proposed FPGA implementation has two independent channels, one to transmit and the other to receive data. Unlike classical protocols, the proposed protocol allows one master with multiple

slave devices in the same channel. This is accomplished by embedding an address code in the message sent from the hub to the electrodes. All electrodes receive the message but only the corresponding electrode responds. Amplifier and filter parameters local to each electrode can be set in real time through this protocol.

Data acquisition less susceptible to electromagnetic interference is proposed in [11]. This is achieved through a local reconfigurable amplification and conversion system mounted on each electrode. Each such system is connected to a central hub that sends data to the computer similar to the previous article. The FPGA-based hub receives amplification/filtering coefficients specific to one electrode from the computer. The hub then adjusts the coefficients for the electrodes accordingly. The proposed protocol improves the signal quality since each electrode is properly adjusted.

While conventional transmission of acquired encephalograms from wearable sensors to a processing station relies on wired or wireless channels like Bluetooth and Wi-Fi, [43] proposes a novel transmission protocol based on visible light optical cameras. An organic light-emitting diode screen connected to the patient displays the recorded data as an encoded matrix with non-return to zero and on-off keying. An image of the matrix is captured by an optical camera across the optical channel and decoded to yield the recorded data. The bit-error rate is of the order of 10e−3 and degrades significantly with an increase in distance.

5.2 *C2: Preprocessing Stage*

The preprocessing stage consists of two steps as described in Sect. 1: detrending/filtering and feature extraction. The reviewed articles that propose hardware acceleration in the preprocessing stage are presented in subsections according to the step they correspond to. An article may appear in both subsections.

5.2.1 C2.1: Detrending/Filtering

A reconfigurable BPF is designed in [3] which is based on the FIR filter architecture. It is used to remove baseline wander and powerline interferences at 50/60 Hz, as a part of a reconfigurable "biopotential processor" which also includes an SA-ADC (see Sect. 5.1.1). The processor is prototyped on a Xilinx Spartan-3E FPGA and implemented as an ASIC using 180 nm CMOS technology, but there is no comparison with related work to quantify the benefits.

[12] uses the LabVIEW FPGA tool to implement a biomedical signal preprocessing architecture on a Xilinx Virtex-5 FPGA. The EEG signal is filtered using a notch filter at 50Hz and a BPF between 1 and 70 Hz. The signal is then amplified by a factor of 10,000 using a gain amplifier.

A large Laplacian spatial filter is used in [13] to reduce SNR from multichanneled EEG data. The filter multiplexes the input to one channel which is then filtered using a single zero-phase BPF unit to optimize resource utilization. The BPF is implemented through four different methods. The method that employs a manually designed and not predesigned FIR filter, using systolic multiply accumulator blocks and taking advantage of the symmetric property of filters, has the least computational delay when implemented on the Xilinx Virtex-4 FPGA. Coefficients close to zero are discarded for an approximate result which decreases resource consumption.

[14] uses LabVIEW to design a preprocessing circuit for EEGs. The acquired signal is first filtered by an HPF with a cutoff frequency of 0.5 Hz. Then it is amplified by an instrumentation amplifier INA114 with high CMRR with a driven-right-leg circuit. The next stage consists of a two-stage amplifier with a gain of 50 dB composed of operational amplifiers. The final stage is an LPF with a cutoff frequency of 40 Hz which is used as an anti-aliasing filter before digital-to-analog conversion.

[15] proposes to remove muscle artifacts from EEGs using an eight-order Butterworth LPF with a cutoff frequency of 35 Hz to preserve the important information which is mostly in the lower frequencies and remove out-band interferences, followed by a least mean square (LMS) adaptive filter to remove the in-band interferences since an EEG is a nonstationary signal. Both the filters are implemented on a Xilinx Spartan-6 FPGA and verified empirically.

A real-time cascaded moving average filter is proposed in [16] to remove baseline wander from EEG signals. An IIR implementation of the filter is preferred over the FIR design for resource optimization. The hardware implementation on a Xilinx Zynq-7000 SoC is validated by comparison with theoretical results with very low root mean square error (RMSE).

[19] presents the adaptive-maximum-mean-minimum (AMaMeMi) filter, which is used to detrend various kinds of biomedical signals. The simple time-domain algorithm is an approximation of an envelope detector. The mean of the envelope is approximated as the noise to be removed from the acquired signal. The approach is verified through implementation on the Xilinx Zynq-7000 SoC and demonstrates low-power and resource utilization. It is one of the few articles with a completely available description of the hardware architecture.

A framework for embedded signal processing, reconfigurable Signal Processing and Classification Environment (reSPACE), is proposed in [39]. The preprocessing flow of this framework is composed of a notch IIR filter to remove the DC offset. An anti-aliasing FIR filter is also used before the decimation of the sampling rate (see Sect. 5.1.4). The architecture is implemented on a Zynq-7000 SoC for validation along with the processing stage (see Sect. 5.3.4).

The same design for a mean-averaged reference filter is used to remove baseline wander in [40, 44]. While in the latter article, it is prototyped on a Xilinx Artix-7 FPGA, in the former, it is implemented on a 180 nm CMOS ASIC. Both are followed by their respective feature extraction stages (see Sect. 5.2.2) and processing stages for emotion (see Sect. 5.3.2) and autism (see Sect. 5.3.3) detection, respectively.

Powerline interference is filtered in [45] using a fifth-order dual-notch LPF. The filter is implemented in a fully differential fashion with 250 nm CMOS technology through a realization of an inductance-capacitance (LC) ladder. The inductance in the ladder is realized using digitally programmable operational transconductance amplifiers (OTAs). The circuit is verified through a PSpice simulation and shows a considerate decrease in power consumption compared to the state of the art.

A compact long-short-term architecture inference kernel (CLINK) is proposed in [46] to approximate conventional filtering functions. A long-short-term memory (LSTM) recurrent neural network based on the CLINK architecture is trained offline using FIR and IIR implementations of the BPF and Hilbert transform in the 4–12 Hz range. LSTM inferencing reduces computational costs by almost 84%. The circuit is constructed using an array of custom processing elements that work in a massively parallel way and a controller. Weights for the neural network are stored in buffers and LUTs in each processing element. The offline-trained floating-point values are converted to fixed point for online reference, which is shown not to affect the accuracy. The neural network architecture is prototyped on various platforms including the Zynq-7030 and Zynq-7045 SoCs, the Xilinx Virtex-690t, and Virtex-VU9 FPGAs. The ASIC implementation using 15 nm CMOS technology demonstrates highly reduced power consumption compared to FPGA and SoC prototypes.

In [47], the blind source separation (BSS) technique based on canonical correlation analysis (CCA) is used to remove muscle and eyeblink artifacts from EEGs. The various stages of BSS-CCA are implemented and parallelized in FPGA hardware wherever possible. A high correlation with clean signals is demonstrated.

The discrete wavelet transform (DWT) is emulated in [48] using the Mallat algorithm to remove noise from acquired EEG signals. Multilevel DWT is implemented with pipelining. The Daubechies Db4 wavelet function is selected for the parameters to decompose into eight levels. The basic brainwaves are observed empirically in some levels. The implementation is targeted at the Intel/Altera Cyclone-3 FPGA.

To divide acquired EEG signals into their different components, [49] proposes a resource-efficient implementation of the online recursive ICA (ORICA). It is composed of a whitening unit to accelerate the training process. A training algorithm takes the whitened signals to produce an independent component and an unmixed weight value, which forms the output. An efficient singular value decomposition (SVD) and a shared matrix multiplication unit ensure low resource utilization. The algorithms are theoretically validated and implemented using 90 nm CMOS technology.

[50] also proposes an efficient eigenvalue decomposition for the FastICA algorithm. The unit uses a modified approximate Jacobi method for efficient computations which are supported by a coordinate rotation digital computer (CORDIC). CORDICs calculate complex trigonometric functions using only addition, shift, and LUTs. A 6× speedup and a reduction in the area of over 77.2% for eigenvalue decomposition computation is observed through implementation on a Xilinx Virtex-2 FPGA.

[51] proposes a FastICA algorithm for the decomposition of EEG signals. The acquired signal goes through various steps including overlapping for smoothening,

garbage (outlier) detection to reduce redundant elements, channel permutation, and the final weight update from the classical ICA. The design is verified theoretically and simulated on 90 nm CMOS.

For removing eyeblink artifacts and filtering EEG signals to be used for movement intention detection, [52] uses a statistical eyeblink artifact detection module and a removal module. Resource sharing is used for all channels to increase hardware efficiency. The architecture is based on the computation and comparison of variances of electrooculograms. A fourth-order Chebyshev_II filter is used to further filter the signals and extract alpha frequencies in the range 8–12 Hz. The entire preprocessing, feature extraction (see Sect. 5.2.2), and detection architecture (see Sect. 5.3.4) are prototyped on the Nallatech BenOne SoC integrated with a Xilinx Virtex-4 FPGA. The validated architecture is implemented on a 180 nm CMOS ASIC which shows a 69% decrease in power consumption and a 74% decrease in the area compared to previous works.

[53] uses a simple fourth-order FIR LPF to smooth the acquired signal by filtering noise above 64 Hz. The filter is implemented along with the feature extraction module (see Sect. 5.2.2) and classification module (see Sect. 5.3.1) on the Xilinx Virtex-5 FPGA of the ML506 SoC.

[54] presents the hardware architecture for a second-order FIR LPF. It consists of Baugh-Wooley multiplier, flip flops, and adders. It proposes to use a 64th-order FIR filter with a similar architecture and implements the filter along with a feature extraction module (see Sect. 5.2.2) and seizure detection/classification module (see Sect. 5.3.1) on a Xilinx Virtex-5 FPGA-based SoC.

An ICA system integrated with multi-instance learning (MIL) is proposed in [55] for only EEG artifact detection and noise estimation. The ICA extracts independent additive components from multichanneled EEG data. The components may be associated with noise. Features are extracted from these components using Symbolic Aggregate Approximations (SAX) which conveys the overall structure of the sampled data. The MIL system is used for the classification of data as noisy or not. The system is implemented on an NVIDIA Jetson TK1 SoC.

[56] uses a depth-wise separable convolutional neural network (CNN) to identify movement artifacts in EEG. All sorts of movement and eyeblink artifacts are identified. Average accuracy of 93% is demonstrated with significantly reduced complexity compared to the state of the art. The model is optimized for hardware implementation by optimizing the number of filters in the first layer and validated on a Xilinx Artix-7 FPGA with 14-channeled data. An ASIC is also designed with 65 nm CMOS technology that demonstrates greater power consumption and latency improvement.

A 64th-order FIR BPF is utilized in [57] and time-shared between different physiological signals to reduce hardware consumption. The BPF is used to extract the delta wave from the acquired EEG. Then, a window-averaging filter is used to obtain the mean amplitude of the filtered signals and avoid instantaneous misclassifications. The architecture is implemented on an Actel IGLOO AGL060 FPGA with a sleep-stage detection stage (see Sect. 5.3.5). [58] presents an optimized implementation of the same filter on the same platform by the same authors. The FIR filter is merged with the window averaging filter using time multiplexing.

The Savitzky-Golay smoothing filter is prototyped in [59] using an FIR filter architecture. The prototype is vigorously compared with a moving average filter for various kinds of synthetic noise EEG and ECG signals. The superior performance of the Savitzky-Golay filter is demonstrated along with lower resource and power consumption (though not significantly low), on a Xilinx Spartan-3E FPGA.

The short-time Fourier transform (STFT) is accelerated in [60]. A window size of 8- and 32-bit precision is used with 15 bits in the integral part and 16 bits in the decimal part. LUTs were used for trigonometric and window functions which reduced the errors. The results are compared with MATLAB-based theoretical results for validation.

[61] uses the Altera Nios SoC with a Stratix-4 FPGA to accelerate the filtering stage. A 500th-order FIR filter is implemented along with a 128th-order IIR filter in parallel to reduce time delay. The filter coefficients are reconfigurable to realize specific filters like equiripple, kaiserwin, Butterworth, and Chebyshev. The filters are designed to be subject-specific for maximum accuracy. The two parallel co-processors accompany the software-based seizure detection in a hardware-software codesign system for maximum efficiency.

While a 5th-order IIR filter has been used in [62] to aid in gustation detection, a 64th-order FIR filter has been prototyped in [63]. The former utilizes a Xilinx Spartan-6 FPGA, while the latter uses a Xilinx Atlys SoC as the prototyping platform. The former also implements a decomposition algorithm on the FPGA for feature extraction (see Sect. 5.2.2). The latter also accelerates feature extraction (see Sect. 5.2.2) and processing (see Sect. 5.3.4) for movement intention detection.

A tenth-order FIR LPF with a cutoff frequency of 40 Hz is implemented for EEG filtering in [64]. The FIR filter is compared with a 10-point moving average filter, and while it consumes more resources, the FIR filter also shows better performance. A wavelet transform is also implemented on the Xilinx Spartan-6 FPGA (see Sect. 5.2.2) for feature extraction.

A very comprehensive analysis of the implementation of the normalized-LMS (NLMS) and its various extensions like set-membership NLMS (SM-NLMS), partial-update NLMS (PU-NLMS), and SM-bi-normalized LMS (SM-BNLMS) is described for removal of powerline interference in [65]. The PU-NLMS is demonstrated to be the most efficient algorithm when implemented on an ASI using 65 nm CMOS technology, while having the least latency of 8 clock cycles and one of the lowest gate counts and chip areas among all algorithms.

[66] presents an active comb filter based on low-transconductance OTAs. The traditional RLC ladder has been implemented using OTAs for a low-power design. OTA-capacitance topology is used to realize notch filters that are cascaded to form the comb filter using 250 nm CMOS technology to demonstrate the lowest power consumption in the state of the art.

[67] is the only article to deal with noise from the electrode-tissue interface. It does so by assuming the spurious multiplicative and additive gains separately from the interface. The additive gain is approximated using windowed averaging and

subtracted from the quantized signal. The multiplicate gain is then approximated by a correlation block that compares the additive-corrected signal with the analog signal to estimate the multiplicative factor and assist with calibration as well. The reference channel goes through the same calibration technique that results in the correct differential EEG signal to be used for further processing. The entire algorithm is prototyped on a Zynq-7000 SoC.

A further-desensitized 12th-order half-band FIR filter is designed in [68] to reduce the sensitivity of the input to quantization error. A modified Russian peasant multiplier block is used for additional optimization of the architecture. An almost 50% reduction in the logic slice consumption of a Xilinx FPGA is observed.

The BSS problem is tackled in [69] by a variable-channel hardware implementation of the FastICA algorithm. The algorithm is optimized using Gram-Schmidt orthonormalization and corresponding whitening which reduces the computational complexity of the hardware. An optimum sample size of 512 samples is empirically selected, as a trade-off between accuracy and memory consumption. The IEEE-754 floating-point representation of data is used in the hardware implementation to maintain accuracy, and this is the only article to do so, in this review. Standard 90 nm CMOS technology is used for designing the ASIC, and the design demonstrates the lowest per-channel energy consumption for 2–16 channels.

[70] proposes to tackle the BSS problem for EEGs using second-order blind identification (SOBI) instead of ICA. Only the correlation unit of the SOBI algorithm is hardware accelerated considering the computationally intensive steps in matrix correlation calculation and the ability of hardware implementations to deliver results with minimum time delay using parallelism.

5.2.2 C2.2: Feature Extraction

A complete data acquisition system is presented in [9] and integrated with a Wi-Fi component to transmit data to a cellular phone. A radix-2 fast Fourier transform (FFT) is implemented on an FPGA which is supported by off-the-shelf ADC and amplifier components. An empirical spectral analysis of transmitted data is performed through a cellular device to verify the proposed system.

A very simple seizure detection algorithm is implemented in [22] based on the acceleration of computation of accumulated energy and coastline features. The advantage is that these are time-domain-based features that are simpler to compute than frequency-domain features. The seizure prediction hardware accelerator based on these features is described in Sect. 5.3.1. Classification using frequency-based features is compared with time-domain features to prove hardware efficiency, while not sacrificing accuracy on a Xilinx Spartan-6 FPGA.

The 95% confidence ellipse area to be used as a feature for detection of epileptic seizures is accelerated in [24]. The second-order difference plot is used to study the variability of the acquired signals. The variability expressed in the plot is quantified through the elliptical area. The computation of the plot and area is implemented on a

Xilinx Artix-7 FPGA. The implementation is validated by comparison with theoretical output, and low resource utilization is demonstrated.

[38] implements a separate feature extraction unit to compute various time- and frequency-domain features. The time-domain features extracted include deceleration capacity, acceleration capacity, sample entropy, Shannon permutation entropy, and detrended fluctuation analysis. The frequency-domain features include spectral energies of the different brainwaves obtained using DWT and the SEF95 frequency, which is the frequency band with 95% of the total energy of the signal. The feature extraction unit is implemented using 180 nm CMOS technology along with the level-of-hypnosis processing unit (see Sect. 5.3.5). The feature extraction unit occupies a separate area of 0.6 mm^2. The architecture is synthesized on the Xilinx Virtex-5 FPGA as well.

In [40] the eight-channeled EEG data is sent to a feature extraction engine in a systolic manner which reduces the pin and resource utilization by 8×. A BPF is implemented using 50th-order Butterworth filters, in hardware. The absolute interhemispheric power spectral difference (IHPD) and interhemispheric power spectral ratio (IHPR) features are extracted because of high classification accuracy and low resource usage for computation of these features. The features are down-scaled with rounding to reduce memory usage. The features are normalized to increase classification accuracy. The entire architecture is implemented using 180 nm CMOS technology, including a down-sampling unit (see Sect. 5.1.4), a filtering unit (see Sect. 5.2.1), and the emotion classification unit (see Sect. 5.3.2).

A feature extraction engine is implemented in [44] to calculate the Katz fractional dimension and kurtosis in the beta band of the acquired EEG signals. Since these features are highly complex and require high memory resources, approximation indicators are used to compute their values instead. The feature extraction engine is implemented in a Xilinx Artix-7 FPGA, with an autism classifier described in Sect. 5.3.3. Hardware savings of 54% and 38% were demonstrated, respectively, for the Katz fractional dimension and kurtosis.

[52] implements a subject-specific mother wavelet by applying a unit-scale wavelet transform with a basis function (extracted from the training phase) of 6-s length. Resource sharing is used to minimize area by using just a multiplier and adder. This continuous wavelet transform (CWT) is prototyped along with a filtering unit (see Sect. 5.2.1) and an adaptive threshold processing unit for detection of movement onset (see Sect. 5.3.4) on the Nallatech BenOne SoC with the Xilinx Virtex-4 FPGA and implemented as an ASIC with 180 nm CMOS technology.

[53] prefers to use the DWT instead of the Fourier transform because of the latter's poor resolution due to a finite length window. The Daubechies4 (Db4) wavelet is used for decomposition. After determination of the wavelet coefficients, the following features are extracted from each brainwave: maximum coefficient, minimum coefficient, mean absolute value, standard deviation, wavelet energy, relative wavelet energy, Hurst exponent, coastline, detrended fluctuation analysis, and skewness. So, there are a total of 5 brainwaves × 10 features = 50 extracted features from the acquired EEG signal. The entire architecture is implemented on a

Xilinx ML506 SoC along with a filtering module (see Sect. 5.2.1) and a seizure classification system (see Sect. 5.3.1).

[54] proposes a DWT algorithm for feature extraction which uses two stages of alternating LPFs and down-samplers. The variance, mean, and frequency features from the different levels of the DWT are calculated and implemented on a Virtex-5 FPGA-based SoC along with a linear seizure detection classifier (see Sect. 5.3.1) and a 64th-order FIR LPF (see Sect. 5.2.1).

An analog approximation of the DWT is proposed in [67]. The analog DWT is emulated by mimicking the filter bank in a typical DWT. Each filter in the analog DWT filter bank is an s-domain filter that is an estimation of the z-domain filters that perform the same filtering operation. In place of down-sampling, the cutoff frequency of each analog-domain filter is scaled by a factor of 2. A Butterworth mother wavelet is used due to its potential to estimate the output from the Daubechies mother wavelets in the analog domain. To achieve low power, g_mC filters are used. A 180nm CMOS technology is used for the design and fabrication of the circuit.

[68] proposes a hardware accelerator for real-time extraction of various graph-theoretic brain-connectivity parameters like clustering coefficient, degree, weighted degree, density, transitivity, characteristic path length, eccentricity, radius, and diameter. All of these features are dependent on the phase lag index (PLI). A separate block calculates the PLI, and the related features are pipelined or parallelized to avoid redundant hardware. The architecture is implemented on Intel/Altera Stratix-4 FPGA and shows considerable resource and power savings when compared to a non-optimized architecture. It is also synthesized using 90nm CMOS technology to demonstrate further power savings.

A circuit to extract the Teager energy operator (TEO) is proposed in [69]. It utilizes a fully CMOS subthreshold multiplier and a programmable subthreshold g_mC Butterworth LPF using OTAs. The LPF is used as a delay element. The work is verified empirically and claims to be one of the lowest power-consuming (53.59 nW on 90 nm CMOS) ASICs for biomedical recording reported.

Like [68, 70] proposes the hardware acceleration of the computation of various graph-theoretic features like the PLI, degree of a node, weighted degree, density, clustering coefficient, and transitivity. Other features like shortest path network matrix, characteristic path length (using Dijkstra's algorithm), global efficiency, eccentricity, radius, and diameter are also calculated parallelly by conversion of the PLI from weights to length units. The computation of all these features is maximally parallelized to reduce time delay. The hardware is implemented on 130 nm CMOS technology and validated by comparison with theoretical results from MATLAB. The high speed of the hardware accelerator against MATLAB is demonstrated as well.

[71] presents an efficient implementation of the Morlet CWT on a Xilinx Spartan-3AN FPGA. The complex convolution in CWT is replaced with a simpler multiplication operation by converting the signal and wavelet into the frequency domain using FFT, which is executed by the Xilinx FFT core. Different brainwaves are extracted on the hardware, and the results are validated by comparison with

software-based theoretical output. A low time delay demonstrates possible use in real-time scenarios.

[72] optimizes the memory utilization of the computation of the Morlet CWT from the previous article by compressing the shape of the wavelet function through zero elimination: all the zeros are eliminated. This results in a lesser number of computations. Furthermore, scale elimination is applied, by ignoring the frequency components of the EEG greater than 60 Hz, since they are weak and don't contain much information. Hence, the lower scales of wavelet can be ignored and need not be stored in the block RAM (BRAM) of the same Spartan-3AN FPGA, increasing memory efficiency. A comparison with theoretical output without optimizations validates the proposed optimizations.

The same authors as the previous two articles propose the acceleration of the calculation of wavelet coherence in [73]. Two complex multipliers are used in the calculation. Frequency smoothing is preferred in this work over temporal smoothing to average the coefficients of the wavelet spectrums. The frequency smoothing algorithm is modified for hardware implementation. It is modified by choosing a divisor as a multiple of two to allow right shifting in hardware. An appropriate word length is chosen to trade-off between resource utilization and accuracy. The algorithm is implemented on the Xilinx Spartan-3AN FPGA.

[74] hardware accelerates the computation of the Hilbert transform and the mean power frequency features for seizure detection. An analytical filter is applied as an approximation of a Hilbert transform as a digital filter. An FIR filter with additional delays is used to realize the analytical filter. The mean power frequency feature is used for feature selection. The feature extraction stage is implemented on a Microsemi IGLOO2 FPGA along with a perceptron classifier (see Sect. 5.3.1) for seizure detection.

The Hanning window function (to remove discontinuities and smooth the signal) is used in [75], followed by the STFT implemented by an FFT block in fixed-point arithmetic to reduce complexity. The real and imaginary outputs are used to compute the power spectrum, followed by normalization. Each of the five common brainwaves is divided into quarters, and the power in each frequency quarter is extracted as one feature. This gives a total of 20 features extracted on the Zynq-7000 SoC platform.

[76] proposes to use an ANN to measure chaotic behavior-related features of the acquired EEG. The bipolar sigmoid activation function is used for the hidden neuron layer, and the ramp activation function is used for the output neurons layer. The Henon map bifurcation and the Lyapunov exponent are calculated using the network. The neural network-based FPGA implementation is compared with the equation-based computational architectures designed using Simulink models and VHDL as well. Significantly higher resource utilization and power consumption than the baseline implementations are reported.

[77] presents a simple dedicated feature extraction hardware design for calculation of the slope and mean energy feature which accompanies a bit-serial neural network for seizure detection, described in Sect. 5.3.1. A picoMips architecture is

used to design the various modules and prototype them on an Altera Cyclone-5 FPGA.

The TEO is extracted in [78] using digital OTA circuits. The delay elements are modeled by LPFs, which are created by cascading two first-order GMC LPFs with cutoff frequencies of 65 Hz. The advantage is that the LPF attenuates the high-frequency components remaining on the signal the input to the circuit. The multipliers are implemented using standard Gilbert cell topology. The median frequency is the frequency at which the signal power spectrum is split into two halves. The FFT is the classical algorithm to calculate the median, but it is very power hungry. To simplify the computation, an analog approach based on a feedback loop that compares the upper and lower power content is proposed. The feature extraction stage is implemented along with the processing stage using 180 nm CMOS technology. The overall system, including a processing stage for level-of-hypnosis estimation (see Sect. 5.3.5), occupies 1.63 mm^2.

[79] presents the multi-scale power spectral density (PSD), which computes the PSD at multiple timescales. It is believed that the dimension reduction in PSDs is not enough. So, the moment of the PSD is utilized instead, for feature extraction. The article presents the first 3 moments of multi-scale PSD which are multi-scale spectral centroid, multi-scale variation coefficient, and multi-scale spectral skew. The feature extraction unit is implemented on the Xilinx Zynq ZCU102 Multi-Processor SoC (MPSoC) with a random forest seizure classifier (see Sect. 5.3.1)

Spectral features are extracted in the range of 3–26 Hz from the 14-channeled EEG in [80]. 0.5–3 Hz frequency is discarded since the spectrum of seizures contains large energy components in this band which suppresses energy in other bands. Eight spectral energy features are extracted from each channel which gives a total of 112 features from 14 channels. The extraction unit is implemented using s simplified form of the FFT, i.e., the real-valued FFT. CORDICs are used for the efficient implementation of trigonometric functions. Circular and linear vectoring are employed to utilize the same CORDIC for spectral energy computation of all features. The extraction unit is implemented on an ASIC using 40 nm CMOS technology, with a seizure detection processing unit (see Sect. 5.3.1).

[81] implements the DWT using cascading HPF-LPF pairs to realize the Daubechies Db2 wavelets since it was found to have the highest accuracy for the seizure detection unit (see Sect. 5.3.1). The variance of the DWT coefficients is used as a feature for the classifier. The architecture is prototyped on the Nexys 7 SoC which is equipped with the Xilinx Artix-7 FPGA.

An optimized variational mode decomposition is implemented in [82]. The most informative and suitable band-limited intrinsic mode functions are selected based on the kurtosis value with the correlation coefficient and mean of selection index. These mode functions are used by the processing stage for seizure detection (see Sect. 5.3.1). The entire architecture is implemented on the Xilinx Virtex-5 FPGA.

[83] uses a reduced-complexity DWT architecture which is based on an only-LPF implementation. The Daubechies Db4 mother wavelet is used to extract seven different levels, which are fed to a quasi-averaging filter that mimics a moving window average filter. The moving average is used to approximately compute the

detail coefficients of the next stage, as an adaption of the Mallat algorithm for DWT. The theoretical soundness of the reduced-complexity architecture is proved, and it is consequently implemented with an adaptive thresholding block (see Sect. 5.3.1) for seizure detection, using 65 nm CMOS technology.

A feature vector comprising 16 features is used in [84], which includes 8 PSDs, 4 IHPDs, and 4 scaled IHPR of the selected electrodes in the beta bands similar to the work in [40]. While [40] utilized a logarithmic divider, the current work utilizes a customized LUT-based divider to reduce hardware costs by almost 47% on 65 nm CMOS along with a linear SVM for emotion classification (see Sect. 5.3.2).

[85] selects the beta ratio, SEF95, and spectral energies of different brainwaves as the most accurate features for depth of anesthesia estimation using the classifier described in Sect. 5.3.5. Parallelism and pipelining are used to optimize latency and resource consumption wherever possible on a 65 nm CMOS ASIC and a Xilinx Virtex-5 FPGA prototype.

[86] proposes approximate feature extraction through genetic programming (GP) to take advantage of the efficiency of fixed-function accelerators and the flexibility of programmable hardware accelerators. Genetic programming is a synthesis framework that takes inputs and the related outputs and learns functions called "GP models" which relate the inputs and outputs. The structure of the GP models is very particular. They are composed of genes, which are computation trees, where the nodes are base functions and edges correspond to intermediate inputs and outputs. The classifier is retrained with features approximated by the GP model. The resulting model is called an *error-aware model*. A four-core feature extraction accelerator based on genetic programming is accompanied by an SVM accelerator for seizure detection (see Sect. 5.3.1). A very significant reduction in the number of cycles to compute features through approximation is reported. A prototype is implemented in 130 nm CMOS technology. Voltage-scaling capabilities are added for energy efficiency using a power management unit. A significant reduction in power consumption is observed as well.

Following the filter described in Sect. 5.2.1, the stationary wavelet transform is preferred in [62] over the DWT for acceleration so that the time invariant of the signal is not lost. An up-sampling method at each of the cascaded levels is used (to preserve the time invariant) in the architecture implemented on a Xilinx Spartan-6 FPGA. The mean and standard deviation from each of the components of the five-level transform are used for gustation classification. The coefficients for the transform are calculated offline.

Haar wavelet decomposition is implemented in [64] after a tenth-order FIR filter (see Sect. 5.2.1), on a Xilinx Spartan-3e-based SoC. The Haar wavelet is preferred because of its simplicity while maintaining similar performance accuracy as other wavelets.

In [91], an FFT architecture based on second-order Butterworth BPFs is used to extract delta (0.5–2.75 Hz), theta (3.5–6.75 Hz), low-alpha (7.5–9.25 Hz), high-alpha (10–11.75 Hz), low-beta (13–16.75 Hz), and high-beta (18–29.75 Hz) brainwaves. After this, the spectral energy of each brainwave is extracted using a simple addition ladder, to result in a six-feature vector that is used by a classifier that is described in Sect. 5.3.2. The entire system, christened XGBoost, is implemented on a Xilinx Artix-7 FPGA.

[92] implements the real-time EEG analysis for seizure event detection (REACT) system on an ASIC with 90 nm CMOS technology. A feature vector comprising 55 features is used. The features are extracted post an FFT of the input data and include the SEF80, SEF90, SEF95, dominant spectral peak, variance of first and second derivatives of PSD, zero crossings of PSD, power in 2 Hz epochs, normalized power in 2 Hz epochs, spectral entropy, number of maxima and minima, line length of PSD, nonlinear energy, wavelet coefficients, Hjorth activity/mobility/complexity, RMS amplitude, spectral power, entropy, Fisher coefficient, skewness, autoregression model fitness, kurtosis, and decomposition entropy.

After a 64th-order filter described in Sect. 5.2.2, [63] implements the 8-level DWT in the standard cascaded architecture on a Xilinx Atlys Soc and extracts the mean, variance, and standard deviation features for a brain-computer interface (see Sect. 5.3.4). [93] uses an optimized version of the DWT where a single LPF-register and HPF-register pair is multiplexed and used to extract all the levels of the required brainwaves from EEG. The LPF and HPF are used from standard Xilinx libraries, and the coefficients are calculated from MATLAB. [94] implements the DWT with Daubechies Db6 wavelet coefficients on a Xilinx Basys-3 SoC. The transformation is used to generate the standard deviation of the D1 and D2 outputs which are the selected features for a linear discriminant analysis classifier for sleep detection (see Sect. 5.3.2).

The sixth-order Gaussian CWT is implemented in [95] using g_mC filters. The filter is based on a leapfrog multiple loop feedback system to synthesize a wide range of arbitrary transfer functions with minimum resources. Other architectures like the LC ladder and leapfrog fail to realize a large class of transfer functions. The coefficients are calculated as part of an optimization problem using various tools. The architecture is implemented on an ASIC using 180 nm CMOS technology to demonstrate picowatt-level power consumption.

[96] proposes an architecture for efficient calculation of the multichanneled FastICA on the Xilinx Virtex-6 FPGA. It is a scalable macro-pipelined structure where a single shared symmetric orthogonalization unit is used for all channels and the memory units are updated online by a control unit as and when required. A highly reduced computation time compared to the state of the art is demonstrated.

[97] utilizes the SpiNNaker platform to implement NeuCube, a spiking neural network to map, learn, and understand spatiotemporal neural data. The EEG data are transformed into corresponding spikes by an encoding module. After a detailed analysis, the step forward spiking algorithm was selected for the transformation. The NeuCube is a three-dimensional matrix of $5 \times 5 \times 5$ (125) neurons modeled after the human brain. The input spike train introduced these map-coordinated neurons simultaneously. This simple model was chosen instead of a complex 1471 neuron structure to reduce classification complexity. Three different feature vectors are extracted from NeuCube for every input: the first is the number of spikes in the input (spike count), the second is the number of spikes that occurred at each reservoir

neuron in the SNN (spikes per neuron), and the third is the final weights of the connections formed during spiking (final weights).

The EMD is accelerated in [98], and a single feature extraction block is reused for all channels using a multi-rate architecture. The amplitude modulation and frequency modulation for each of the five levels of the EMD output are computed for the feature vector. This is done using a simplified version of the Hilbert-Huang transform. Hence, each feature vector contains 10 features, and there are 23 such feature vectors for 23-channeled input. The architecture is implemented on a Xilinx Virtex-7 FPGA.

[99] presents an efficient architecture for the DWT based on a lifting-scheme structure based on the Haar and Daubechies Db4 and Db6 wavelets which are more efficient than the classical HPF-LPF filter bank structure. The power spectral density and corresponding band energies were computed using the FFT. Other extracted features include the correlation, zero-crossing histogram, autoregressive coefficients, and phase synchronization coefficient. The feature extraction architecture is prototyped on the Xilinx Virtex-7 FPGA.

5.3 *C3: Processing Stage*

The processing stage can be subcategorized according to the application domain as follows: seizure detection, emotion classification, autism detection, movement intention/motor imagery detection, sleep detection/classification, and speech reconstruction. The articles that propose hardware acceleration in the processing stage are presented in this section according to these subcategories. An article may appear in more than one subcategory.

5.3.1 C3.1: Seizure Detection

[22] presents a multi-window event-based counting threshold based on the accumulated energy and coastline features. The accelerator is implemented on a Xilinx Spartan-6 FPGA with the feature extraction unit (see Sect. 5.2.2). The average accuracy of the proposed algorithm is shown to be ~97%. Low power consumption compared to the state of the art is also demonstrated.

[53] uses a population-based optimization method, the binary particle swarm optimization (BPSO) to select the most optimal features. BPSO can escape from local optima, has an easier implementation, and requires the adjustment of just a few parameters which makes it advantageous to be utilized. An Optimized Adaptive Neuro Fuzzy Inference System (OANFIS), a fuzzy inference system, is used for seizure detection. The Xilinx ML506 SoC is used for implementing the OANFIS classifier. The preprocessing algorithm (see Sect. 5.2.1) and a wide array of 50-feature extraction unit (see Sect. 5.2.2) are also implemented. A minor improvement in hardware resource utilization and power consumption is reported, and a

significant improvement in the performance of the algorithms because of the optimization is observed.

[54] uses a linear SVM to classify acquired EEG as seizure or non-seizure. The algorithm is reportedly implemented on a Virtex-5-based SoC with the feature extraction unit (see Sect. 5.2.2) and a 64th-order FIR low-pass filter (see Sect. 5.2.1).

A multilayer perceptron artificial neural network is developed in [78] for seizure detection. The network was trained offline with backpropagation and Adam optimization algorithm. The optimal weights were computed and supplied to a Microsemi IGLOO2 FPGA implementation for online classification. The implementation achieved an accuracy of 100% with the mean power frequency and Hilbert transform features which were also implemented on the platform (see Sect. 5.2.2). 159.7 mW power utilization was reported on the selected platform.

[81] presents a neural network using bit-serial hardware architecture which depends on a bit-serial data processing unit. The unit implements the functionality of a neuron and uses bit-serial arithmetic. Arrays of these units are controlled by a finite-state machine. A significant reduction in the number of logic elements required to implement the proposed neural network compared to the baseline is reported with an implementation on an Altera Cyclone-5 FPGA. The hardware implementation was trained offline, and MATLAB and the coefficients were then supplied for online classification. When integrated with the hardware-based slope and mean energy feature extraction units (see Sect. 5.2.2), an accuracy of 90% is reported.

The random forest classifier is used in [83] to detect seizures. A random forest contains several independent decision trees which operate as an ensemble to solve any classification problem. The classifier is selected because of its resistivity to noise, capability to handle high-dimensional data, and parallelizability making it computationally less costly. The highest reported accuracy of the algorithm is 100% using multi-scale features (see Sect. 5.2.2) based on the multi-scale power spectral density. The classification algorithm is implemented on Xilinx ZYNQ Ultrascale+ MPSoC. Two different execution schemas are reported: parallel and serial, and it is reported that the parallel execution has more than 10× lower latency while the resource utilization is scaled up similarly.

A nonlinear SVM for seizure detection using spectral energy features in the 326 Hz range (see Sect. 5.2.2) is proposed in [84]. Highly parallel computing is enabled by an alternating direction method of multipliers. Minimum redundancy maximum relevance (mRMR) feature selection algorithm and rank approximation are used to reduce the memory storage during matrix inversion in the training stage without impacting accuracy. The complexity is reduced by 99.4%, whereas memory utilization is reduced by 90.4%. A customized floating-point arithmetic unit is used where the input is converted to fixed-point representation and the output is converted back to the floating-point which gives 20% area savings and 11% power savings. An array of CORDIC-based processing elements is used to execute various matrix manipulations. The approximate Jacobi method is used for efficient eigenvalue decomposition. With all of these optimizations, the area (2.5 mm^2) is reduced by 87% and the latency is reduced by 99%. The architecture is implemented in a 40 nm CMOS technology for verification with 1.4 mW power consumption.

The approach in [85] is an end-to-end description of the hardware implementation of a simple SVM classifier, with principal component analysis to identify patterns in data. The features and wavelets are selected offline to extract required coefficients and keep the hardware architecture simple. The Daubechies Db2 wavelet is selected. The architecture is implemented on a Nexys A7 SoC. A theoretical evaluation yielded 100% accuracy of the algorithm, while the hardware implementation, with accelerated feature extraction (see Sect. 5.2.2), has an accuracy near 98.7%.

[86] proposes to combine a reduced deep convolutional neural network with a multi-kernel random vector functional link network for epileptic seizure detection. The selected link network for classification is advantageous due to its less computationally complex single-layer feed-forward network, linear framework, and the well-known kernel recursive least squares. A 100% accuracy is obtained through MATLAB simulations. Consequently, the architecture is verified by implementation on Xilinx Virtex-5 FPGA, along with an optimized version of mode decomposition described in Sect. 5.2.2.

A thresholding algorithm with reconfigurable coefficients is used in [87] to detect the onset of seizures with patient-specific data. The algorithm shows an accuracy of up to 99% in some patients. The training phase consumes more power than the online phase which consumes power in the range of tenths of microvolts when implemented with 65 nm CMOS technology with the feature extraction module which is an LPF-only DWT algorithm (see Sect. 5.2.2).

The real-time EEG analysis for event detection (REACT) algorithm is implemented using a coarse-grained reconfigurable array architecture in [92]. An SVM classifier with a Gaussian kernel is used for the detection of seizures. The output of the SVM is converted into a probability of containment of seizure activity in the corresponding epoch. Probabilities from successive epochs are smoothed using a moving average filter and transformed into binary yes/no answers. Low power and resource consumption using the SYSCORE architectural elements is demonstrated.

[100] presents a unique reconfigurable 64-core platform, where each core is routed in a hierarchical topology. Programs are loaded onto each of the five-stage RISC-based cores to configure the platform. A locally synchronous, globally asynchronous clocking approach is utilized to reduce power consumption compared to a common global clock routing. Cores are configured individually by loading corresponding programs. The proposed platform is compared with traditional digital signal processing (DSP) slices. The prototype platform was implemented in 65 nm CMOS technology, and a seizure prediction algorithm was mapped to the cores and consumed ~17 mW at 400 MHz.

[101] presents the same 64-core architecture as the previous article. Instead of a five-stage RISC processor, a six-stage pipeline is used that is augmented with a loop control block that exploits the highly repetitive nature of DSP applications and a small dedicated FFT computation block that uses the radix-2 Cooley-Tukey algorithm for FFT computations. Pointer support is also added. Each processor was placed and routed in a 65 nm CMOS process and occupied 0.07 mm^2 area. The seizure detection is mapped onto 60/64 cores to verify the processor.

A neuromemristive echo-state network reservoir computing architecture is designed in [102]. The network is described in a fully parallel and not partially sequential fashion, to solve latency problems. Digital design specifications for the neurons and synapses of the reservoir are described. Two topologies of the echo-state network for seizure detection are reported—ring and hybrid. It is observed that the accuracy increases as the reservoir size increases. The maximum accuracy achieved is 90% with hybrid topology with a little more than 100 neurons. The article proves that a scalable neuromemristive echo-state network architecture with power constraints is feasible. The network with 30 neurons is synthesized on three different FPGA platforms for verification and it achieved 67.4% accuracy.

[103] advocates for the stochastic implementations of two popular classifiers—the linear SVM and the artificial neural network. Stochastic computing requires fewer logic gates and is inherently fault-tolerant. A linear transformation for translating binary data to stochastic data is proposed. Likewise, stochastic arithmetic is used in the classifiers. Both the classifiers are designed using 32 nm libraries and simulated using Synopsys Design Compiler. The accuracy of the neural network is increased by 5.89% through stochastic optimization. The power consumption of the linear SVM and artificial neural network is reportedly decreased by 74% and 88%, respectively.

A Naive Bayes' classifier is implemented in [104] on the Xilinx Kintex-7 FPGA for binary classification of acquired EEG data into seizure and non-seizure classes. Floating-point data is converted to fixed-point data to ensure the scalability of the architecture. The algorithm is first verified theoretically using MATLAB, and the resource utilization is presented after implementation. Accuracy above 95% is reported.

[105] describes a reduced deep convolutional stack autoencoder (RDCSAE) integrated with an improved kernel random vector functional network (IKRVFLN). The method is first simulated on MATLAB to verify the accuracy and then implemented on Xilinx Virtex-5 FPGA-based ML506 SoC. The reported accuracy is 99.96% which outperforms the state of the art like deep convolutional neural networks.

A single-input rule model fuzzy classifier is demonstrated in [106] for the classification of epilepsy risk level in diabetic patients. Various tuned and untuned fuzzy systems are analyzed, and the five-rule tuned fuzzy model is selected for its highest accuracy. The system is prototyped on various FPGAs as the Xilinx Spartan-2, Virtex-2, and Virtex-2P.

5.3.2 C3.2: Emotion/Vigilance Classification

[40] implements two linear SVMs to detect the valence and arousal of the subject which are used as decoding signals for quaternary detection of the emotion of the subject. Average accuracy of 70% is reported. The entire implementation including the feature extraction engine (see Sect. 5.2.2) is implemented on a 5.4 mm^2 area chip manufactured using a 180 nm CMOS process.

[88] uses emotion classification to detect early signs of autism spectrum disorder. A custom LUT-based divider specifically for emotion classification is developed which consumes approximately 47% lower power. A patient-specific linear SVM is utilized to separately and parallelly detect the valence and arousal of the subject, and these two factors are decoded to classify the emotion, similar to the work in [40]. The accuracy of the approach is approximately 63%. It is implemented using 65 nm CMOS technology and consumed 12.7 uW at 1 kHz frequency. The approach uses features from a specifically selected set of 8 channels, compared to other works that use up to 64 channels, to reduce the hardware resources utilized, while minimally impacting the accuracy, as described in Sect. 5.2.2.

The two-class linear discriminant analysis is used in [94] for attention/vigilance detection. The classifier is selected based on its simplicity for a two-class classification and similar accuracy to complex classifiers. A projection vector between the two classes is calculated in the training phase which provides the highest amount of discrimination between the two. Distances based on this projection vector are calculated in the detection phase for classification.

[107] proposes to tackle the emotion classification problem using a structural radial basis function neural network (SRBFNN). The SRBFNN is specifically designed for analyzing nonlinear behavior. It temporally observes nodes and edges. The fitness function used is a hybrid grey wolf ant lion optimization. The approach is theoretically validated on software and then implemented on the Xilinx PYNQ Z2 SoC with an Intel Movidius neuromorphic chip, being one of the few reviewed articles to have utilized a neuromorphic chip for implementing the statistics-based processing stage. An accuracy greater than 99% is consistently achieved during testing.

[108] implements a multiphase convolutional neural network on a custom 28 nm CMOS chip integrated with Xilinx Spartan-3 FPGA for binary emotion classification and a standalone Xilinx Virtex-7 FPGA for quaternary classification. The multistage pipelined implementation enabled optimized memory usage since the entire batch input need not be stored on the FPGA. The network is composed of various layers and units like the softmax layer, forward propagation layer, a backpropagation unit, and a learning engine. Mean accuracy of 77% is achieved which is consistent with the single-phase results. Power consumption of 77 mW is observed at 70 MHz operating frequency.

[109] implements a convolutional neural network composed of multiple layers for EEG and ECG/PPG separately. A max pooling layer follows every batch normalization layer to reduce the dimensions. A softmax layer gives the final classification one-dimensional output. The output from both layers is fed into a fuzzy-logic decision-maker module to output the emotion with the highest confidence level. An accuracy of 100% was achieved in some cases while there were cases with less than 50% accuracy attributed to noise. Average accuracy of 77% is reported. The classification stage is implemented on a Xilinx Spartan-6 FPGA for verification.

The vigilance of a subject is classified in [110] using a learning vector quantization neural network. The algorithm architecture adequacy methodology is employed for optimized hardware implementation of the ANN. Since the methodology does not account for fixed-point precision limitation, an analysis of the quantization error is integrated into the methodology. Parallelizing optimizations are explored, and corresponding VHDL code is generated and implemented on the Xilinx Virtex-4 and Virtex-5 FPGAs. Various topologies consisting of different numbers of inputs and different numbers of neurons are tested for accuracy, resource consumption, and latency, and the trade-offs are discussed in detail.

5.3.3 C3.3: Autism/Anomaly Detection

[44] describes a four-channel EEG-based ASD detection processing system that utilizes a shallow neural network. The channel selection is performed offline to avoid the discomfort that comes with wearing high-density EEG headsets. The shallow neural network employs a sigmoid function which is implemented in hardware in a customized fully pipelined fashion. The classification accuracy is shown to be ~85% which is respectable. The classifier and feature extraction engine (see Sect. 5.2.2) are implemented on a Xilinx Artix-7 FPGA which consumes 8.62 uW power at a standard 100MHz clock frequency.

5.3.4 C3.4: Intention/Imagery/Gesture/Speech Recognition

An architecture called reSPACE is proposed in [39] for the development of FPGA-based signal processing systems. The first step in classification is the XDAWN spatial filter for feature selection followed by further dimensional reduction. The resulting feature vector is normalized, and a linear SVM performs the final classification which is followed by a final thresholding optimization step. The architecture is used for an ERP3000 oddball classification experiment. The flow for various stages, including preprocessing (see Sect. 5.2.1) and acquisition (see Sect. 5.1.4), is discussed and implemented on the Zynq-7000 SoC FPGA with a classification accuracy of 89%.

[52] aims to shift the complexity of the classification for movement intention detection from the online phase to the offline phase, to reduce the complexity and utilization of hardware. A simple "spiky peak" detector instead of the classical peak detector is used for classification based on the continuous wavelet transform features (see Sect. 5.2.2) and a Laplacian filter (see Sect. 5.2.1). The spike parameters are supplied manually. A 66% reduction in power consumption and a 74% reduction in the area were achieved through the optimizations in implementations on the Virtex-4 FPGA and the XtreameDSP Nallatech BenOne SoC. An ASIC is designed as well, using 180nm CMOS technology.

The Xilinx Spartan-6 FPGA is used to accelerate an SVM for detection of the user's intention of moving the left or right hand in [63]. This system is implemented together with a feature extraction module (see Sect. 5.2.2) for a brain-computer interface system.

[93] uses the particle swarm algorithm, a nature-inspired optimization algorithm, for arm and finger movement intention detection. In this algorithm, each particle of the swarm searches the space for the solution of the given problem. All particles share and compare their local solutions, and the nearest one to the optimum is defined as the global position of the swarm. With each trial, the particles update their velocity and position and move nearer to the optimum. The algorithm is implemented in a Xilinx MicroBlaze soft-core processor on an FPGA, and average accuracy of around 90% is observed.

A low-complexity XNOR-logic-based stochastic SVM is proposed in [111] for real-time gesture recognition. Parallelizing optimization is used to reduce latency, while trading off with hardware resource utilization. In stochastic computing, a number is represented with a random bit-stream, and complex operations can be realized with simple logic. The design is implemented in 65 nm CMOS technology for a single thread and eight parallel classification threads. A higher accuracy (73%) and better latency are observed in the parallel implementation, while power and resource utilization increase.

Motor imagery is detected in [112] using spiking neural networks implemented on SPiNNaker neuromorphic hardware that supports massively parallel architectures. The spiking model is inspired by an insect olfactory system. The accuracy of the three-layer classifier is reported as 75%. The advantage of the implementation is the massively parallel architecture that enables low latency and high throughput.

[113] presents a novel $Pr_{0.7}Ca_{0.3}MnO_3$ (PCMO) oxide-based analog synapse array for neuromorphic computing. Analog synapse characteristics are obtained by optimizing the redox reaction at the metal/oxide interface of a PCMO resistive switching device. A high-density 8Kbit cross-point synapse device array in a quad flat package is fabricated in the same fashion. For verification, the array is connected to an FPGA, along with a CMOS neuron. An artificial neural network was implemented on the FPGA software component in tandem with the developed neuromorphic hardware. The network was able to recognize imaginary vowels with an accuracy of 80%. The network was also used to classify a rat's fear state with accuracy near 90%.

The *Darwin Neural Processing Unit* is proposed in [114]. The unit is based on the leaky integrate and fire spike neural network model implemented with digital logic. The unit supports eight physical neurons on chip. The unit supports more logical neurons through reconfigurable time multiplexing. It is implemented using Verilog and fabricated with 180 nm CMOS technology on a 25 mm^2 chip. It consumes 0.84 mW/MHz. A RISC-based core is integrated on the chip which acts as an interface for system configuration and communication. The prototype is implemented on a Xilinx Spartan-6 FPGA for prototyping in the EEG motor imagery decoding scenario. The reported classification accuracy is 92.7%.

[115] proposes a neural network architecture composed of a common spatial pattern encoder layer and an attention layer. During the training phase, the encoder layer learns to map the input EEG sequence to feature vectors. A segmentation module is used to correct the classifier from variation of EEG temporal patterns similar to a sliding kernel filtering approach. Each segment is fed into a different

encoding pipeline. The training phase consists of a forward pass and a backward pass. To demonstrate efficiency, the classifier is implemented on several FPGAs: Xilinx Artix-7, Kintex-7, and Kintex-7 UltraScale families are used. Very low power consumption and latency in comparison to software implementation on a CPU are reported.

Movement intention is detected in [116] using a convolutional neural network. An accuracy of 80% is demonstrated. The Theano deep learning framework is used for the design and implementation of the classifier. Sixteen-bit fixed-point representation is used for data representation on the Xilinx PYNQ platform because it shows no difference in accuracy from a 32-bit floating-point representation.

5.3.5 C3.5: Sleep/Hypnosis-Level Classification

A bagged tree ML (BTML) classifier is implemented in [38] to identify the level of hypnosis of patients. Such algorithms are usually very costly for hardware, but the proposed approach utilizes (1) sequential decision at node level for each tree and (2) serialize the class decision in each of the 15 trees to reduce the area by more than 56%. It uses a LUT-based approach in which a single comparator is used for every comparison across any node in the same tree and across different ensemble model trees. The algorithm is implemented using 180 nm CMOS technology and occupies an area of 0.3 mm^2 on the chip. It has a reported accuracy of 95.4% and power consumption of 1.6 mW for the entire system which also consists of an analog front end for acquisition (see Sect. 5.1.4) and feature extraction (see Sect. 5.2.2). A Xilinx FPGA is used just for data storage and upload but not for hardware acceleration.

[57] presents the VLSI architecture for a sleep-stage classification algorithm. A thresholding stage classifies the different sleep stages between random-eye-movement, non-random-eye-movement, and awake classes. The architecture is implemented in an Actel $ProASIC_3$ FPGA with an average power consumption of 3.6 mW including preprocessing unit (see Sect. 5.2.1). An accuracy of 82% is achieved through this algorithm.

A patient anesthetic drug-independent algorithm is proposed in [89] to measure the depth of anesthesia. A fine decision tree classifier is used to give a four-state output. The inputs to the classifier consist of spectral features derived using the FFT (see Sect. 5.2.2). The tree is trained using a Gini index as a splitting criterion. The algorithm gives superior accuracy compared to other algorithms. It is implemented with 65 nm CMOS technology and occupies 0.375 μm^2. The reported accuracy is 92%.

Using 180 nm CMOS technology, the implementation of a sleep-stage classifier based on the XGBoost methodology is described in [117]. The depth of a classification tree is chosen as 3, and 250 such tree layers are used in the architecture for fine-tuning the classification accuracy for a five-level sleep-stag classification problem.

6 Results

6.1 RQ1: What Hardware Accelerators Exist?

Hardware acceleration has been embraced by the neurodiagnostic research community as the answer to developing low-cost, low-power, wearable devices for accurate diagnosis. As mentioned in Sect. 1, hardware acceleration offers various advantages that a traditional software implementation is inherently not capable of. In this review, we presented a stage-by-stage analysis of the various hardware accelerators proposed in the last 10 years by the neurodiagnostic research community. Over 72 different articles were studied, with many of them introducing hardware acceleration in more than one of the three stages of computer-aided neurological diagnosis: acquisition, preprocessing, and processing. Hardware acceleration is not just limited to application-specific integrated circuits, but various designs have been proposed which have been implemented on FPGAs and neuromorphic hardware as well. Hardware acceleration is also not limited to simple time-domain algorithms but has been implemented for complex frequency-domain and statistical algorithms as well. In the case of statistical algorithms, often the training is performed offline, and the obtained parameters are stored in the accelerator, to simplify the implementation of the classifier.

6.2 RQ2: Which Algorithms Are Implemented?

Various time-domain, frequency-domain, time-frequency-domain, and statistical algorithms have been implemented as hardware accelerators in each stage of the neurodiagnostic process. We presented a stage-by-stage analysis of the various proposed hardware accelerators in Sect. 5. A brief list of the algorithms encountered in this SLR is given below:

Acquisition: Band-pass filter [2, 3], successive approximation register analog-to-digital converter [1, 3], fast Fourier transform [9], Log2 sub-band compression [41], data compression [5, 6], ADC downsizing [7], bidirectional communication [10], and optical communication [43].

Detrending: Gain amplification [2, 14], low-pass filter [12, 14, 15, 45, 53], high-pass filter [14], LSTM neural network [46], blind source separation [47], least mean square adaptive filter [15], Laplacian spatial filter [13], discrete wavelet transform [48, 71], adaptive-maximum-mean-minimum filter [19], independent component analysis [49–51], moving average filter [16], statistical detection [52], and depth-wise separable convolutional neural network [56].

Feature extraction: Variance/standard deviation [53, 54], mean [54], accumulated energy [22], coastline [22, 53], 95% confidence ellipse area [24], max/min DWT coefficient [53], mean absolute value [53], wavelet energy/power [38, 53, 79], relative wavelet energy [53], Hurst exponent [53], detrended fluctuation analysis

[38, 53], skewness [53], clustering coefficient [72, 74], degree [72, 74], weighted degree [72, 74], density [72, 74], transitivity [72, 74], characteristic path length [72, 74], wavelet coherence [77], Morlet continuous wavelet transform [75, 76], eccentricity [72, 74], radius [72, 74], diameter [72, 74], Henon map bifurcation [80], Lyapunov exponent [80, 118], symbolic aggregate approximations [55], Teager energy operator [73, 82], phase lag index [72, 74], absolute interhemispheric power spectral difference (AIHPD) [40], interhemispheric power spectral ratio (IHPR) [40], Katz fractional dimension [44], kurtosis [44], mean power frequency [78], Hilbert transform [78], acceleration/deceleration capacity [38], sample entropy [38], Shannon entropy [38], SEF95 [38], multi-scale spectral centroid [83], multi-scale variation coefficient [83], and multi-scale spectral skew [83].

Processing: Thresholding [22, 57, 87], perceptron artificial neural network [78], nonlinear support vector machine [84], linear support vector machine [39, 40, 54, 85, 88, 90, 103], minimum redundance maximum relevance feature selection [81], neuromemristive echo-state network [102], bit-serial neural network [81], artificial neural network [103, 113, 115], Naïve Bayes' classifier [104], reduced deep convolutional stack autoencoder (RDCSAE) with improved kernel random vector functional network (IKRVFLN) [86, 105], random forest classifier [83], structural radial basis function neural network (SRBFNN) [107], convolutional neural network [81, 82], shallow neural network [44, 112, 114], spiky peak detector [52], stochastic support vector machine [111], bagged tree machine learning classifier [38], and fine decision tree classifier [89].

6.3 RQ3: Which Technology/Platform Is Preferred?

Hardware accelerators can be implemented on four different platforms: application-specific discrete circuits (ASDCs), application-specific integrated circuits (ASICs), field-programmable gate arrays (FPGAs), and neuromorphic chips. ASICs are manufactured for a customized single use; they have a single simple/composed function to perform which is generally defined by the hardware design. This is the class of the classical rigidly hardware-oriented accelerators. While on the one hand, they offer unmatched resource, power, weight, and area efficiency, on the other hand, they lack the customizability desired in today's fast-changing multi-paradigmatic world. Various articles proposing ASIC-based implementations have been studied in this SLR. They range from single-function ASICs to accelerate a specific stage of the neurodiagnostic process to dedicated processing arrays for a wide range of efficient neurodiagnostic computing. Figure 5 describes the various CMOS technologies used for the design of these ASICs. Technologies ranging from 15 to 250 nm have been used for neurodiagnostic hardware acceleration. Each of these nanometer-scale technology comes with its own set of challenges and techniques. As can be seen, the 180 and 65 nm CMOS technologies are the most popular among researchers. As the technology is scaled down, the power consumption and time

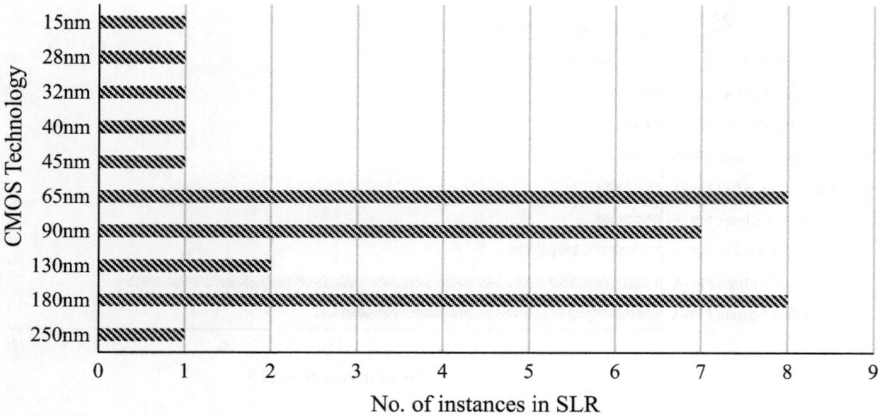

Fig. 5 Distribution of various CMOS technologies used in this SLR

delay are also reduced, but at the cost of increasing complexity in the design and testing phase. This is probably a reason why higher-scale technologies are preferred.

As described in Sect. 1, FPGAs are composed of an array of programmable logic blocks and reconfigurable interconnects that allows a user to program the desired hardware architecture to be emulated by a corresponding setting of the FPGA blocks and interconnects. With the integration of processors, FPGAs allow the synergistic execution of instructions on hardware and software, in tandem, that allows the user to extract benefits from both paradigms. The only limitation is the cost of manufacturing an FPGA. Evaluation platforms allow the testing and verification of various system architectures before they are finally realized as an ASIC, saving significant financial and temporal overheads in the development process. Due to this reason, FPGAs are the most popular hardware platform for acceleration. While standalone FPGAs have been used by researchers for acceleration, FPGA-based systems-on-chip (SoCs) provide the synergistic benefit of on-the-fly coupling with a software-based processor. Hence, researchers also utilize the SoC platforms. Figure 6 describes the various FPGA products available in the market and the number of instances that they were encountered in this SLR. As can be seen, Xilinx FPGAs are by far the most popular FPGA manufacturer among researchers. Similarly, Xilinx SoCs are the most popular SoC platforms, with the Xilinx Zynq SoC series being the most used.

Neuromorphic chips are the latest in the line of hardware platforms. Having been developed on a commercial scale as late as 2014, they are inspired by the mathematical model of neurons, where biological neurons are mimicked in silicon chips. The spiking of the neurons can be represented by a unit as small as a bit, where the timing of the spike from the silicon-based artificial neuron can convey important information as well, which unlocks faster data communication. They are being increasingly used for implementing neural networks as researchers still try to identify the true potential of neuromorphic computing. Three neuromorphic

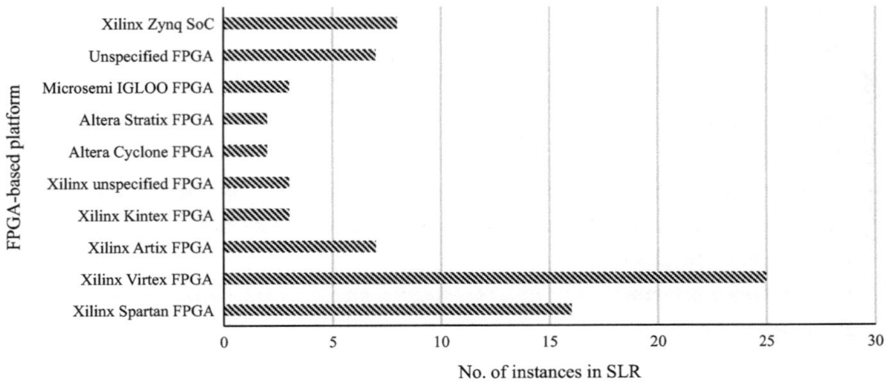

Fig. 6 Distribution of various FPGA platforms used in this SLR

platforms have been reported, showing that technology is still in its infancy. The Intel Movidius neuromorphic package has been reported in [107]. SpiNNaker is a massively parallel, many-core supercomputer architecture developed at the University of Manchester. The platform is based on spiking neural networks and is one of the components of the Human Brain Project. It surpassed its goal of simulating up to a billion neurons, in real-time, in 2018. [97, 112] report studies that have been implemented on the SpiNNaker platform. A custom PCMO-based neuromorphic platform has also been proposed in [113].

Application-specific discrete circuits (ASDCs) are the simplest platform of all and are only used in very small-scale laboratory settings. They are typically developed using breadboards and discrete circuit components.

All the articles studied in this SLR have at least one subsystem implemented or simulated on one or a combination of these four platforms. The distribution of implementations is described graphically in Fig. 7. As can be seen, the FPGA-based implementations are the most popular: FPGAs and FPGA-based SoCs were used 78 times, ASICs were used 32 times, whereas ASDCs and neuromorphic platforms were used 4 times each for hardware acceleration.

6.4 RQ4: Which Stages Are Hardware Accelerated?

All stages of the neurological diagnostics described in Sect. 1 have been hardware accelerated. This SLR categorizes the articles in four different stages from raw data to diagnostic knowledge: data acquisition, filtering, feature extraction, and processing. The category-wise distribution of the reviewed articles is described graphically in Fig. 8. From the reviewed articles, 16 hardwares accelerated the acquisition stage, 36 articles accelerated the filtering stage, 41 articles accelerated the feature extraction stage, and 43 articles accelerated the processing stage.

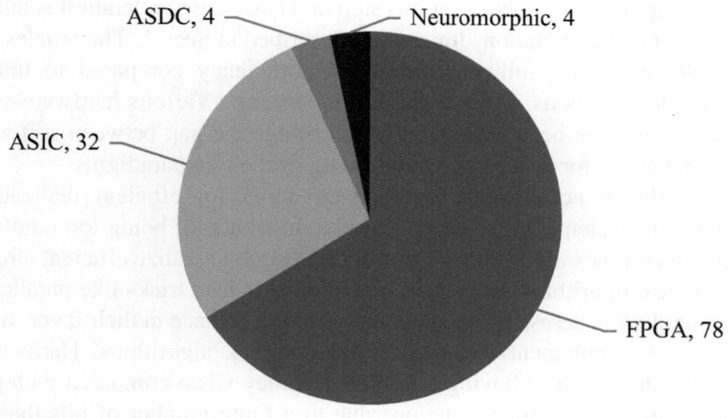

Fig. 7 Distribution of various hardware acceleration platforms in this SLR

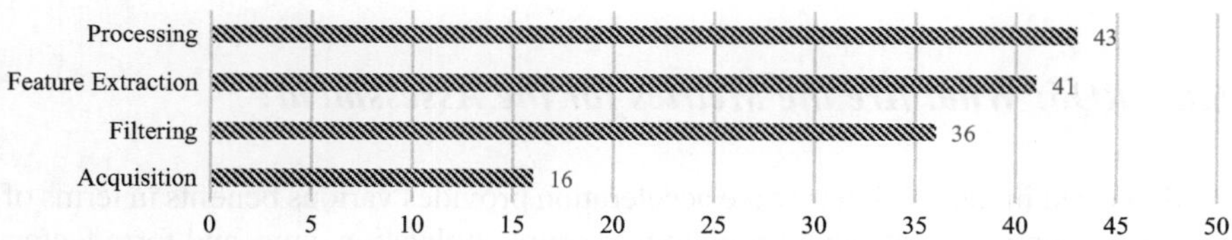

Fig. 8 Category-wise distribution of reviewed articles

6.5 RQ5: What Are the Advantages and Disadvantages?

As has already been described, a purely software-based implementation of neurodiagnostic systems poses several disadvantages. A software-based system inherently consumes more power than is truly required to execute an algorithm, since it is composed of various simultaneously necessary utilitarian programs that are required for the smooth and effective functioning of the required algorithm. This poses several problems:

1. Execution of more programs requires higher power consumption.
2. Execution of more programs leads to higher latency and lower throughput.

These disadvantages result in a device with a higher form factor than that required to execute just the diagnostic algorithm, through an increase in the area of the processing chip, increase in number of electronic components required, and higher volume of batteries. These disadvantages are essentially grave in the domain of wearable devices, an industry that is growing at an exceptional rate today. A wearable device has to meet various constraints including low power consumption, low cost, low form factor, low weight, low latency for real-time results, and an ever-increasing need for higher computing power. A purely software-based system cannot meet these growing requirements.

This is where hardware accelerations step in. Hardware acceleration is inherently able to overcome the limitations for reasons described in Sect. 1. The articles studied in this SLR report a significant increase in efficiency compared to traditional software implementations of neurodiagnostic systems. Various hardware-software implementations have been suggested which bridge the gap between software and hardware and allow for synergistic functioning of the two paradigms.

While hardware acceleration has opened doors for efficient realizations of neurodiagnostic systems, they suffer from the drawback of being too cumbersome to implement complex algorithms. It becomes difficult to realize efficient circuits for highly complex algorithms. Expert designers possess neat tricks like parallelization and resource sharing across many channels to tilt the balance in their favor, but often it is difficult to implement these tricks for complex algorithms. Hardware also possesses the drawback of having a limited accuracy when compared with theoretically correct software implementations due to a finite number of bits that can be contained in a channel.

6.6 RQ6: What Are the Metrics for the Assessment?

As described in Sect. 6.5, hardware acceleration provides various benefits in terms of power consumption, latency, throughput, resource utilization, area, and form factor. Consequently, the articles analyzed in this SLR use the same metrics for the assessment of their proposed hardware accelerators. Since hardware accelerators have the drawback of lower accuracy as compared to theoretically correct software implementations, most analyzed studies also study the correlation between a hardware accelerator and its equivalent theoretically correct software implementation to verify the correctness of the hardware implementation with a small margin of error. Metrics like root mean square error (RMSE) and correlation are reported to verify that the hardware accelerator works as expected theoretically.

Metrics like accuracy and sensitivity are also reported to prove the efficiency of statistical ML and ANN-based algorithms.

7 Discussion

Today's wearable device industry is faced with newer and greater challenges by the day. Wearable devices are required to have lower form factors and ultralow power consumption which enables long-term recording. Long-term recording is essential to unlocking new information about the inner workings of physiology. Wearable diagnostic devices also have to meet stringent real-time requirements, since the time to the reception of preventive care is a critical factor in saving a patient's life from adverse occurrences like seizures.

Hardware acceleration provides the key to unlocking the true potential of the wearable device industry. In this SLR, we have seen a wide range of hardware accelerators proposed for every stage of computer-aided neurological diagnosis. Researchers have proposed many innovative hardware accelerators from code compression to the final neurological diagnosis of seizure onsets and sleep-stage levels. The articles which propose such solutions were queried from reputed databases, analyzed, and also assigned a quality score from 0 to 8. While the mode of quality scores is 7, most papers could not reach a perfect quality score of 8. This is because most articles are not accompanied by completely specified hardware design or simulation files. This greatly reduces the reusability of the proposed accelerators, which is a significant obstruction toward the advancement of the domain. Researchers in the circuits and systems domain are strongly encouraged to make their complete hardware files available to give their work more reach and make it more reusable.

8 Conclusion

In this chapter, we conducted a detailed systematic literature review (SLR) of the hardware-accelerated techniques reported in the literature for computer-aided diagnosis of neurological disorders. The largest curated digital libraries were queried with topic-relevant phrases to acquire the initial pool of 1273 articles which were filtered down to 99 relevant articles. These 99 articles were assessed for quality and assigned quality scores in the range of 0–8. The average quality score of reviewed articles is 5.63/8. Consequently, each article was studied to answer the formal research question. A trend of increasing popularity is found in the domain of hardware-accelerated neurological diagnostics, given its advantages in power consumption, area, and latency. The chapter aims to serve as a guide for the scientist/developer designing their hardware-accelerated neurodiagnostic system, allowing the reader to browse through the recent trends and proposals in the domain, specific to each stage in the neurological diagnosis domain. All the data used in the SLR is available in a Zenodo repository with DOI: 10.5281/zenodo.5592811.

References

1. Nazzal TB, Mahmoud SA, Shaker MO (2016) A 200-NW 7.6-ENOB 10-KS/s SAR ADC in 90-Nm CMOS for Portable Biomedical Applications. Microelectron J 56:81–96. doi: https://doi.org/10.1016/j.mejo.2016.08.004
2. Kumar SCP, Chandrasekar A, Nagaraj A, et al (2016) Design of an ElectroEncephaloGram (EEG) amplification circuit for neonates. In: 2016 International Conference on Communication and Signal Processing (ICCSP). pp 2069–2072

3. Senapati B, Kumar MGL, Ray KB (2017) High resolution reconfigurable bio-potential processor for portable biomedical application. In: 2017 Devices for Integrated Circuit (DevIC). pp 517–521
4. Dai C, Bailey C (2015) A lossless data reduction technique for wireless EEG recorders and its use in selective data filtering for seizure monitoring. Annual International Conference of the IEEE Engineering in Medicine and Biology Society IEEE Engineering in Medicine and Biology Society Annual International Conference 2015:6186–6189. doi: https://doi.org/10.1109/EMBC.2015.7319805
5. Liu D, Wang Q, Zhang Y, et al (2019) FPGA-based real-time compressed sensing of multichannel EEG signals for wireless body area networks. Biomedical Signal Processing and Control 49:221–230. doi: https://doi.org/10.1016/j.bspc.2018.12.019
6. Ranjith J, Muniraj NJR (2012) VLSI implementation of efficient code compressor for bio-signal processors. Procedia Engineering 30:953–962. doi: https://doi.org/10.1016/j.proeng.2012.01.950
7. Truong ND, Kavehei O (2019) Low Precision Electroencephalogram for Seizure Detection with Convolutional Neural Network. In: 2019 IEEE International Conference on Artificial Intelligence Circuits and Systems (AICAS). pp 299–301
8. Praveen GB, Agrawal A, Shah J, Prince A (2019) Medicloud: Cloud-based solution to patient's medical records. In: Lecture Notes in Computational Vision and Biomechanics
9. Lee J, Kim S, Lee Y (2015) A potable platform to analyze EEG signal for transmitting to mobile device via Wi-Fi. In: 2015 International Symposium on Consumer Electronics (ISCE). pp 1–2
10. Rodrigues AZ, Moreno RL, Crepaldi PC, Pimenta TC (2017) Implementation of a bidirectional serial communication protocol using shared channel. In: 2017 29th International Conference on Microelectronics (ICM). pp 1–4
11. Onofre TB, Moreno RL, Crepaldi PC, Pimenta TC (2014) A bidirectional hub for a programmable gain/filtering data acquisition of a low interference electroencephalogram. In: 2014 International Caribbean Conference on Devices, Circuits and Systems (ICCDCS). pp 1–4
12. Chabchoub S, Mansouri S, Ben Salah R (2015) Biomedical monitoring system using LabVIEW FPGA. In: 2015 World Congress on Information Technology and Computer Applications (WCITCA). pp 1–5
13. Ghani U, Wasim M, Khan US, et al (2018) Efficient FIR Filter Implementations for Multichannel BCIs Using Xilinx System Generator. BioMed research international 2018: 9861350
14. Chandrasiri ME, Dhanapala RMTM, Kumari WGKG, Ranaweera R (2013) PC based Electroencephalogram system. In: 2013 IEEE 8th International Conference on Industrial and Information Systems. pp 116–121
15. Dutande PV, Nalbalwar SL, Khobragade S V (2018) FPGA Implementation of Filters for Removing Muscle Artefacts from EEG Signals. In: 2018 Second International Conference on Intelligent Computing and Control Systems (ICICCS). pp 728–732
16. Sreekrishna RR, Nalband S, Prince AA (2016) Real time cascaded moving average filter for detrending of electroencephalogram signals. In: 2016 International Conference on Communication and Signal Processing (ICCSP). pp 745–750
17. Prince AA, Ganesh S, Verma PK, et al (2016) Efficient implementation of empirical mode decomposition in FPGA Using Xilinx System Generator. In: IECON Proceedings (Industrial Electronics Conference)
18. Nalband S, Valliappan CA, Prince AA, Agrawal A (2018) Time-frequency based feature extraction for the analysis of vibroarthographic signals. Computers and Electrical Engineering. doi: https://doi.org/10.1016/j.compeleceng.2018.02.046
19. Mittal R, Prince AA, Nalband S, et al (2021) Low-Power Hardware Accelerator for Detrending Measured Biopotential Data. IEEE Transactions on Instrumentation and Measurement 70:1–9. doi: https://doi.org/10.1109/TIM.2020.3018235

20. Mittal R, Prince AA, Nalband S, et al (2021) Modified-MaMeMi filter bank for efficient extraction of brainwaves from electroencephalograms. Biomedical Signal Processing and Control. doi: https://doi.org/10.1016/j.bspc.2021.102927
21. Prince AA, Verma PK, Jayakumar C, Raju D (2015) Efficient architecture for real time implementation of Hilbert Transform in FPGA. In: Proceedings of 2015 IEEE International Conference on Electrical, Computer and Communication Technologies, ICECCT 2015
22. Helal KA, Elmkarem AYA, Refaat A-MB, et al (2017) Low-power high-accuracy seizure detection algorithms for neural implantable platforms. In: 2017 29th International Conference on Microelectronics (ICM). pp 1–4
23. Bajpai R, Yuvaraj R, Prince AA (2021) Automated EEG pathology detection based on different convolutional neural network models: Deep learning approach. Computers in Biology and Medicine. doi: https://doi.org/10.1016/j.compbiomed.2021.104434
24. Singh A, Prince AA (2015) FPGA implementation of second-order difference plot for epileptic seizure detection in EEG signals. In: 2015 Annual IEEE India Conference (INDICON). pp 1–5
25. Nalband S, Prince A, Agrawal A (2018) Entropy-based feature extraction and classification of vibroarthographic signal using complete ensemble empirical mode decomposition with adaptive noise. IET Science, Measurement and Technology. doi: https://doi.org/10.1049/iet-smt.2017.0284
26. Nalband S, Sundar A, Prince AA, Agarwal A (2016) Feature selection and classification methodology for the detection of knee-joint disorders. Computer Methods and Programs in Biomedicine. doi: https://doi.org/10.1016/j.cmpb.2016.01.020
27. George GC, Prince AA, Buch JJU, Pathak SK (2019) Design of configurable multi-mode trigger unit. Measurement: Journal of the International Measurement Confederation. doi: https://doi.org/10.1016/j.measurement.2019.03.028
28. Rathod A, Thakker R, Prince AA (2021) Accelerating Parameter Extraction of PSP MOSFET Model on SoC Platform. doi: https://doi.org/10.1142/S0218126621502479
29. Prince AA, Mishra S (2015) Multi-mode electronic stethoscope implementation and evaluation using Dynamic Reconfigurable Design. Souvenir of the 2015 IEEE International Advance Computing Conference, IACC 2015 228–232. doi: https://doi.org/10.1109/IADCC.2015.7154703
30. George GC, Moitra A, Caculo S, Prince AA (2018) Efficient Architecture for Implementation of Hermite Interpolation on FPGA. In: Conference on Design and Architectures for Signal and Image Processing, DASIP
31. Rathod A, Thakker R, Prince AA (2020) Parameter Extraction of PSP MOSFET Model in Multi-core Zynq SoC Platform. Procedia Computer Science 171:1027–1036. doi: https://doi.org/10.1016/J.PROCS.2020.04.110
32. George GC, Moitra A, Caculo S, et al (2020) A Novel and Efficient Hardware Accelerator Architecture for Signal Normalization. Circuits, Systems, and Signal Processing. doi: https://doi.org/10.1007/s00034-019-01262-3
33. Prince AA, Kartha V (2015) A framework for remote and adaptive partial reconfiguration of SoC based data acquisition systems under Linux. 10th International Symposium on Reconfigurable and Communication-centric Systems-on-Chip, ReCoSoC 2015. doi: https://doi.org/10.1109/RECOSOC.2015.7238091
34. Page MJ, McKenzie JE, Bossuyt PM, et al (2021) The PRISMA 2020 statement: An updated guideline for reporting systematic reviews. PLoS Medicine
35. Dybå T, Dingsøyr T (2008) Strength of Evidence in Systematic Reviews in Software Engineering. In: ESEM'08: Proceedings of the 2008 ACM-IEEE International Symposium on Empirical Software Engineering and Measurement
36. Chuang K-C, Lin Y-P (2019) Cost-Efficient, Portable, and Custom Multi-Subject Electroencephalogram Recording System. IEEE Access 7:56760–56769. doi: https://doi.org/10.1109/ACCESS.2019.2914088

37. Pengju Z, Dezhi Z, Shuailei Z, Kai H (2017) Digital EEG signal acquiring system based on FPGA. In: 2017 13th IEEE International Conference on Electronic Measurement Instruments (ICEMI). pp 528–533
38. Khan FH, Saadeh W (2021) An EEG-Based Hypnotic State Monitor for Patients During General Anesthesia. IEEE Transactions on Very Large Scale Integration (VLSI) Systems 29: 950–961. doi: https://doi.org/10.1109/TVLSI.2021.3058047
39. Woehrle H, Teiwes J, Kirchner E, Kirchner F (2013) A Framework for High Performance Embedded Signal Processing and Classification of Psychophysiological Data. APCBEE Procedia 7:60–66. doi: https://doi.org/10.1016/j.apcbee.2013.08.013
40. Aslam AR, Altaf MA Bin (2020) An On-Chip Processor for Chronic Neurological Disorders Assistance Using Negative Affectivity Classification. IEEE Transactions on Biomedical Circuits and Systems 14:838–851. doi: https://doi.org/10.1109/TBCAS.2020.3008766
41. Dai C, Bailey C (2015) Power analysis of a lossless data compression technique for wireless wearable biometric devices. In: 2015 11th Conference on Ph.D. Research in Microelectronics and Electronics (PRIME). pp 97–100
42. Mavinkattimath SG, Khanai R (2021) Simulation of Hardware Accelerator for Wireless Body Sensor Network. Wireless Personal Communications 2021 1–11. doi: https://doi.org/10.1007/S11277-021-08908-Z
43. Aggarwal G, Dai X, Binns R, et al (2018) Wireless EEG Signal Transmission Using Visible Light Optical Camera Communication. Communications in Computer and Information Science 899:152–161. doi: https://doi.org/10.1007/978-981-13-2035-4_14
44. Aslam AR, Hafeez N, Heidari H, Altaf MA Bin (2021) An 8.62 μ W Processor for Autism Spectrum Disorder Classification using Shallow Neural Network. In: 2021 IEEE 3rd International Conference on Artificial Intelligence Circuits and Systems (AICAS). pp 1–4
45. Alhammadi AA, Mahmoud SA (2016) Fully Differential Fifth-Order Dual-Notch Powerline Interference Filter Oriented to EEG Detection System with Low Pass Feature. Microelectron J 56:122–133. doi: https://doi.org/10.1016/j.mejo.2016.08.014
46. Chen Z, Howe A, Blair HT, Cong J (2018) CLINK: Compact LSTM Inference Kernel for Energy Efficient Neurofeedback Devices. In: Proceedings of the International Symposium on Low Power Electronics and Design. Association for Computing Machinery, New York, NY, USA
47. Chou C-C, Chen T-Y, Fang W-C (2016) FPGA implementation of EEG system-on-chip with automatic artifacts removal based on BSS-CCA method. In: 2016 IEEE Biomedical Circuits and Systems Conference (BioCAS). pp 224–227
48. Li N, Nie Y, Zhu W (2012) The Application of FPGA-based Discrete Wavelet Transform System in EEG Analysis. In: 2012 Second International Conference on Intelligent System Design and Engineering Application. pp 1306–1309
49. Shih W-Y, Liao J-C, Huang K-J, et al (2013) An efficient VLSI implementation of on-line recursive ICA processor for real-time multi-channel EEG signal separation. In: 2013 35th Annual International Conference of the IEEE Engineering in Medicine and Biology Society (EMBC). pp 6808–6811
50. Shih Y-H, Chen T-J, Yang C-H, Chiueh H (2012) Hardware-efficient EVD processor architecture in FastICA for epileptic seizure detection. In: Proceedings of The 2012 Asia Pacific Signal and Information Processing Association Annual Summit and Conference. pp 1–4
51. Van L-D, Lu T-C, Jung T-P, Wang J-F (2019) Hardware-oriented Memory-limited Online Fastica Algorithm and Hardware Architecture for Signal Separation. In: ICASSP 2019 - 2019 IEEE International Conference on Acoustics, Speech and Signal Processing (ICASSP). pp 1438–1442
52. Chamanzar A, Shabany M, Malekmohammadi A, Mohammadinejad S (2017) Efficient Hardware Implementation of Real-Time Low-Power Movement Intention Detector System Using FFT and Adaptive Wavelet Transform. IEEE transactions on biomedical circuits and systems 11:585–596

53. Indira PB, Krishna RD (2021) Optimized adaptive neuro fuzzy inference system (OANFIS) based EEG signal analysis for seizure recognition on FPGA. Biomedical Signal Processing and Control 66:102484. doi: https://doi.org/10.1016/j.bspc.2021.102484
54. Jacob R, Menon KP (2017) Implementation of EEG feature extractor and classifier for seizure detection on FPGA. In: 2017 International Conference on Intelligent Computing and Control Systems (ICICCS). pp 307–310
55. Jafari A, Gandhi S, Konuru SH, et al (2017) An EEG artifact identification embedded system using ICA and multi-instance learning. In: 2017 IEEE International Symposium on Circuits and Systems (ISCAS). pp 1–4
56. Khatwani M, Rashid H-A, Paneliya H, et al (2021) A Flexible Multichannel EEG Artifact Identification Processor Using Depthwise-Separable Convolutional Neural Networks. J Emerg Technol Comput Syst 17:. doi: https://doi.org/10.1145/3427471
57. Kassiri H, Chemparathy A, Salam MT, et al (2017) Electronic Sleep Stage Classifiers: A Survey and VLSI Design Methodology. IEEE transactions on biomedical circuits and systems 11:177–188
58. Li PZX, Kassiri H, Genov R (2016) A compact low-power VLSI architecture for real-time sleep stage classification. Proceedings - IEEE International Symposium on Circuits and Systems 2016-July:1314–1317. doi: https://doi.org/10.1109/ISCAS.2016.7527490
59. Agarwal S, Rani A, Singh V, Mittal AP (2015) Performance Evaluation and Implementation of FPGA Based SGSF in Smart Diagnostic Applications. Journal of Medical Systems 2015 40:3 40:1–15. doi: https://doi.org/10.1007/S10916-015-0404-2
60. Beeraka SM, Kumar A, Sameer M, et al (2021) Accuracy Enhancement of Epileptic Seizure Detection: A Deep Learning Approach with Hardware Realization of STFT. Circuits, Systems, and Signal Processing 2021 1–24. doi: https://doi.org/10.1007/S00034-021-01789-4
61. Belwafi K, Ghaffari F, Djemal R, Romain O (2016) A Hardware/Software Prototype of EEG-based BCI System for Home Device Control. Journal of Signal Processing Systems 2016 89:2 89:263–279. doi: https://doi.org/10.1007/S11265-016-1192-8
62. Chandran KS, Perumalsamy M (2018) EEG Based Strategies for Human Gustation Classification Using Spartan—6 FPGA. Wireless Personal Communications 2018 103:4 103:3041–3053. doi: https://doi.org/10.1007/S11277-018-5993-X
63. Aravind M, Suresh Babu S (2017) Embedded implementation of brain computer interface using FPGA. Proceedings of IEEE International Conference on Emerging Technological Trends in Computing, Communications and Electrical Engineering, ICETT 2016. doi: https://doi.org/10.1109/ICETT.2016.7873633
64. Kumbhar S, Darji AD, Singapuri HM (2021) FPGA-Based Implementation of Artifact Suppression and Feature Extraction. Lecture Notes in Electrical Engineering 676:131–143. doi: https://doi.org/10.1007/978-981-15-6229-7_10
65. la Rosa AB, Pereira PTL, Ücker P, et al (2021) Exploring NLMS-Based Adaptive Filter Hardware Architectures for Eliminating Power Line Interference in EEG Signals. Circuits, Systems, and Signal Processing 2021 40:7 40:3305–3337. doi: https://doi.org/10.1007/S00034-020-01620-6
66. Sahu AK, Sahu AK (2018) Low Transconductance OTA Based Active Comb Filter for Biomedical Applications. Communications in Computer and Information Science 956:127–137. doi: https://doi.org/10.1007/978-981-13-3143-5_12
67. Song J, Wang Y (2017) Systematic design and implementation of a high-robust adaptive calibration technique for ETI-induced analog front end circuits in EEG systems. Analog Integrated Circuits and Signal Processing 2017 91:1 91:63–72. doi: https://doi.org/10.1007/S10470-017-0925-3
68. Uthaya Kumar C, Kamalraj S (2019) Ambient intelligence architecture of MRPM context based 12-tap further desensitized half band FIR filter for EEG signal. Journal of Ambient Intelligence and Humanized Computing 2019 11:4 11:1459–1466. doi: https://doi.org/10.1007/S12652-019-01237-X

69. Van L-D, Huang P-Y, Lu T-C (2015) Cost-Effective and Variable-Channel FastICA Hardware Architecture and Implementation for EEG Signal Processing. Journal of Signal Processing Systems 2015 82:1 82:91–113. doi: https://doi.org/10.1007/S11265-015-0988-2
70. Zhang X, Vialatte F-B, Chen C, et al (2014) Embedded Implementation of Second-Order Blind Identification (SOBI) for Real-Time Applications in Neuroscience. Cognitive Computation 2014 7:1 7:56–63. doi: https://doi.org/10.1007/S12559-014-9282-Z
71. Casson AJ (2015) An Analog Circuit Approximation of the Discrete Wavelet Transform for Ultra Low Power Signal Processing in Wearable Sensor Nodes. Sensors (Basel, Switzerland) 15:31914–31929
72. Gutierrez Nuno RA, Chung CHR, Maharatna K (2020) Hardware architecture for real-time EEG-based functional brain connectivity parameter extraction. Journal of neural engineering
73. Khan SI, Diab MS, Mahmoud SA (2020) Design of low power Teager Energy Operator circuit for Sleep Spindle and K-Complex extraction. Microelectronics Journal 100:104785. doi: https://doi.org/10.1016/j.mejo.2020.104785
74. Pal C, Biswas D, Maharatna K, Chakrabarti A (2017) Architecture for complex network measures of brain connectivity. In: 2017 IEEE International Symposium on Circuits and Systems (ISCAS). pp 1–4
75. Qassim YT, Cutmore T, James D, Rowlands D (2012) FPGA implementation of Morlet continuous wavelet transform for EEG analysis. In: 2012 International Conference on Computer and Communication Engineering (ICCCE). pp 59–64
76. Qassim YT, Cutmore TRH, Rowlands DD (2016) Optimized FPGA Based Continuous Wavelet Transform. Comput Electr Eng 49:84–94. doi: https://doi.org/10.1016/j.compeleceng.2014.11.012
77. Qassim YT, Cutmore TRH, Rowlands DD (2017) FPGA implementation of wavelet coherence for EEG and ERP signals. Microprocessors and Microsystems 51:356–365. doi: https://doi.org/10.1016/j.micpro.2017.01.001
78. Daoud HG, Abdelhameed AM, Bayoumi M (2018) FPGA Implementation of High Accuracy Automatic Epileptic Seizure Detection System. In: 2018 IEEE 61st International Midwest Symposium on Circuits and Systems (MWSCAS). pp 407–410
79. Zhang L (2017) Real-time feature extraction for multi-channel EEG signals time-frequency analysis. In: 2017 8th International IEEE/EMBS Conference on Neural Engineering (NER). pp 493–496
80. Zhang L (2017) Artificial neural network model-based design and fixed-point FPGA implementation of hénon map chaotic system for brain research. In: 2017 IEEE XXIV International Conference on Electronics, Electrical Engineering and Computing (INTERCON). pp 1–4
81. Kueh SM, Kazmierski TJ (2018) Low-Power and Low-Cost Dedicated Bit-Serial Hardware Neural Network for Epileptic Seizure Prediction System. IEEE journal of translational engineering in health and medicine 6:2000209. doi: https://doi.org/10.1109/JTEHM.2018.2867864
82. Iranmanesh S, Rodriguez-Villegas E (2017) An Ultralow-Power Sleep Spindle Detection System on Chip. IEEE Transactions on Biomedical Circuits and Systems 11:858–866. doi: https://doi.org/10.1109/TBCAS.2017.2690908
83. Sukriti, Chakraborty M, Mitra D (2021) A computationally efficient automated seizure detection method based on the novel idea of multiscale spectral features. Biomedical Signal Processing and Control 70:102990. doi: https://doi.org/10.1016/j.bspc.2021.102990
84. Huang S-A, Chang K-C, Liou H-H, Yang C-H (2020) A 1.9-mW SVM Processor With On-Chip Active Learning for Epileptic Seizure Control. IEEE Journal of Solid-State Circuits 55:452–464. doi: https://doi.org/10.1109/JSSC.2019.2954775
85. Meddah K, Zairi H, Bessekri B, et al (2020) FPGA implementation of Epileptic Seizure detection based on DWT, PCA and Support Vector Machine. In: 2020 Second International Conference on Embedded Distributed Systems (EDiS). pp 141–146

86. Sahani M, Rout SK, Dash PK (2021) FPGA implementation of epileptic seizure detection using semisupervised reduced deep convolutional neural network. Applied Soft Computing 110:107639. doi: https://doi.org/10.1016/j.asoc.2021.107639
87. Sharad M, Gupta SK, Raghunathan S, et al (2012) Low-Power Architecture for Epileptic Seizure Detection Based on Reduced Complexity DWT. J Emerg Technol Comput Syst 8:. doi: https://doi.org/10.1145/2180878.2180882
88. Aslam AR, Altaf MA Bin (2019) An 8 Channel Patient Specific Neuromorphic Processor for the Early Screening of Autistic Children through Emotion Detection. In: 2019 IEEE International Symposium on Circuits and Systems (ISCAS). pp 1–5
89. Saadeh W, Khann FH, Altaf MA Bin (2020) Design and Implementation of a Machine Learning Based EEG Processor for Accurate Estimation of Depth of Anesthesia. In: 2020 IEEE International Symposium on Circuits and Systems (ISCAS). p 1
90. Jia H, Verma N (2018) Exploiting Approximate Feature Extraction via Genetic Programming for Hardware Acceleration in a Heterogeneous Microprocessor. IEEE Journal of Solid-State Circuits 53:1016–1027. doi: https://doi.org/10.1109/JSSC.2017.2787762
91. Liao Y, Zhang M, Wang Z, Xie X (2019) Design and {FPGA} {Implementation} of an {High} {Efficient} {XGBoost} {Based} {Sleep} {Staging} {Algorithm} {Using} {Single} {Channel} {EEG}. In: Sun F, Liu H, Hu D (eds) Cognitive {Systems} and {Signal} {Processing}. Springer Singapore, Singapore, pp 294–303
92. Patel K, Bleakley CJ (2016) Coarse {Grained} {Reconfigurable} {Array} {Based} {Architecture} for {Low} {Power} {Real}-{Time} {Seizure} {Detection}. Journal of Signal Processing Systems 82:55–68. doi: https://doi.org/10.1007/s11265-015-0981-9
93. Wafeek N, Elbably ME, Mubarak RI (2021) FPGA Implementation of EEG Classification System for Arm and Fingers Movements Based on Particle Swarm Algorithm. 335–345. doi: https://doi.org/10.1007/978-3-030-76346-6_31
94. Ellawala N, Thayaparan S (2019) Hardware Implementation of EEG Classifier Using LDA. Proceedings - 2019 2nd International Conference on Bioinformatics, Biotechnology and Biomedical Engineering—Bioinformatics and Biomedical Engineering, BioMIC 2019. doi: https://doi.org/10.1109/BIOMIC48413.2019.9034742
95. Zhang Y, Zhao W (2021) A 392-pW 42.7-dB Gm-C wavelet filter for low-frequency feature extraction used for wearable sensor. Analog Integrated Circuits and Signal Processing 2021 109:2 109:335–344. doi: https://doi.org/10.1007/S10470-021-01909-9
96. Zhao D, Jiang J, Wang C, et al (2014) FPGA Implementation of FastICA Algorithm for On-line EEG Signal Separation. Communications in Computer and Information Science 491: 59–68. doi: https://doi.org/10.1007/978-3-662-45815-0_6
97. Budhraja S, Sen Bhattacharya B, Durrant S, et al (2020) Sleep Stage Classification using NeuCube on SpiNNaker: A Preliminary Study. Proceedings of the International Joint Conference on Neural Networks. doi: https://doi.org/10.1109/IJCNN48605.2020.9207369
98. Wickramasuriya DS, Wijesinghe LP, Mallawaarachchi S (2016) Seizure prediction using Hilbert Huang Transform on field programmable gate array. 2015 IEEE Global Conference on Signal and Information Processing, GlobalSIP 2015 933–937. doi: https://doi.org/10.1109/GLOBALSIP.2015.7418334
99. Wijesinghe LP, Wickramasuriya DS, Pasqual AA (2014) A generalized preprocessing and feature extraction platform for scalp EEG signals on FPGA. IECBES 2014, Conference Proceedings - 2014 IEEE Conference on Biomedical Engineering and Sciences: "Miri, Where Engineering in Medicine and Biology and Humanity Meet" 137–142. doi: https://doi.org/10.1109/IECBES.2014.7047472
100. Bisasky J, Chandler D, Mohsenin T (2012) A many-core platform implemented for multichannel seizure detection. In: 2012 IEEE International Symposium on Circuits and Systems (ISCAS). pp 564–567
101. Bisasky J, Homayoun H, Yazdani F, Mohsenin T (2013) A 64-core platform for biomedical signal processing. In: International Symposium on Quality Electronic Design (ISQED). pp 368–372

102. Kudithipudi D, Saleh Q, Merkel C, et al (2015) Design and Analysis of a Neuromemristive Reservoir Computing Architecture for Biosignal Processing. Frontiers in neuroscience 9:502
103. Liu Y, Venkataraman H, Zhang Z, Parhi KK (2016) Machine learning classifiers using stochastic logic. In: 2016 IEEE 34th International Conference on Computer Design (ICCD). pp 408–411
104. Memon N, Zaidi SSH, Suleman M (2021) The Prognosis of Epilepsy with Naive Bayes Classifier On FPGA Using HDL Coder. In: 2020 IEEE-EMBS Conference on Biomedical Engineering and Sciences (IECBES). pp 18–23
105. Sahani M, Rout SK, Dash PK (2021) Epileptic Seizure Recognition Using Reduced Deep Convolutional Stack Autoencoder and Improved Kernel RVFLN From EEG Signals. IEEE transactions on biomedical circuits and systems 15:595–605
106. Balamurugan NB, Jothi M, Harikumar R (2013) FPGA Synthesis of SIRM Fuzzy System-Classification of Diabetic Epilepsy Risk Levels from EEG Signal Parameters and CBF. Lecture Notes in Electrical Engineering 221 LNEE:313–322. doi: https://doi.org/10.1007/978-81-322-0997-3_29
107. K AT, S.P K, Poddar S, Rahaman H (2020) FPGA based Structural Radial Basis Function Neural Network with Hybrid Optimization for Neural Activity. In: 2020 IEEE 7th Uttar Pradesh Section International Conference on Electrical, Electronics and Computer Engineering (UPCON). pp 1–6
108. Fang W-C, Wang K-Y, Fahier N, et al (2019) Development and Validation of an EEG-Based Real-Time Emotion Recognition System Using Edge AI Computing Platform With Convolutional Neural Network System-on-Chip Design. IEEE Journal on Emerging and Selected Topics in Circuits and Systems 9:645–657. doi: https://doi.org/10.1109/JETCAS.2019.2951232
109. Yang C-J, Fahier N, He C-Y, et al (2020) An AI-Edge Platform with Multimodal Wearable Physiological Signals Monitoring Sensors for Affective Computing Applications. In: 2020 IEEE International Symposium on Circuits and Systems (ISCAS). pp 1–5
110. Blaiech AG, ben Khalifa K, Boubaker M, Bedoui MH (2016) LVQ neural network optimized implementation on FPGA devices with multiple-wordlength operations for real-time systems. Neural Computing and Applications 2016 29:2 29:509–528. doi: 10.1007/S00521-016-2465-7
111. Han K, Wang J, Xiong X, et al (2020) A Low Complexity SVM Classifier for EEG Based Gesture Recognition using Stochastic Computing. In: 2020 IEEE International Symposium on Circuits and Systems (ISCAS). pp 1–5
112. Tayeb Z, Erçelik E, Conradt J (2017) Decoding of motor imagery movements from EEG signals using SpiNNaker neuromorphic hardware. In: 2017 8th International IEEE/EMBS Conference on Neural Engineering (NER). pp 263–266
113. Lee D, Park J, Moon K, et al (2015) Oxide based nanoscale analog synapse device for neural signal recognition system. In: 2015 IEEE International Electron Devices Meeting (IEDM). pp 4.7.1-4.7.4
114. Ma D, Shen J, Gu Z, et al (2017) Darwin: A neuromorphic hardware co-processor based on spiking neural networks. Journal of Systems Architecture 77:43–51. doi: https://doi.org/10.1016/j.sysarc.2017.01.003
115. Nikitakis A, Makantasis K, Tampouratzis N, Papaefstathiou I (2019) A Unified Novel Neural Network Approach and a Prototype Hardware Implementation for Ultra-Low Power EEG Classification. IEEE transactions on biomedical circuits and systems 13:670–681
116. Ma X, Zheng W, Peng Z, Yang J (2019) FPGA-Based Rapid Electroencephalography Signal Classification System. 2019 IEEE 11th International Conference on Advanced Infocomm Technology, ICAIT 2019 223–227. doi: https://doi.org/10.1109/ICAIT.2019.8935935
117. Liao Y, Zhang M, Wang Z, Xie X (2018) Design and FPGA Implementation of an High Efficient XGBoost Based Sleep Staging Algorithm Using Single Channel EEG. Communications in Computer and Information Science 1005:294–303. doi: https://doi.org/10.1007/978-981-13-7983-3_26

Wearable Real-Time Epileptic Seizure Detection and Warning System

Muhammad E. H. Chowdhury, Amith Khandakar, Khawla Alzoubi, Aisha Mohammed, Safaa Taha, Aya Omar, Khandaker R. Islam, Tawsifur Rahman, Md. Shafayet Hossain, Mohammad T. Islam, and Mamun Bin Ibne Reaz

Abstract Epilepsy is an unpredictable neuronal brain disorder, affecting approximately 70 million people. It is characterized by seizures resulting in the patient losing the capability of controlling his/her actions. Besides being prone to injuries, losing consciousness, and control of the body, patients have a higher risk of experiencing sudden unexpected death in epilepsy (SUDEP). Therefore, continuous multimodal monitoring using electrodermal activity (EDA) and accelerometer (ACC) sensors can help in detecting epileptic seizures based on the seizure types and symptoms. EDA was used to detect emotional activities, while ACC detected physical activities. This paper describes the design and implementation of a real-time wearable epileptic seizure detection and warning system to monitor 12 epilepsy patients during their daily activities at home. The acquired EDA and ACC signals from the patients' body were sent continuously to the detection and warning subsystem, where it was continuously processed and analyzed. The later block can also automatically alert the parent/caregiver of the patient over the cellular network

M. E. H. Chowdhury (✉) · A. Khandakar · A. Mohammed · S. Taha · A. Omar
Department of Electrical Engineering, Qatar University, Doha, Qatar
e-mail: mchowdhury@qu.edu.qa; amitk@qu.edu.qa; am1202703@student.qu.edu.qa; st1306687@student.qu.edu.qa; ao1302928@student.qu.edu.qa

K. Alzoubi
Engineering Technology, Department Community College of Qatar, Doha, Qatar
e-mail: Khawla.Alzoubi@ccq.edu.qa

K. R. Islam
Department of Orthodontics, Bangabandhu Sheikh Mujib Medical University, Dhaka, Bangladesh

T. Rahman
Department of Biomedical Physics & Technology, University of Dhaka, Dhaka, Bangladesh
e-mail: tawsifurrahman@bmpt.du.ac.bd

M. S. Hossain · M. T. Islam · M. B. I. Reaz
Department of Electrical, Electronics & Systems Engineering, Universiti Kebangsaan Malaysia, Bangi, Selangor, Malaysia
e-mail: tariqul@ukm.edu.my; mamun@ukm.edu.my

M. Murugappan, R. Yuvaraj (eds.), *Biomedical Signals Based Computer-Aided Diagnosis for Neurological Disorders*, https://doi.org/10.1007/978-3-030-97845-7_11

in case of a seizure event. Among the various machine learning algorithms, support vector machine (SVM) and bagged decision tree classifiers yielded the highest accuracy of 86.9% and 90.7%, respectively, for ACC and EDA data individually. However, for fused ACC-EDA data, the bagged decision tree showed the highest accuracy of 96.7% in detecting epileptic seizures. It was found that fused ACC data with EDA helped to distinguish epileptic onset from daily activities reliably with a very low false alarm rate.

Keywords Wearable sensors · Epileptic seizure detection · Electrodermal activity (EDA) · Accelerometer (ACC) · Machine learning algorithm

1 Introduction

Epileptic seizures are neural system disorders resulting from multiple sets of recurrent and unpredictable interruptions in the brain area leading to excessive or synchronous neuronal activity [1]. Epilepsy is characterized by recurrent seizures that originate without any prior warning or symptoms and, therefore, result in the brief suspension of awareness or loss of consciousness as well as violent convulsions. Epileptic seizures are widespread complications of epilepsy affecting 70 million people in the world [2, 3]. Epilepsy is the second leading cause of death among neural disorders [4]. An epileptic seizure can also be categorized into two main types of seizures: focal and generalized seizures. A focal or partial seizure can make the patient see, hear, or feel things that are not there. A generalized seizure occurs when the nerve cells on both sides of the patient's brain misfire. This type of seizure can make the muscles of the patient spasms, fall, or blackout and sometimes even leading to loss of consciousness. The severity of the disease demands immediate response to avoid further consequences. The advances in the field of biomedicine and machine learning can assist physicians to track early symptoms and take proactive action [5, 6]. COVID-19 pandemic has significantly affected the healthcare system of the world and has increased the demand for remote health monitoring of patients and the use of artificial intelligence. Despite the technological advances for the early detection of epilepsy and associated seizures, there is still an adequate scope for further improvement in the hardware and decision-making system to get a real-time warning system as the disease is severe and death chances are eminent [7–9]. Therefore, there was an urgent need to develop a system that can monitor the onset of epilepsy in a home setting and notify the caregiver in real-time.

The epileptic seizures are typically detected by EEG and some newer techniques such as magnetic resonance imaging (MRI), functional magnetic resonance imaging (fMRI), computerized tomography (CT), and positron emission tomography (PET) [10, 11]. Blood tests require blood samples, and others require radiations to record the brain signals of epilepsy patients. These methods suffer from disadvantages such as the invasive nature of blood collection, side effects caused by radiations, and cumbersome data collection techniques. With the emergence of real-time detection systems such as electroencephalogram (EEG) and electrocardiogram (ECG) as well as video detection systems, the monitoring of epileptic seizures became comparatively easier to manage [12]. Moreover, electrodermal activity (EDA) or galvanic skin response (GSR) and accelerometers (ACC) were recently used for real-time seizure detection [13, 14]. However, EEG and ECG systems are more suitable for hospital and clinical setups, and EDA and ACC require further research to make them reliable and suitable for clinical application [15].

The EDA sensor can detect focal complex seizures and generalized tonic-clonic seizures (GTCS) [16]. The GSR complex has two main components, which are the general tonic-level GSR and phasic skin conductance response (SCR). The phasic SCR is related to the faster-changing element of the EDA signal that results from the sympathetic neuronal activity [17]. To measure the electrical conductance of the skin, the EDA electrodes are placed over the skin. This measures the electrical conductance of the skin, which is directly proportional to the mental activity of the person. The EDA signal remains elevated for a longer time in the GTCS compared to the complex partial seizure. Moreover, the results obtained from the EDA sensors might be erroneous due to pressure or motion artifacts [18]. Therefore, it is better to integrate it with other sensors to make a multimodal system for reliable epileptic seizure detection. A respiration sensor and ACC sensor can be used for this purpose [19]. An ACC sensor is an electromechanical device that is used to measure the dynamic forces caused by any movement of the limb where it is placed [20]. The ACC is used to detect the changes in motion along the *x*-, *y*-, and *z*-axis. Thus, ACC is capable of detecting rhythmic repetitive movement of an extremity. During seizures, the patient might experience different abnormal physical movements (e.g., twitching in the arms and legs, the inability of controlling the muscles, or sudden falling), which can be detected by the ACC sensor.

Many wearable epileptic seizure detection systems were reported in the literature. Jerger et al. [21] proposed a continuous EEG measurement system for epileptic patients to predict and ultimately control the seizure activity using a control system. Different linear and nonlinear methods such as analysis of power spectra, cross-correlation, principal components, phase, wavelets, correlation integral, and mutual prediction were proposed in their work. Bandarabadi et al. [22] proposed a novel approach using the intracranial electroencephalogram (iEEG) electrodes placed over the foci for automated early seizure detection with a sensitivity of 86.9 % and a false detection rate of 1.4 per day. Ganesan et al. [23] reported a wireless electronic diagnosing system that combines EDA, a global positioning system (GPS), and a global system for mobile (GSM) techniques. The location of the epileptic patient was updated through GPS technology, and a warning was sent to the epileptic patient's caregiver. However, the accuracy of the system on real patients was not reported. A novel method for continuous monitoring of the sympathetic nervous system activity during epilepsy was reported using EDA and tri-axis ACC sensors [24]. The recordings of the ictal video-EEG were reviewed by a team of experts with the EDA and ACC data. Neither machine learning algorithm was applied for detection, nor there was any result reported by this group to show how effective the system was in epilepsy detection. Van de Vel et al. [25] used an accelerometer, video, and radar-induced activity recorder for detecting tonic-clonic and clonic seizures. The test was performed on two patients for 1 month in home environment. The mean sensitivity was reported to 66.87% and the false alarm rate was 1.16 per night. The accuracy of the epileptic seizure detection algorithm mentioned above can be further improved with the help of machine learning techniques using the fusion of EDA and ACC data.

Epileptic seizure detection techniques using machine learning (ML) algorithms were also reported in the literature. Heldberg et al. [26] showed an overall sensitivity of 89.1% and specificity of 93.1% using the k-nearest neighbor (kNN) classifier, for seizures without motor activity using EDA and ACC sensors. The performance of this system was promising; however the precision was low, which could be improved by optimizing the classifier. This also motivated the authors to investigate other machine learning techniques with a new set of features to achieve a better detection performance. A computerized seizure-onset detection method based on EEG data was developed to identify the onset of a seizure using an ML algorithm that showed 96% accuracy while trained on 163 seizures from 844 h of continuous EEG signal from 23 epilepsy patients [27]. But the prototype made use of EEG signals which are not suitable for the home environment. Gouravajhala et al. [28] showed a multimodal system using EDA, temperature, and rotational position sensors to send data to the emergency center via a cellular network. The result showed an accuracy of 90% with a less complex system in comparison to EEG. However, misclassification of seizure cases should be reduced as much as possible to get the maximum benefit from such wearable systems. Furthermore, long-term epileptic seizure detection system based on wearable ECG, hospital ECG, and photoplethysmography (PPG) devices was used to evaluate the seizure detection algorithm proposed in [29]. This algorithm categorized the seizure based on the heart rate features and was applied to 11 patients in a hospital setting to capture 47 temporal lobe seizures. It was reported that the sensitivity of the wearable PPG, wearable ECG, and the hospital ECG system was 32%, 70%, and 57%, respectively. It is reported that PPG has the lowest sensitivity compared to the hospital and the wearable ECG [30]. In all the aforementioned works, the accuracy could be improved by the fine-tuning machine learning algorithm and fusing with other modalities.

Some recent works showed the use of the convolutional neural network (CNN) in detecting real-time epileptic seizures using EEG data [31]. It was reported that epileptic seizures can be detected using the EEG with very high accuracy of around 93.80%, 99.83%, and even 100%; however, the hardware is very complex and impractical to set up for home users for continuous monitoring [32]. Moreover, the number of channels in the clinical EEG cap is 21, which can go up to 256 for the EEG cap used for research. Therefore, practical usage of such a system in a home setting is not yet technically feasible. Several multimodal techniques were reported recently [33–36] for the detection of epileptic seizures, and the overall accuracy reported in those works were 94.55%, 90%, and 95.23% respectively. Picard et al. (2017) reported a multisite clinical evaluation of the machine learning classifier embedded in the Embrace system, which was validated with 94.5% sensitivity in GTCSs detection with a false alarm rate lower than 0.2/day [35]. In a very recent work by Merve and Seda in [37], a detailed investigation was done on multiple features extracted from EEG signals, and feature selection and classification using the SVM method result in an accuracy of 98.8%. Similar high accuracy was reported in [38–41] using deep learning techniques with EEG signals. It can be noticed that EEG-based techniques reached the highest accuracy; however, it is still not practical as a wearable system for daily usage, whereas other multimodal techniques are feasible for home users but the accuracy should be improved further to reduce false alarms. Moreover, the EEG-based technique suffers from motion artifacts due to movement (particularly in children) and thus will be difficult to detect epileptic seizures without the video recording along with the EEG recording. Even tough EEG is considered as gold standard for epileptic seizure detection with or without multimodal system, and therefore, a significant research is carried out to improve EEG quality to reduce false alarm when EEG is used alone [42]. Therefore, even though EEG-based deep learning techniques are useful for the hospital setting and research applications, still they are not suitable for patient monitoring in the home setting. In this work, an extensive investigation of the recent works was carried out, and systematic design of wearable signal acquisition and smart decision-making systems were proposed with the following key contributions:

1. Development of a prototype wearable system using the EDA and ACC sensors as a wristband to be worn by the user like a watch.
2. Several machine learning algorithms were investigated to evaluate their accuracy in detecting epileptic seizures by EDA and ACC individually and in combination.
3. Development of the machine learning models using EDA and ACC data from a public database.
4. ML implementation on a Raspberry Pi 3 for real-time epileptic seizure detection and alerting using GSM service.
5. A limited clinical trial on 12 epileptic patients was carried out to show the efficacy of the proposed system.

This type of system will be particularly useful for monitoring patients at home during their daily activities. Their parents or caregivers can feel safe, while the system is monitoring the health status of the patient. The rest of the paper is divided into five sections: Sect. 2 discusses the research methodology and the details of the studies done in this work; Sect. 3 discusses the analysis of the data, while Sect. 4 reports the results of this work and discusses them; and finally, the manuscript is concluded in Sect. 5.

2 Materials and Methods

The proposed system consists of two subsystems: a wearable sensor subsystem and an intelligent seizure detection and warning subsystem. The two subsystems were able to communicate wirelessly using Bluetooth Low Energy (BLE) technology. The sensor subsystem consists of two EDA electrodes, which were used to measure the electrical conductance of the skin, along with an ACC sensor to measure the physical activities. The raw signals were acquired from the patient's body and sent continuously to the detection and warning subsystem, where they are continuously

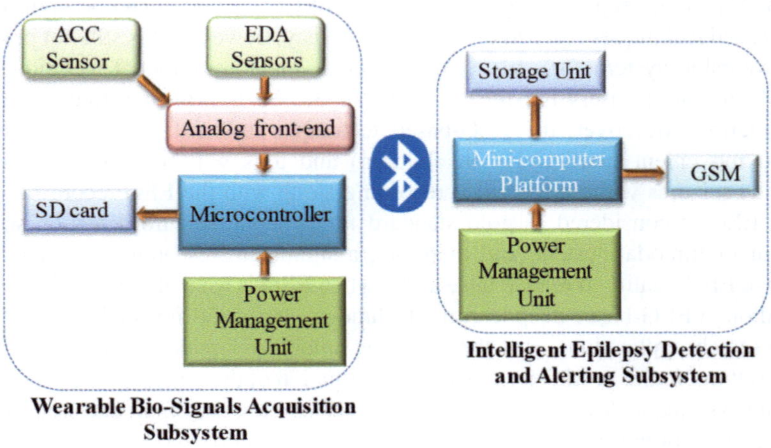

Fig. 1 Overall system block diagram: Wearable bio-signal acquisition subsystem (acquiring data, forming packets, and continuously sending these data using Bluetooth to the other subsystem) and intelligent epilepsy detection and alerting subsystem (data preprocessing, feature extraction, and classification)

processed and classified to make a decision and alert the caregiver. The implementation of the detection and alerting subsystem was accomplished in two phases. In the first phase, the signal processing, segmentation, feature extraction, and training and testing of the best-performing machine learning algorithms were implemented on a personal computer (PC) using MATLAB software. In the second phase, the detection and alerting subsystem were implemented in the Raspberry Pi 3 (RPi3) using Python to make the system standalone and self-operating without the need for user intervention.

The block diagram provides the detailed methodology of the study and also the details of each component of the system (Fig. 1).

2.1 Block Diagram of the System

Figure 1 shows the wearable subsystem consists of an ACC and EDA sensor, RFduino microcontroller (MCU), and a power management unit, and the intelligent subsystem consists of a minicomputer platform (Raspberry Pi 3) and a GSM system. The intelligent system is designed to process the data acquired from the wearable system along with an SD card storage unit to store the data and a power management unit to deliver power to this module. The Raspberry Pi 3 (RPi3) was designed to alert the caregiver using the GSM calling and messaging service.

RFdunio is the smallest microcontroller with built-in BLE capability in a fingertip-sized module suitable for a compact design and can be programmed using Arduino IDE, and the user can utilize all the available resources of Arduino.

RPi 3 was chosen as a minicomputer platform because of three reasons: (i) open source with plenty of online resources, (ii) built-in BLE and easy to develop Python-based machine learning algorithm for real-time application, and (iii) off-the-self GSM shield available for the RPi3 which was used for the alerting service.

2.2 *Wearable Bio-signal Acquisition Subsystem*

This subsystem is a real-time data acquisition system, which includes dry electrodes, EDA sensor, analog front end (AFE), ACC module, RFduino microcontroller, and power management unit. The dry electrodes were used to acquire the skin conductance, and AFE was used for amplification and filtering, and then the signal was digitized and transmitted to the decision-making subsystem.

Figure 2a, b shows the wearable wristband subsystem with embedded dry metal electrodes (EDA sensors, positive and negative electrodes, Fig. 2b). The circuit board was printed to host the AFE, ACC module, RFduino microcontroller, the power management module, and the rechargeable battery. The power management module (PMM) is responsible for feeding power to the system. The longer the battery can last, the more convenient and suitable for the user it would be, as the system was designed for continuous monitoring. However, the system should not be bulky for the user. The system has an undervoltage lockout facility, which was to avoid the battery drain completely when the battery voltage reaches a value below 2.6 V.

2.2.1 Electrodermal Activity (EDA) Sensor

EDA module was designed to measure the changes in the skin conductance of the subject at different emotional situations such as laughing, breathing, and screaming.

Fig. 2 Wearable bio-signal acquisition subsystem (**a**) to be worn as a wristband by a patient, (**b**) back view of the hardware with the electrodes, (**c**) decision-making, and alerting system without and with (**d**) 3D printed packaging

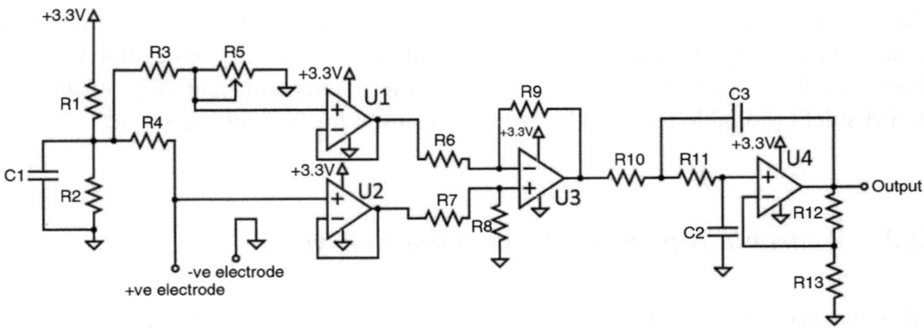

Fig. 3 Circuit diagram of EDA module

The module allows spotting strong emotions by attaching two dry electrodes to the wrist of the patients. Dry metal electrodes were utilized for a long time monitoring as their impedances improve over time. Moreover, the metal electrodes are mechanically strong and are not prone to corrosion [43], which make them suitable for real-time continuous monitoring devices. Moreover, dry electrodes are comfortable and do not cause irritations to the skin compared to wet electrodes [44]. The complete AFE for the EDA module was constructed using rail-to-rail low-power op-amp MCP6022 [45] (as shown in Fig. 3). The AFE consists of two buffer amplifiers, a differential amplifier, and a second-order Butterworth low-pass filter with a cutoff frequency of 1.6 Hz. U1 and U2 were designed as a buffer amplifier, where U1 was providing a buffered reference voltage and U2 was providing a buffered EDA electrode voltage. The differential amplifier, U3, was used to amplify the difference outputs of U1 and U2 amplifiers. The active low-pass filter was designed around the U4 amplifier. EDA electrodes are connected between the positive input of U2 op-amp and Gnd, as shown in Fig. 3. The circuit has an adjustable sensitivity via potentiometer (R5) to deal with the intersubject variability. The circuit was tested whether it can capture EDA signal during emotional changes or not.

To calculate the skin conductance, G, from the circuit shown in Fig. 3, the following mathematical equation was derived from the circuit parameters:

$$G = \frac{V_{0(U4)}(\text{without electrode}) - V_{0(U4)}}{\left(V_{0(U4)} + 1.175\right) \times 74.8\ \text{K}\Omega} \tag{1}$$

The above equation was used in the RFduino to convert the analog voltage read at the input of the microcontroller to skin conductance. To generate the above equation, which can be used to calculate skin conductance, a commercial EDA signal measurement system (wearable wireless BioRadio™ (Great Lakes NeuroTechnologies, OH, USA)) [30] was used to calibrate the designed wearable system. EDA electrodes of the commercial system and prototype system were placed on the same pair of fingers simultaneously and skin conductance from the BioRadio system was measured. To obtain the same skin conductance from the prototype system, parameters of the prototype circuit (reference voltage, gain, etc.) were adjusted. The range of the skin conductance measured by the prototype system was 2–100 microsiemens (μS).

The skin conductance can be best measured on the palmar sites [46]. The recording site can be in either the inner or the outer wrist (forearm) or on the surface of the palm [47]. Moreover, it can also be measured on the fingertips [16]. These sites are known for their high ability to measure skin conductance because of the availability of several sweat glands. The belt of the module should be worn in such a way that the electrodes are touching the skin directly, refer to Fig. 2; otherwise, the EDA signal will be noisy. The wrist was chosen for electrode placement, similar to some of the commercially available devices (e.g., Empatica) [48], and was also compared with the EDA recorded from fingertips.

2.2.2 Accelerometer (ACC) Module

MMA8452Q, a low-power-consuming high-resolution ACC module, was used to track the 3 degrees of freedom (DOF) motion of a subject in the x-, y-, and z-planes, and the range of the ACC module was selected to be $\pm 8g$ [49]. ACC sensor can help in detecting motor-type seizures. ACC module was interfaced with the RFduino microcontroller using the I^2C protocol. RFduino was acquiring the EDA and ACC data, which is formed into packets and sent to the decision-making module over the BLE interface.

2.2.3 Wireless Communication over BLE

RFduino uses BLE protocol to transmit data through a generic attribute (GATT) profile to RPi3. Once the connection was initiated, the peripheral device (RFduino) was connected to the central device (RPi3). In our system, RFduino was GATT master, which holds the GATT service and characteristic, and GATT Client (RPi3) was responsible for sending the requests and receiving the responses. To ensure reliable data transmission to RPi3 without missing any data packet, notification of incoming data packet reception in the BLE buffer was used. Moreover, to increase the sampling frequency of the RFduino to 100 Hz in data acquisition, EDA and ACC data packets were buffered before transmission, and after every 50 ms, a buffered frame was sent to RPi3. From the authors' previous works in [1–3], it is found that 50 ms interval is suitable with reliable signal transmission and reasonable battery life. Thus, the 50 ms helped to ensure low-power consumption of the wearable system, while keeping high-frequency sampling for reliable acquisition. RFduino timer interrupt was used to ensure 10 ms interrupt-driven data acquisition for guaranteeing 100 Hz sampling frequency. In particular, accelerometer data cannot be sampled using a 1-s sampling interval to reliably track human movements, especially during seizure periods. However, we will consider in the next version of the wearable hardware to collect data for 1 s and buffer in a new microcontroller with larger memory and transmit every 1 s, which we believe will significantly increase battery life without losing information.

2.3 Intelligent Epilepsy Detection and Alerting Subsystem

This is a decision-making subsystem to detect the event of the epileptic seizure in real time depending on the acquired EDA and motion signals. Therefore, it is considered as the brain of the whole system. This subsystem incorporates SIM 908 GSM/GPRS module interfaced directly to RPi3 using the cooking-hack GSM/GPRS shield (Fig. 2b), and the system was packaged using 3D-printed packaging (Fig. 2c). Initial signal processing was accomplished in the personal computer (PC) (Intel Core i7, 8 GB RAM, Windows 7 ×64) environment using MATLAB 2018a. In the MATLAB-based implementation, the EDA and ACC signals were segmented to a 10 s window (1000 samples) with 50% overlapping. Therefore, the features were extracted for this window. The classification was accomplished using the statistics and machine learning toolbox in the MATLAB in PC-based implementation. In the RPi3-based implementation, we used sliding windows of 10 s (1000 samples) to include seizure and motion events. The sliding window continuously moves to the next interval, overlapping half of the interval. The features for EDA and ACC were extracted from each sliding window. Python 2.7-based machine learning algorithm was implemented using NumPy (v1.13.3), Matplotlib (v3.0.2), PyBrain (v0.31), and Scikit-learn (v0.20) libraries, and the multi-threaded application was used for real-time preprocessing and classification of an epileptic seizure. In-house built C++ program was used to initiate an emergency call and text short message service using the GSM/GPRS module to a predefined caregiver number.

2.3.1 Database Description

Empatica [33] is the only Food and Drug Administration (FDA)-approved multi-modal research system available in the market for epilepsy monitoring. Empatica has shared a database upon the request of the authors for this study. In the Empatica dataset, EDA and ACC along with PPG, heart rate, and temperature data were recorded for 24 h for 14 days from 5 epileptic patients (2 males, 3 female patients; age, 27–57 years) comprised of normal and epileptic seizure events. There were 41 absence seizures, and 7 GTCS seizure events lasting from a few seconds to approximately 2 min were recorded in the dataset. This dataset was recorded while the users were wearing Empatica wristbands during their daily activities; however, an event monitoring by video monitoring and event-logging diary by the caregiver was used for labeling the dataset by the experts. Besides, an ACC and EDA database (similar to the Empatica dataset) was created using our wristband prototype system from 12 epileptic patients (6 male and 6 female subjects; age, 20–47 years; mean

Table 1 Details of the patients in this study

Patient number	Age (years)	Sex	Weight (kg)	Height (cm)	Type of seizures	Seizure frequency (/month)
01	20	Male	69	175	Focal and GTC seizures	15
02	24	Male	71	165	Focal and GTC seizures	10
03	47	Male	66	170	GTCS	25
04	22	Male	69	178	GTCS	30
05	26	Male	63	167	GTCS	25
06	34	Male	64	171	GTCS	35
07	25	Female	64	170	Focal and GTC seizures	15
08	23	Female	58	160	GTCS	15
09	24	Female	63	150	Focal and GTC seizures	25
10	20	Female	62	171	GTCS	40
11	24	Female	60	162	Focal and GTC seizures	25
12	23	Female	59	159	GTCS	25

age, 26 years; mean weight male, 67 kg, and female, 61 kg; mean height male, 171 cm; mean female height, 162 cm) (Table 1).

The EDA and ACC data acquisition from 12 epileptic patients using the prototype system were approved by the Hamad General Hospital and Qatar University Institutional Review Board (QU-IRB) ethical committee. Each subject has been introduced to the study they are going to take part, and written consent was taken from each participant before participating in the study. The machine learning algorithm was initially trained, validated, and tested on the labeled data from the Empatica database [33]. Later on, patients' data were recorded using the prototype system and added with the Empatica database utilizing the same sampling frequency and normalization technique. In this new database, created with Empatica data and the patients acquired using the prototype system, the ML model was trained, validated, and tested. Finally, the trained ML model was deployed in the prototype decision-making and alerting system.

For acquiring the ACC data, the subjects were asked to do normal daily activities like laying down, sitting, standing, eating, walking, and running. Each activity was recorded for 60 s with a sampling rate of 100 Hz for six trials in the regular home setting with the subject being asked to pretend to do their regular activities at a regular pace. Therefore, several traces were obtained for each ACC component from each subject with 10-s trace duration. It might be useful to note that these activities represent the most regular activities done by the subjects on daily basis; therefore distinguishing these signals would help in detecting epileptic seizures from daily activities. RPi3-based real-time system was evaluated on these 12 epileptic patients, and the subjects continued to use the system for 1 month to evaluate whether the prototype system can identify the seizure event reliably or not. In the dataset created using the prototype system, EDA and ACC data were recorded for 24 h, which was made up of normal and epileptic seizure events. There were 65 generalized tonic-clonic (GTC) and 160 absent seizure events lasting from a few seconds to a few minutes that were recorded in the dataset created using the prototype device. The events of GTCS and their timestamps were confirmed with the help of patients' caregivers' diaries. Since the absence of seizures cannot be confirmed by the caregiver as the accurate timing of absence seizure is difficult to identify even by the patients, the ACC and EDA data were investigated using the model trained on Empatica database, and the events of sudden EDA change where the change of ACC was not shown noticeable were considered as the absence seizures. Moreover, it was observed that most of the absence seizures in the patients' data from the Empatica database have the same pattern. There is a chance of incorrect absence seizure labeling or missed absence seizures in the collected data from the 12 patients; however, in the Empatica database, the absence seizures were labeled properly using the ground truth EEG signals (which were confirmed by the clinician for the presence of a seizure), which was used for the initial ML model development. This initial model was used for identifying the potential absence seizures along with visual inspection on EDA and ACC to distinguish the absence seizures reliably. It is therefore expected that most of the absence seizures were detected reliably in the labeling process.

3 Analysis

The EDA and ACC data undergone several steps: preprocessing, feature extraction, and feature reduction techniques before being used for training, validation, and testing using machine learning models (which are explained in detail below). Figure 4 shows the steps of the best-performing machine learning model selection.

3.1 Data Preprocessing

The in-house built MATLAB code was used to preprocess the ACC data during different activities. The ACC data undergone digital band-pass filtering to remove

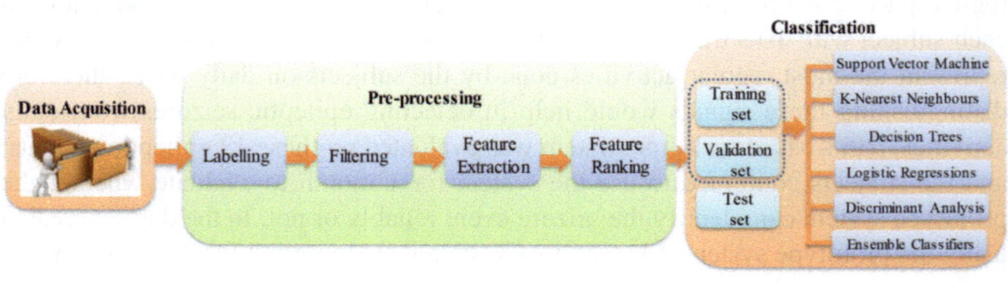

Fig. 4 Stages of data preprocessing and the application of ML models

any unwanted spikes or noise from the data. Figure 5 shows the labeling of the z-component of the raw and high-pass filtered data recorded during the normal activities of the subject.

3.2 Feature Extraction

Several time-domain, statistical, and frequency-domain features were investigated to observe the difference between the EDA and ACC signals during different activities. There are small variations in EDA noticed during regular activities; therefore, ACC was utilized to distinguish them.

Different statistical parameters were used to visualize the raw values with respect to their position in time, which can help in the process of easily separating the activities from each other such as walking and sitting. However, when dealing with signals, the statistical analysis seems to be insufficient because it can be hard to distinguish some of the activities. Thus, the time-domain analysis must be taken into consideration to study the variation of the signal over time (Fig. 5). The ACC signal obtained can be classified into fast variation over time due to the body movements of the subject and slow variation over time due to the position of the body with respect to the vertical gravitational field. Therefore, the time-domain analysis on the effect of the dynamic activities should be the main interest of this analysis. Moreover, power spectral density (PSD) is plotted to observe any potential variation in the signal obtained from each activity. Power spectral density shows the strength and the

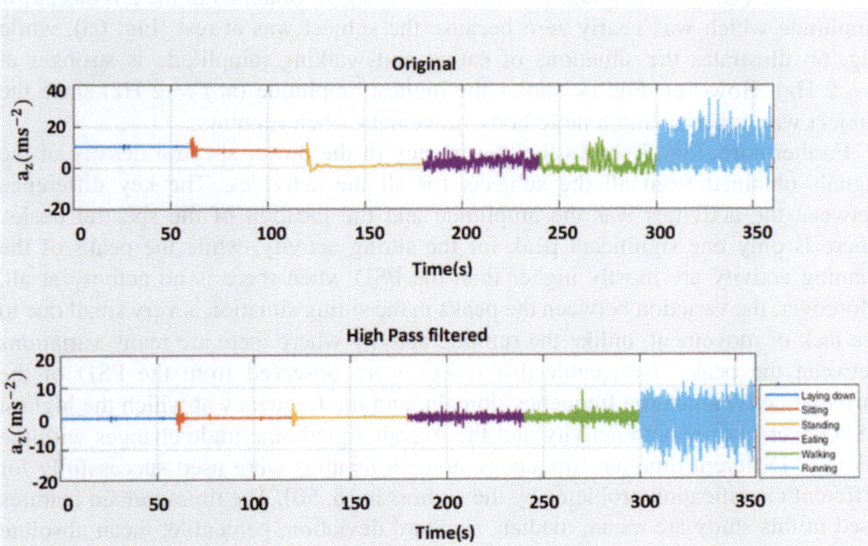

Fig. 5 Original vs high-pass filtered ACC signal

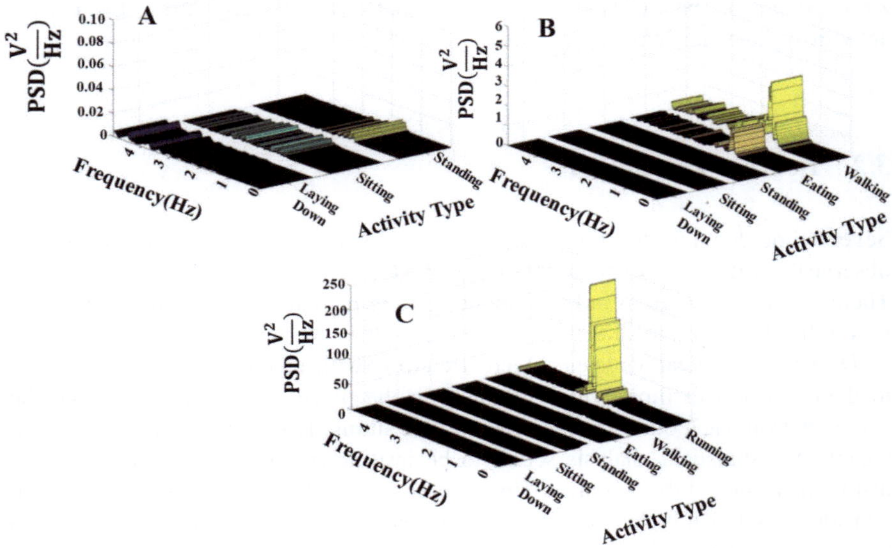

Fig. 6 Power spectral density (PSD) of different stationary (**a**) and nonstationary activities (**b** and **c**) of a user

variation of the signal as a function of frequency (Fig. 6) for a test subject. The x-axis represents the activity type, the y-axis is the frequency spectrum of the signal, and the z-axis is the PSD of the signal. For instance, the signal obtained in the left has a low amplitude which was nearly zero because the subject was at rest (Fig. 6a), while Fig. 6b illustrates the situations of eating and walking (amplitude is stronger at $f = 2$ Hz). However, Fig. 6c shows the highest amplitude (at $f = 2$ Hz) since the subject was experiencing a large body movement when running.

Furthermore, Fig. 7 shows the consistency of the power spectral density of the signals obtained from all the subjects for all the activities. The key difference between the activities was the amplitude and the location of the spectral peaks. There is only one significant peak for the sitting activity, while the peaks of the running activity are mostly higher than the PSD when there is no activity at all. Moreover, the variation between the peaks in the sitting situation is very small due to the lack of movement, unlike the running activity where there are many variations between the peaks. Noticeable differences were observed from the PSD of the different activities in the frequency domain, and the frequency at which the highest PSD occurs varies with activity and the overall signal amplitude changes with the activity. Different time and frequency domain features were used successfully for different classification problems by the authors in [6, 50]. The time-domain features used in this study are mean, median, standard deviation, percentile, mean absolute deviation, interquartile range, skewness, kurtosis, Shannon's entropy, and fractal dimension using Higuchi fractal dimension (HFD) [3]; and the frequency-domain

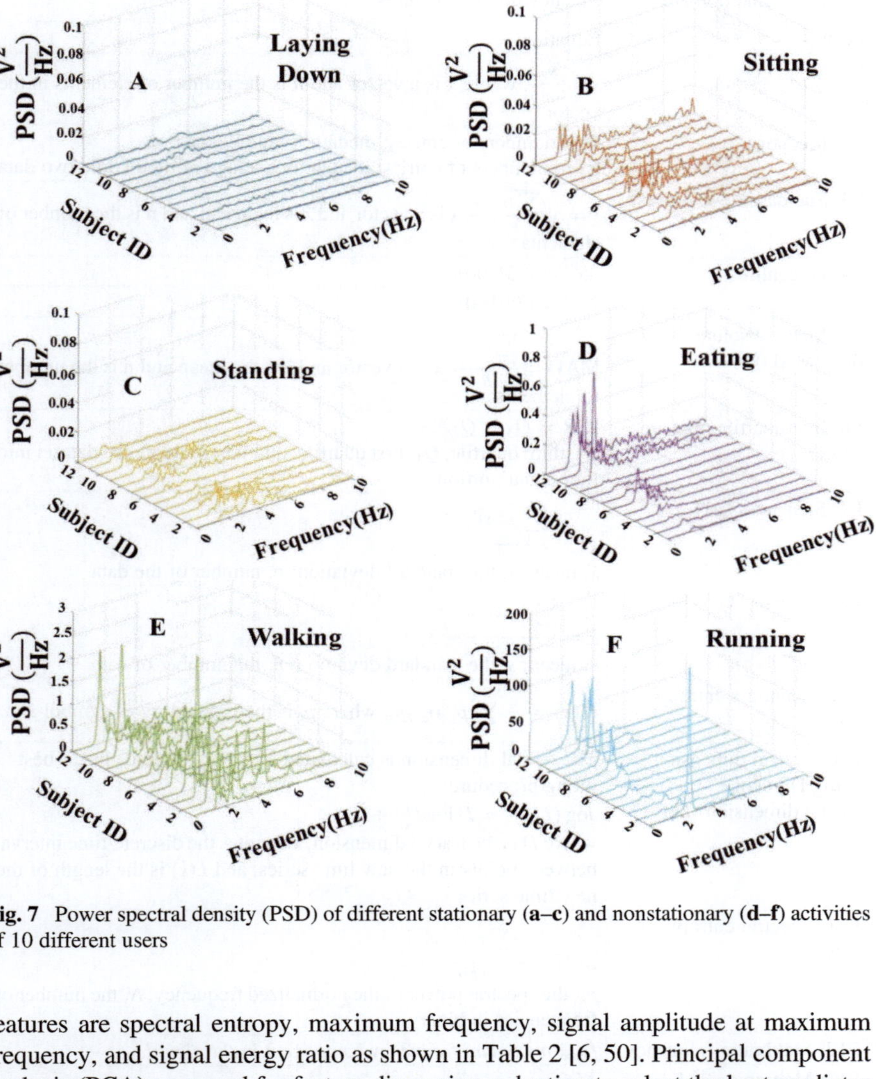

Fig. 7 Power spectral density (PSD) of different stationary (**a–c**) and nonstationary (**d–f**) activities of 10 different users

features are spectral entropy, maximum frequency, signal amplitude at maximum frequency, and signal energy ratio as shown in Table 2 [6, 50]. Principal component analysis (PCA) was used for feature dimension reduction to select the best predictor variables that provide the best predictive power in modeling the data. It was observed that the best-selected features were suitable to characterize the difference of different activities and the nature of movements during epileptic seizures using EDA and ACC.

Table 2 Time-domain and frequency-domain features

Features	Equation
F1: Mean ($\bar{x}$)	$\bar{x} = \frac{\sum x}{n}$, where x is a vector and n is the number of elements in the vector
F2: Median	Odd numbers of entries: median = middle data Even numbers of entries: median = average of the middle two data
F3: Standard deviation (s)	$s = \sqrt{\frac{\sum (x-\bar{x})^2}{n-1}}$, x is a vector and $\bar{x}$ is the mean and n is the number of elements
F4: Percentile	$25^{th} = (25/100)n$ $75^{th} = (75/100)n$
F5: Mean absolute deviation (MAD)	$\text{MAD} = \frac{\sum_{i=1}^{n} \lvert x_i - \bar{x} \rvert}{n}$, x is a vector and $\bar{x}$ is the mean and n is the number of elements
F6: Interquartile range (IQR)	$\text{IQR} = Q_3 - Q_1$ Q_3, third quartile; Q_1, first quartile; quartile, dividing the dataset into four equal portions
F7: Skewness ($g1$)	$g1 = \frac{\sum_{i=1}^{N} (x_i - \bar{x})^3 / n}{s^3}$ $\bar{x}$, mean; s, the standard deviation; n, number of the data
F8: Kurtosis (k)	$k = \frac{\sum_{i=1}^{N} (x_i - \bar{x})^4 / n}{s^4} - 3$ $\bar{x}$, mean; s, the standard deviation; n, the number of data
F9: Shannon's entropy ($H(x)$)	$H(x) = -\sum_{i=0}^{N-1} p_i \log_2 p_i$, where p_i is the probability of symbol i
F10: Fractal dimension using Higuchi fractal dimension (D_f)	The fractal dimension is calculated by a least-squares linear best-fitting procedure: $\log(L(k)) = D \log\left(\frac{1}{k}\right) + \overline{b}$ where D is the fractal dimension, k denotes the discrete-time interval between points in the new time series, and $L(k)$ is the length of the new time series
F11: Spectral entropy (SEN)	$\text{SEN} = \frac{-\sum_{i=0}^{N-1} p_k \log_2 p_k}{\log N}$ p_k, the spectral power of the normalized frequency; N, the number of frequencies in binary
F12: Maximum frequency	f_{max}, *maximum frequency component in the signal*
F13: Magnitude at $F_{\max}$	$X(f_{max})$
F14: Ratio of signal energy	$X(fmax \pm \Delta f) / \sum_{i=0}^{N-1} X_i(f)$

3.3 *Training, Validation, and Testing*

Several machine learning algorithms using the classification learner app from the statistics and machine learning toolbox of MATLAB were trained using 80% of the dataset, where 20% of the training data were used for validation and the remaining 20% of the dataset were used for testing. Fivefold cross-validations (patient-wise stratification was done) were used for training and validation of different machine learning algorithms to identify the best algorithm for classifying the EDA and ACC data separately and combined to detect an epileptic seizure. Several classification models including support vector machine (SVM), k-nearest neighbors (kNN), decision trees, logistic regression, discriminant analysis, and ensemble classifier were trained, validated, and tested. Performance matrixes were calculated to compare the performance of different algorithms.

4 Results and Discussion

In the following section, results from different experiments are reported and discussed.

4.1 *Performance Evaluation of EDA and ACC Sensor*

As shown in Fig. 8, the sensor module is capable of acquiring an EDA signal with good quality and low noise in the presence of an active second-order Butterworth filter and reasonable signal amplitude with an increased overall gain of the circuit. Figure 8 shows the EDA signal acquired while the subject was asked to breathe deep and release it frequently.

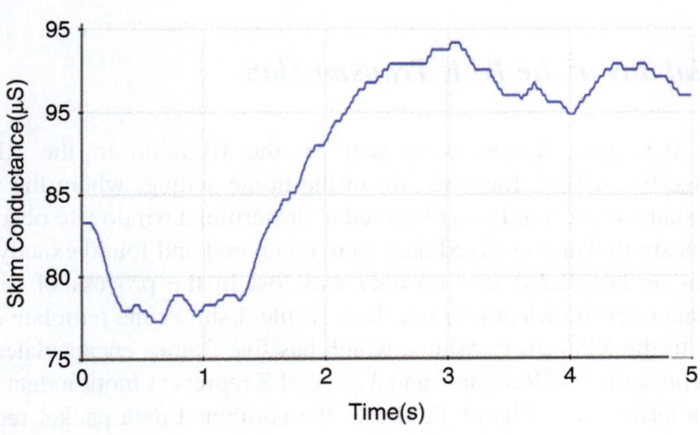

Fig. 8 EDA signal during deep breathing

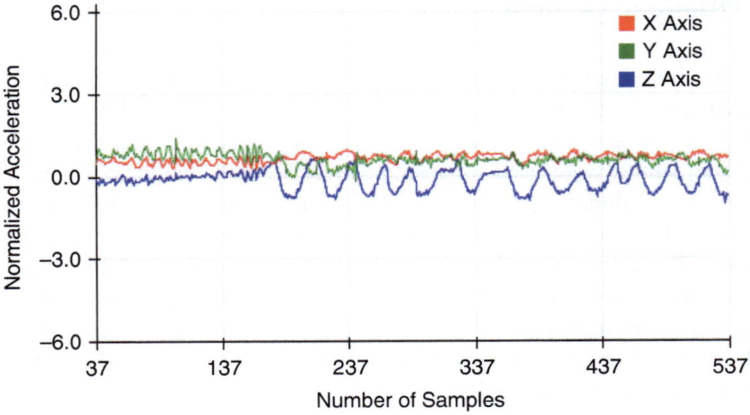

Fig. 9 Sample ACC data (x-axis, blue; y-axis, red; z-axis, green) acquired by the wearable system

ACC data were acquired over the I^2C interface to the RFduino microcontroller and sent to the RPi3 for decision-making. Three degrees of freedom motion data along the three axes (normalized by $g = 9.8\ ms^{-2}$) acquired by the wearable system were shown in Fig. 9. These EDA and ACC signals were compared with the signal acquired by the commercial BioRadio system and were found consistent.

Figure 10 shows the results of the experiments to compare the effect of the placement of the dry electrodes for the EDA module at the wrist and the fingertip. The EDA signals were recorded from the wrist and the fingers for different types of situations like deep breathing in and out, sudden shock, and laughing. This is apparent from Fig. 10 that the signals in the wrist and fingers were similar, and therefore using the EDA sensor in the wrist should still reliably record the EDA signal.

4.2 Reliability of the BLE Transmission

EDA and ACC data frames were sent by the RFduino to the RPi3 using. Acknowledge-based BLE transmission in the home setting, where the number of the packets sent and received was observed in the terminal window to observe packet loss. The transmitted and received data were compared and found exactly the same. Thus, it can be concluded that no data was lost in the process of sending and receiving data over the Bluetooth interface. Table 3 shows the template of the data packet sent to the RPi3 every 50 ms, which has five frames encapsulated together, and EDA represents "GSR signal" and *X*, *Y*, and *Z* represent motion data from three different Cartesian axes. Figure 11 shows the combined data packet received and plotted in RPi3.

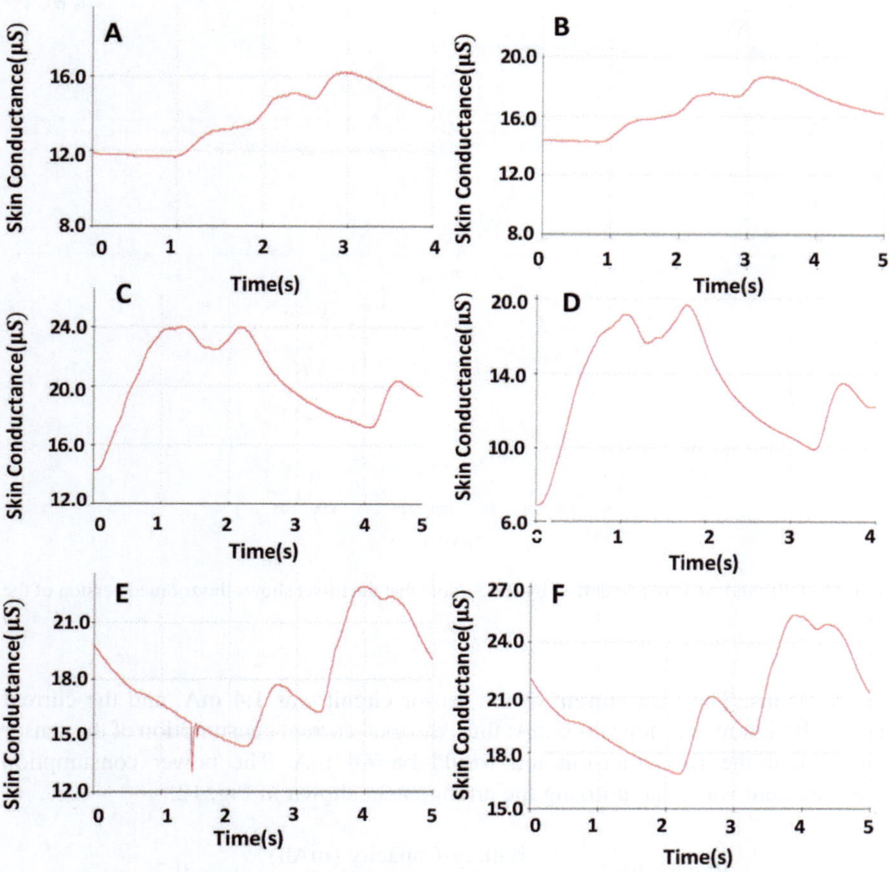

Fig. 10 GSR data was acquired when the subject was asked to breathe in and out at a fast pace, during sudden shock, and when the subject was laughing. Note that (**a**, **c**, and **e**) and (**b**, **d**, and **f**) are showing EDA data acquired from wrist and fingers, respectively

Table 3 Data frame format

Header	,	EDA	,	*X*	,	*Y*	,	*Z*	;	Header	…………	;

4.3 Power Consumption of the Wearable Module

The wearable subsystem was powered by a Li-Po battery using a power management module that is directly connected to the RFduino, the AFE of the EDA module, and the ACC module. The power module converts the battery voltage to 3.3 V, which was the operating voltage of the RFduino, ACC module, and EDA module. The overall power consumption was tested when the RFduino was acquiring data from both the EDA and ACC modules in addition to continuous BLE transmission at

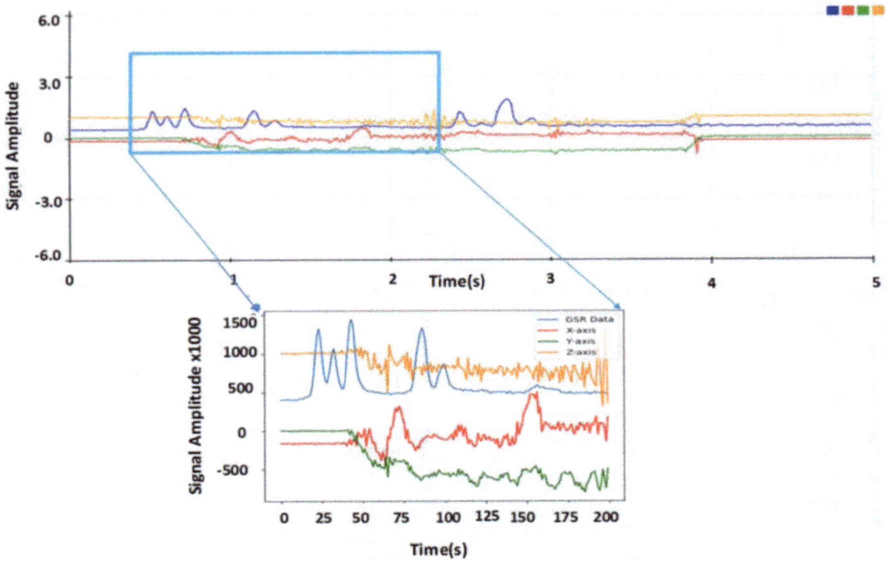

Fig. 11 GSR and ACC data received by RPi3. Note that the in-set shows the zoomed version of the signals

every 50 ms. The load current of the sensor circuit was 1.4 mA, and the current during BLE transmission was 6 mA; thus, the total current consumption of the sensor circuit and the BLE transmission would be 7.4 mA. The power consumption measurement was done utilizing the arrangement shown in Fig. 12.

$$\text{Battery life} = \frac{\text{Battery Capacity (mAh)}}{\text{Current Consumed by load (mA)}} \times 0.7$$

The battery life can be found using the following equation, where the factor 0.70 (https://www.digikey.com/en/resources/conversion-calculators/conversion-calculator-battery-life) was an allowance for nonlinearity effects of battery and external factors that can affect the battery life.

The rating of the battery was 300 mAh, and the battery life should be 41 h [(300 mAh/7.4 mA) × 0.7]. Therefore, the system should run continuously for around 1.5 days from that battery if we consider the BLE is running continuously. However, in practice, data were sent over BLE every 50 ms and the average current was 1.75 mA. Therefore, the battery life should be 120 h (i.e., 5 days) [(300 mAh/1.75 mA) × 0.7]. The system was allowed to discharge entirely after charging fully, and it was noticed that the system can last for 6 days and 2 h, which was even greater than the estimated duration.

$$Battery\ \ life = \frac{Battery\ \ Capacity(mAh)}{Current\ \ Consumed\ \ by\ \ load(mA)} \times 0.7$$

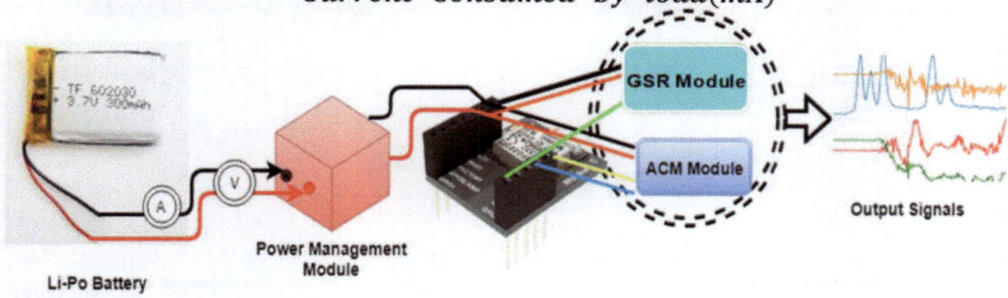

Fig. 12 Setup for the power consumption test

4.4 Classification Results for ACC Data Alone

ACC data collected from 12 users were used to train, validate, and test the different machine learning algorithms to reliably classify the daily activities. ACC data segmentation and feature extraction (features are reported in Table 2) were carried out in MATLAB. Classification learner was used to perform interactively exploring data, data partitioning, training models, and obtaining the validation evaluation matrix. Fifty trials of 12 patients for each activity were used for the training, validation, and testing of the ML models.

Figure 13 shows the scatter plot of ACC *x*-component vs *y*-component for six different activities for SVM, K-nearest neighbor (kNN), and bagged decision tree model which was identified as the three best classifiers for classifying different daily activities. The scatter plot is showing different colors for different activities, where dot (.) represents correctly classified components and cross (×) represents an incorrectly classified component (Fig. 13). The scatter plot clearly shows that different daily activity data patterns were distinguishable and how well it could distinguish different activities and this reflects the reason for higher accuracy for some activity classes. Moreover, Fig. 13a–c also reflects that different models have differing accuracy for different activity detection. It can be noticed from Fig. 13 that there are some overlapping activity data points, which explain the higher misclassifications for some classes as shown in Fig. 14. SVM was the best classifier in detecting daily activities and was able to detect most of the human activities with overall 94.7% accuracy; however, some of the activities were classified with much higher accuracy, while some were detected with lower accuracy of 90%. The other two best-performing algorithms were kNN and bagged decision trees with an overall accuracy of 93.9% and 93.8%, respectively.

Table 4 shows the percentage accuracy of classifying each activity using three different classifiers. In each case, eating and running showed the least accuracy because of the potential randomness in the data. Higher misclassifications were observed therefore at eating and running classes as shown in Fig. 14. This six-

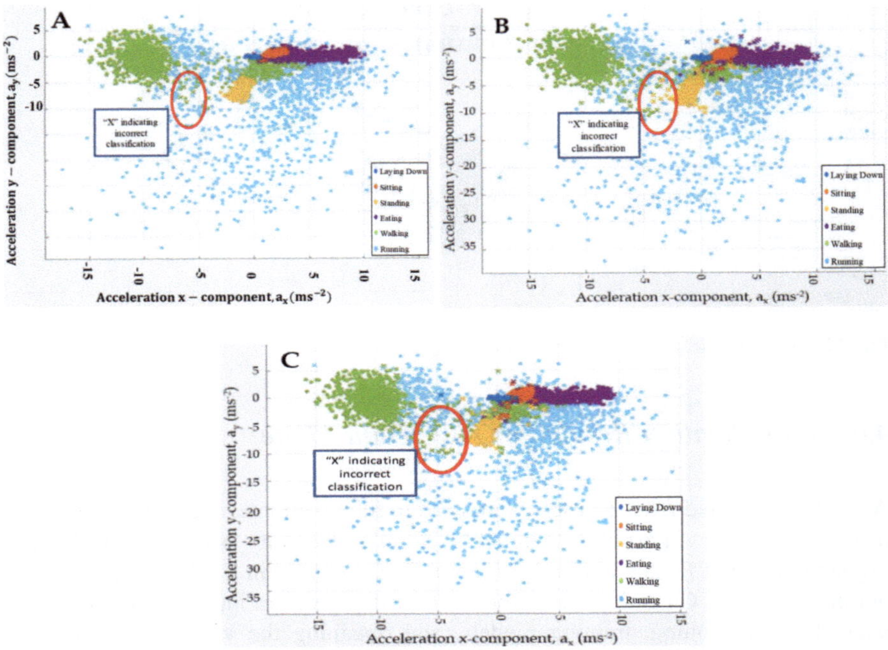

Fig. 13 Scatter plot of acceleration *x*- and *y*-component relationship in identifying different activities using three different algorithms: support vector machine (**a**), K-nearest neighbor (**b**), and bagged tree (**c**) model

class-based activity recognition best-performing model was then selected for training the ACC data into two classes: normal activity (non-seizure) and abnormal activity (seizure) while using ACC data alone for classifying epileptic seizures.

4.5 EDA and ACC Data from Empatica and Prototype System

ACC and EDA data from the Empatica were first segmented into a seizure and non-seizure data with the ground-truth label (from certified clinicians) from the database. These segmented seizure and non-seizure data were combined to train the machine learning algorithms to classify the data into two classes: non-seizure and seizure events.

The window length for EDA and ACC data was 10 s (1000 samples) with 50% overlapping, and the features were extracted for this window. Figure 15 shows the EDA and ACC data that were labeled with seizure and without a seizure, depending on the seizure onset. To observe the difference between ACC during a seizure and non-seizure daily activities, seizure events are plotted in Fig. 16a and non-seizure

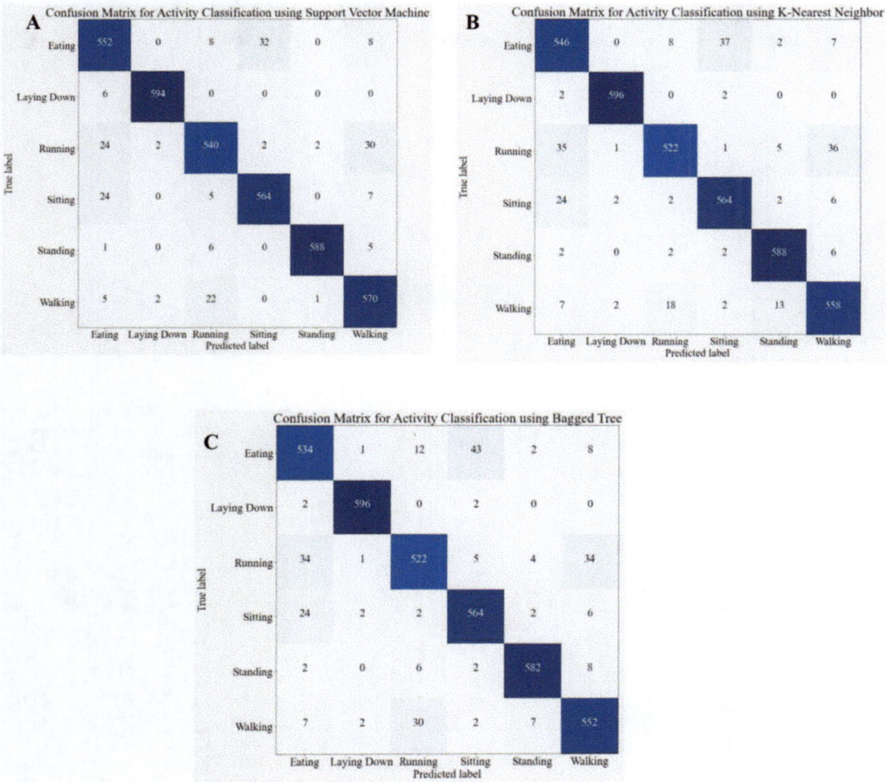

Fig. 14 Confusion matrix for (**a**) support vector machine, (**b**) K-nearest neighbor, and (**c**) bagged tree model

Table 4 Comparison of performance between the best machine learning algorithms

Technique activity	Support vector machine (SVM)	k-nearest neighbors (kNN)	Bagged decision tree
Laying down	99%	>99%	>99%
Sitting	94%	94%	94%
Standing	98%	98%	97%
Eating	92%	91%	89%
Walking	95%	93%	92%
Running	90%	87%	87%

daily activity in Fig. 16b. This is clearly visible from Fig. 16 that the daily activity (for a very small duration) shows some repetitive regular patterns (for a repetitive activity), and the signal amplitude is smaller in the ACC data, whereas for the clonic-tonic seizures, the ACC data appears like a burst, and its amplitude varies in a very

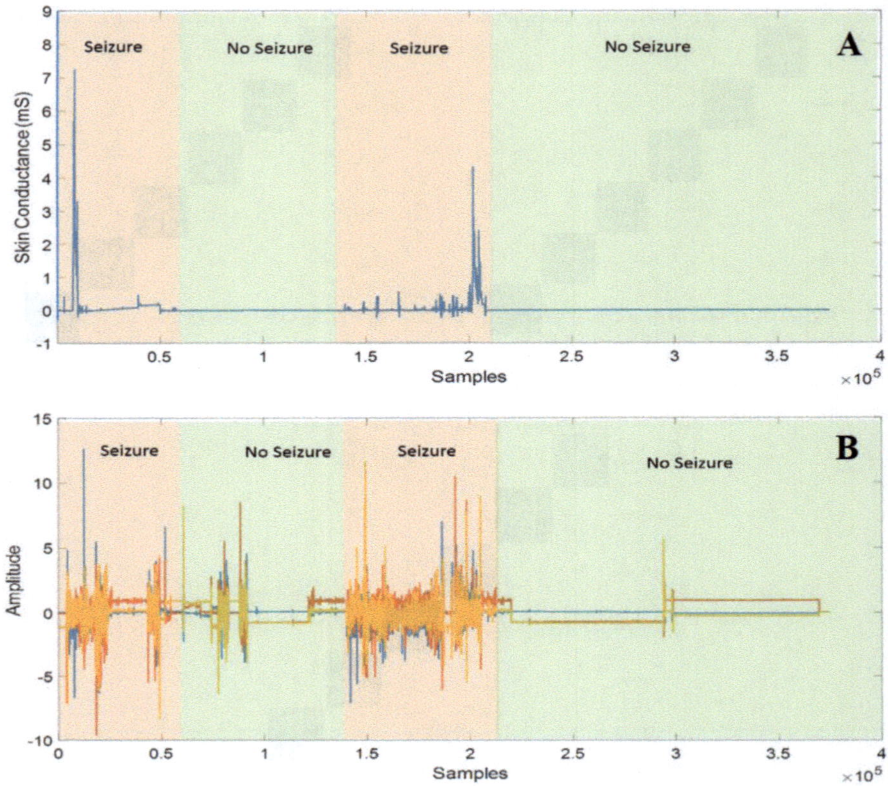

Fig. 15 (**a**) EDA and (**b**) ACC data from an epilepsy patient from the Empatica database

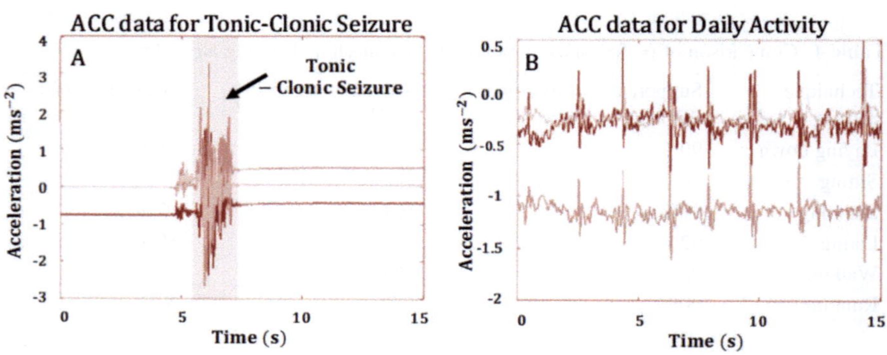

Fig. 16 ACC data showing the (**a**) tonic-clonic seizure events and (**b**) daily activity

short duration and the amplitude of the signal is quite high. However, longer duration of daily activity will not be like Fig. 16; rather it will look like the signal shown in Fig. 15.

4.6 *Classification of Seizure Using EDA Alone, ACC Alone, and Fused EDA-ACC*

It can be visualized from Fig. 15 that the EDA signal showed high activity during the epileptic seizures and the ACC signals that were synchronously recorded with EDA showed high activity during that time slots (0–50 s and 140–220 s). However, the ACC showed high activity during the 50–100 s time slot, which could lead to misclassification of seizure if ACC signals were used alone. Moreover, there is a chance of misclassification of seizures during emotional arousal from the EDA data alone, but those can be filtered out by the use of ACC data. Therefore, the multi-modal approach can significantly improve the accuracy of classification in different scenarios.

ML model on combined Empatica and prototype system data. The EDA and ACC signals from the Empatica database were used to train, validate, and test the best-performing ML model which was used to label the absence seizure of the patients from the prototype system along with the visual inspection verified by a certified neurologist. The EDA and ACC signals from Empatica and prototype systems were then categorized into two classes (seizure and non-seizure) and segmented into equal-length segments (depending on the different recording duration for each patient) each with a specific window (1000 data points). These segmented seizures and non-seizure data were partitioned into two sets: training (80%) and testing (20%) sets. Twenty percent of the training set was used for validation to identify the best validation model for the features of EDA alone, ACC alone, and when they are fused. Table 5 summarizes the highest-ranked features, while ACC and EDA were tested alone and when they were fused.

Table 6 summarizes the accuracy (A), recall (R), precision (P), F1-score (F), and specificity(S) of ACC, EDA alone, and fused EDA-ACC. ACC performed well in classifying non-seizure events with very high accuracy however failed to identify

Table 5 Listing of ranked features for different modalities

EDA-ACC	EDA	ACC
F1	F1	F2
F2	F2	F3
F5	F3	F4
F6	F4	F5
F7	F5	F8
F8	F6	F9
F12	F7	F11
F13	F8	F14

Table 6 Comparison of classification performance metrics with different modalities

	Bagged decision tree					Bagged decision tree					Support vector machine (SVM)				
Technique class	EDA-ACC					EDA					ACC				
Performance metrics	A (%)	R. (%)	P (%)	F (%)	S (%)	A (%)	R (%)	P (%)	F (%)	S (%)	A (%)	R (%)	P (%)	F (%)	S (%)
Seizure	97.3	97.2	96.1	96.7	96.1	98.4	98.3	98.4	98.3	83	75.2	75.1	98.1	85.1	98.6
No seizure	96.1	96.1	97.1	96.6	97.3	82.9	83.5	83.3	83.3	98.1	98.6	98.5	80	88.3	75.2
Overall accuracy	96.7	96.6	96.6	96.6	96.7	90.7	90.9	90.8	90.8	90.5	86.9	86.8	89.0	86.7	86.9

reliably the seizure events, and overall accuracy is 86.9%. However, the EDA alone can identify the seizure events with very high accuracy (98.4%); however, non-seizure events were not classified with reasonable accuracy (82.9%). Therefore, there is a high chance of considering non-seizure events as seizures by the EDA. Interestingly, when both the modalities were combined, the classification accuracy of the seizure (97.3%) and non-seizure (96.1%) classes significantly improved to an overall accuracy of 96.7%. Therefore, it validates that adding different modules together (like ACC and EDA rather than EDA alone) would increase the accuracy of the epileptic seizure detection and reduce the false alarm. It was observed that the proposed system was more accurate (98.3%) in detecting the absence of seizures than the GTCS (95.1%). During GTCS, the epilepsy patient undergoes mechanical movements along with the skin-resistance changes; therefore both EDA and ACC signals were altered during GTCS. On the other hand, during the absence seizure, there was no mechanical movement of the patients observed; however, their skin resistance has altered; i.e., there was no change in the ACC signal, but the EDA signal was changed. Therefore, detection of absence seizure was comparatively easier for the algorithm as EDA was changing, while no change in ACC (in most of the cases). However, this is true when the patient is in rest or sleeping. If the patient is involving any activity and the absence seizure occurred during that, the system might not identify that reliably because of the ground truth that EEG was not used to label the data and there is no way to be confident in the current system that the absence seizure labels are the correct seizure activities. On the other hand, changes in both EDA and ACC were considered as GTCS which can be misclassification of regular activity (ACC) with emotional arousal (EDA) of the patients. It should be noted that the number of non-seizure and seizure observations for both the training and testing datasets is unequal or the dataset is imbalanced. Moreover, misclassifying observations of seizure class have more severe consequences than misclassifying observations of non-seizure class. Since the classes are adequately represented in the training data but we have to treat them asymmetrically, the cost of classes was made different. Since we want to classify patients with non-seizure and seizure, failure to identify a seizure class (false negative) has far more severe consequences than misidentifying a non-seizure class as a seizure (false positive). We have assigned ten times more cost to misidentifying seizure as non-seizure and low cost to misidentifying non-seizure as a seizure. We have trained the best-performing classification algorithm with unequal classification costs to increase accuracy for seizure with higher accuracy by partially scarifying the accuracy of the non-seizure classification. This partial reduction in the accuracy in non-seizure classification should be acceptable, as this will act as a false alarm.

The best-performing bagged tree model was implemented in the RPi3 for real-time epileptic seizure detection. The RPi3-based system was evaluated on 12 epileptic patients, and the system was used to monitor the patients for 24 h for an additional 1-month duration. Epileptic seizure events were recorded by the video monitor and the caregiver in the patients' daily notes, and the EDA and ACC data were recorded in the RPI3 system. The performance of the implemented system was evaluated based on the alarms generated by the system and the reference seizure identified in the video monitor and note of the caregiver. Interestingly, the system has identified several events where the patient has undergone seizures but their caregivers did not notice that as seizures due to sleep time or absence seizures; however the system managed to track most of them, which was confirmed by the clinicians. However, there were some false alarms created by the system when the patients were involved in some physical activities but the system considered those as seizure events. Table 7 summarizes the real and false seizure alarms generated by the prototype system for different subjects for the EDA-ACC combined technique. On average, few false alarms were recorded in the system which is quite reasonable if it is considered that the system missed very few GTC seizure events. The absence of seizures missed by the system was not considered in the false alarm calculation due to the lack of ground truth which can be done by an EEG-based monitoring system. This reflects the fact that the system was capable to identify the GTCS seizure event reliably and the reliability of absence seizure detection should be further investigated with the video-EEG system. The number of these events are reported in Table 7. The inference time for the machine learning model is less than 1 ms, which made it possible to detect any seizure event as soon as it starts, and therefore, the system cannot be called as early detection system; however, it is certainly capable to detect seizure in real-time, which will significantly help in patient management.

5 Conclusions

This work summarizes the implementation of an epileptic seizure detection system using EDA and ACC sensors in a wearable subsystem and a decision-making subsystem using RPi3, which is suitable for home setting. BLE communication was used between the two subsystems. This device is capable to monitor the occurrence of epileptic seizures and warn the caregiver. The wearable system was implemented as a wristband to be worn by the patient, and it was continuously monitoring the patient and sending the data to the RPi3. The decision-making system, after successful reception of the signal, preprocessed and segmented the signal and extracted the relevant features and classified them to seizure/non-seizure events by the pretrained model, and if any seizure event is detected, it notifies the caregiver. A GSM module attached to the RPi3 was used to call and send alerting text messages to the people responsible for patients' healthcare. The RPi3-based implementation was used with the best-performing fused algorithm to test 12 epileptic patients in the home setting for a month. Since the absence of seizures cannot be confirmed by the caregiver as the accurate timing of absence seizure is difficult to identify even by the patients, the ACC and EDA data were investigated using the model trained on Empatica database, and the events of sudden EDA change, where the change of ACC was not shown noticeable, were considered as the absence seizures. As can be seen from Table 8, the work proposed in the paper is of comparable performance with many state-of-the-art work done using wearable devices. It was observed that EDA alone can provide an overall accuracy of

Table 7 Comparison of classification accuracy for different subjects using the fused technique

	Sub01	Sub02	Sub03	Sub04	Sub05	Sub06	Sub07	Sub08	Sub09	Sub10	Sub11	Sub12
Real alarm	11	9	21	23	18	26	9	13	23	32	19	21
False alarm	2	2	3	2	3	2	1	2	3	3	2	3
Potential absence seizures (missed)	1	3	4	5	3	3	1	1	3	3	1	3
False alarm rate (per day at night)	0.15	0.18	0.13	0.08	0.14	0.07	0.1	0.13	0.12	0.09	0.09	0.13

Table 8 Comparison of classification accuracy reported in the study with other similar works

Studies	Technique	Performance (false alarm rate)
Kramer et al. in [51]	Wrist 3D accelerometer	0.11 (per day at night) with 15 patients
Meritam et al. in [52]	Using wrist 3D accelerometer (Epi-Care)	0.1 (per day at night) with 71 patients
Beniczky et al. in [53]	Using wrist 3D accelerometer (Epi-Care)	0.2 (per day at night) with 20 patients
This study	Developed prototype with a trained network using EDA-ACC data combined	0.11 (per day at night) with 12 patients

90.7%, while ACC alone can provide 86.9%; however, fusing them can provide 96.7% overall accuracy. It was observed that seizure events were reliably detected with a reasonably low false alarm rate for epileptic patients. Therefore, this system can be implemented at a low cost and can be used to monitor patients easily in home settings. In this work, we have discussed the systematic system design, signal conditioning, feature extraction, and optimized best-performing classification algorithm, whereas Empatica has used a proprietary algorithm for classification, and the research community will not be benefited from this unless it is reported or made open source. Moreover, the data from this device could help the patient's healthcare center to remotely monitor the patients. Since different human activity recognition system uses both accelerometer and gyroscope, in this work, our focus was on ACC; however, in future work, ACC and gyroscope can be combined to check whether inertial measurement units (IMU) can provide better accuracy than ACC alone or not. In the future, it could warn the medical caregivers about the location of the patient in case of emergency and send an ambulance if needed. To investigate the pre-ictal seizure events and reliable absence seizure detection, the authors are preparing to carry out a new extension of this work. We have applied for an IRB approval where the proposed system would be evaluated along with the EEG signal and video recordings. The comparison of the EEG and video recordings with the proposed system of EDA and ACC signals will be used to identify pre-ictal EDA and ACC responses to see whether the proposed multimodal system is suitable for early prediction or not. This will also help in the assessment of the sensitivity and specificity of the seizure prediction system as per the guidelines provided by Mormann et al. in [54].

Acknowledgments Authors would like to thank Engr. Ahmed Zahran, electrical engineering, Qatar University, for helping with the 3D printing of the system.

Funding: This research was partially funded by Qatar National Research Foundation (QNRF), grant number UREP19-069-2-031 and UREP23-027-2-012, and Research University Grant DIP-2018-017.

References

1. Y. H. Yin, N. Ahmad, and M. Makmor-Bakry, "Pathogenesis of epilepsy: challenges in animal models," *Iranian journal of basic medical sciences,* vol. 16 (11), p. 1119(2013)
2. N. Haddad, G. Melikyan, H. Al Hail, A. Al Jurdi, F. Aqeel, A. Elzafarany, *et al.*, "Epilepsy in Qatar: causes, treatment, and outcome," *Epilepsy & Behavior,* vol. 63 pp. 98-102(2016)
3. R. D. Thijs, R. Surges, T. J. O'Brien, and J. W. Sander, "Epilepsy in adults," *The Lancet,* vol. 393 (10172), pp. 689-701(2019)
4. R. Hanna, "The contribution of medical conditions to passenger vehicle crashes," 2009.
5. M. E. Chowdhury, A. Khandakar, Y. Qiblawey, M. B. I. Reaz, M. T. Islam, and F. Touati, "Machine learning in wearable biomedical systems," in *Sports Science and Human Health-Different Approaches*, ed: IntechOpen, 2020.
6. M. E. Chowdhury, K. Alzoubi, A. Khandakar, R. Khallifa, R. Abouhasera, S. Koubaa, *et al.*, "Wearable real-time heart attack detection and warning system to reduce road accidents," *Sensors,* vol. 19 (12), p. 2780(2019)
7. Z. Chen, G. Lu, Z. Xie, and W. Shang, "A Unified Framework and Method for EEG-Based Early Epileptic Seizure Detection and Epilepsy Diagnosis," *IEEE Access,* vol. 8 pp. 20080-20092(2020)
8. S. Deivasigamani, C. Senthilpari, and W. H. Yong, "Machine learning method based detection and diagnosis for epilepsy in EEG signal," *Journal of Ambient Intelligence and Humanized Computing,* pp. 1-7(2020)
9. A. Shoeibi, N. Ghassemi, M. Khodatars, M. Jafari, S. Hussain, R. Alizadehsani, *et al.*, "Epileptic seizure detection using deep learning techniques: A Review," *arXiv preprint arXiv:2007.01276,* 2020)
10. M.Walker, "Diagnosing Epilepsy.," 2014,1 October)
11. M. N. Lava, "Epilepsy and Blood Testing," 2016)
12. A. Ulate-Campos, F. Coughlin, M. Gaínza-Lein, I. S. Fernández, P. Pearl, and T. Loddenkemper, "Automated seizure detection systems and their effectiveness for each type of seizure," *Seizure,* vol. 40 pp. 88-101(2016)
13. G. Iadarola, A. Poli, and S. Spinsante, "Analysis of Galvanic Skin Response to Acoustic Stimuli by Wearable Devices," in *2021 IEEE International Symposium on Medical Measurements and Applications (MeMeA)*, 2021, pp. 1-6.
14. Y. S. Can, N. Chalabianloo, D. Ekiz, and C. Ersoy, "Continuous stress detection using wearable sensors in real life: Algorithmic programming contest case study," *Sensors,* vol. 19 (8), p. 1849 (2019)
15. C. Meisel, R. El Atrache, M. Jackson, S. Schubach, C. Ufongene, and T. Loddenkemper, "Machine learning from wristband sensor data for wearable, noninvasive seizure forecasting," *Epilepsia,* 2020)
16. W. Boucsein, *Electrodermal activity*: Springer Science & Business Media, 2012.
17. J. J. Braithwaite, D. G. Watson, R. Jones, and M. Rowe, "A guide for analysing electrodermal activity (EDA) & skin conductance responses (SCRs) for psychological experiments," *Psychophysiology,* vol. 49 (1), pp. 1017-1034(2013)
18. M. Amengual-Gual, A. Ulate-Campos, and T. Loddenkemper, "Status epilepticus prevention, ambulatory monitoring, early seizure detection and prediction in at-risk patients," *Seizure,* vol. 68 pp. 31-37(2019)
19. A. Sano and R. W. Picard, "Stress recognition using wearable sensors and mobile phones," in *2013 Humaine Association Conference on Affective Computing and Intelligent Interaction*, 2013, pp. 671-676.
20. R. Selvaraju, M. Cogswell, A. Das, R. Vedantam, D. Parikh, and D. Batra, "Visual explanations from deep networks via gradient-based localization," in *Proceedings of the IEEE International Conference on Computer Vision*, pp. 618-626.
21. K. K. Jerger, T. I. Netoff, J. T. Francis, T. Sauer, L. Pecora, S. L. Weinstein, *et al.*, "Early seizure detection," *Journal of Clinical Neurophysiology,* vol. 18 (3), pp. 259-268(2001)

22. M. Bandarabadi, J. Rasekhi, C. A. Teixeira, T. I. Netoff, K. K. Parhi, and A. Dourado, "Early seizure detection using neuronal potential similarity: A generalized low-complexity and robust measure," *International journal of neural systems,* vol. 25 (05), p. 1550019(2015)
23. S. Ganesan, T. A. A. Victoire, and R. Ganesan, "EDA based automatic detection of epileptic seizures using wireless system," in *2011 International Conference on Electronics, Communication and Computing Technologies*, 2011, pp. 47-52.
24. M.-Z. Poh, T. Loddenkemper, N. C. Swenson, S. Goyal, J. R. Madsen, and R. W. Picard, "Continuous monitoring of electrodermal activity during epileptic seizures using a wearable sensor," in *2010 Annual International Conference of the IEEE Engineering in Medicine and Biology*, 2010, pp. 4415-4418.
25. A. Van de Vel, M. Milosevic, B. Bonroy, K. Cuppens, L. Lagae, B. Vanrumste, *et al.*, "Long-term accelerometry-triggered video monitoring and detection of tonic–clonic and clonic seizures in a home environment: Pilot study," *Epilepsy & behavior case reports,* vol. 5 pp. 66-71 (2016)
26. B. E. Heldberg, T. Kautz, H. Leutheuser, R. Hopfengärtner, B. S. Kasper, and B. M. Eskofier, "Using wearable sensors for semiology-independent seizure detection-towards ambulatory monitoring of epilepsy," in *2015 37th Annual International Conference of the IEEE Engineering in Medicine and Biology Society (EMBC)*, 2015, pp. 5593-5596.
27. A. H. Shoeb and J. V. Guttag, "Application of machine learning to epileptic seizure detection," in *Proceedings of the 27th International Conference on Machine Learning (ICML-10)*, 2010, pp. 975-982.
28. S. R. Gouravajhala and L. Khuon, "A multi-modality sensor platform approach to detect epileptic seizure activity," in *2012 38th Annual Northeast Bioengineering Conference (NEBEC)*, 2012, pp. 233-234.
29. G. T. Borujeny, M. Yazdi, A. Keshavarz-Haddad, and A. R. Borujeny, "Detection of epileptic seizure using wireless sensor networks," *Journal of medical signals and sensors,* vol. 3 (2), p. 63(2013)
30. K. Vandecasteele, T. De Cooman, Y. Gu, E. Cleeren, K. Claes, W. Paesschen, *et al.*, "Automated epileptic seizure detection based on wearable ECG and PPG in a hospital environment," *Sensors,* vol. 17 (10), p. 2338(2017)
31. F. Achilles, F. Tombari, V. Belagiannis, A. M. Loesch, S. Noachtar, and N. Navab, "Convolutional neural networks for real-time epileptic seizure detection," *Computer Methods in Biomechanics and Biomedical Engineering: Imaging & Visualization,* vol. 6 (3), pp. 264-269 (2018)
32. A. Subasi, J. Kevric, and M. A. Canbaz, "Epileptic seizure detection using hybrid machine learning methods," *Neural Computing and Applications,* vol. 31 (1), pp. 317-325(2019)
33. S. Raghu, N. Sriraam, G. P. Kumar, and A. S. Hegde, "A novel approach for real-time recognition of epileptic seizures using minimum variance modified fuzzy entropy," *IEEE Transactions on Biomedical Engineering,* vol. 65 (11), pp. 2612-2621(2018)
34. D. Sopic, A. Aminifar, and D. Atienza, "e-glass: A wearable system for real-time detection of epileptic seizures," in *2018 IEEE International Symposium on Circuits and Systems (ISCAS)*, 2018, pp. 1-5.
35. R. W. Picard, M. Migliorini, C. Caborni, F. Onorati, G. Regalia, D. Friedman, *et al.*, "Wrist sensor reveals sympathetic hyperactivity and hypoventilation before probable SUDEP," *Neurology,* vol. 89 (6), pp. 633-635(2017)
36. Y. Liu, S. Sivathamboo, P. Goodin, P. Bonnington, P. Kwan, L. Kuhlmann, *et al.*, "Epileptic Seizure Detection Using Convolutional Neural Network: A Multi-Biosignal study," in *Proceedings of the Australasian Computer Science Week Multiconference*, 2020, pp. 1-8.
37. M. Açıkoğlu and S. A. Tuncer, "Incorporating feature selection methods into a machine learning-based neonatal seizure diagnosis," *Medical hypotheses,* vol. 135 p. 109464(2020)
38. A. Emami, N. Kunii, T. Matsuo, T. Shinozaki, K. Kawai, and H. Takahashi, "Seizure detection by convolutional neural network-based analysis of scalp electroencephalography plot images," *NeuroImage: Clinical,* vol. 22 p. 101684(2019)

39. R. Hussein, H. Palangi, R. Ward, and Z. J. Wang, "Epileptic seizure detection: A deep learning approach," *arXiv preprint arXiv:1803.09848,* 2018)
40. Y. Cao, Y. Guo, H. Yu, and X. Yu, "Epileptic seizure auto-detection using deep learning method," in *2017 4th International Conference on Systems and Informatics (ICSAI)*, 2017, pp. 1076-1081.
41. R. Yuvaraj, J. Thomas, T. Kluge, and J. Dauwels, "A deep learning scheme for automatic seizure detection from long-term scalp EEG," in *2018 52nd Asilomar Conference on Signals, Systems, and Computers*, 2018, pp. 368-372.
42. A. Rahman, M. E. Chowdhury, A. Khandakar, S. Kiranyaz, K. S. Zaman, M. B. I. Reaz, *et al.*, "Multimodal EEG and keystroke dynamics based biometric system using machine learning algorithms," *IEEE Access,* vol. 9 pp. 94625-94643(2021)
43. W. Boucsein, D. C. Fowles, S. Grimnes, G. Ben-Shakhar, W. T. Roth, M. E. Dawson, *et al.*, "Society for psychophysiological research ad hoc committee on electrodermal measures. publication recommendations for electrodermal measurements," *Psychophysiology,* vol. 49 (8), pp. 1017-1034(2012)
44. L. Xie, G. Yang, L. Xu, F. Seoane, Q. Chen, and L. Zheng, "Characterization of dry biopotential electrodes," in *2013 35th Annual International Conference of the IEEE Engineering in Medicine and Biology Society (EMBC)*, 2013, pp. 1478-1481.
45. (2020). *MCP6022-Rail-to-Rail Input/Output, 10 MHz Op Amps* [Online]. Available: https://www.microchip.com/wwwproducts/en/MCP6022. [Accessed on: 06/06/2020]
46. J. Malmivuo and R. Plonsey, "Bioelectromagnetism: Principles and Applications of Bioelectric and Biomagnetic Fields. Bioelectromagnetism: Principles and Applications of Bioelectric and Biomagnetic Fields," ed: Oxford University Press, New York, 2012.
47. A. Sano, R. W. Picard, and R. Stickgold, "Quantitative analysis of wrist electrodermal activity during sleep," *International Journal of Psychophysiology,* vol. 94 (3), pp. 382-389(2014)
48. G. Regalia, F. Onorati, M. Lai, C. Caborni, and R. W. Picard, "Multimodal wrist-worn devices for seizure detection and advancing research: Focus on the Empatica wristbands," *Epilepsy research,* 2019)
49. (2020). *MMA8452Q 3 axix, 12-bit/8-bit digital accelerometer* [Online]. Available: https://www.nxp.com/docs/en/data-sheet/MMA8452Q.pdf. [Access on: 06/06/2020]
50. M. E. Chowdhury, A. Khandakar, K. Alzoubi, S. Mansoor, A. M Tahir, M. B. I. Reaz, *et al.*, "Real-Time Smart-Digital stethoscope system for heart diseases monitoring," *Sensors,* vol. 19 (12), p. 2781(2019)
51. U. Kramer, S. Kipervasser, A. Shlitner, and R. Kuzniecky, "A novel portable seizure detection alarm system: preliminary results," *Journal of Clinical Neurophysiology,* vol. 28 (1), pp. 36-38 (2011)
52. P. Meritam, P. Ryvlin, and S. Beniczky, "User-based evaluation of applicability and usability of a wearable accelerometer device for detecting bilateral tonic–clonic seizures: a field study," *Epilepsia,* vol. 59 pp. 48-52(2018)
53. S. Beniczky, S. Wiebe, J. Jeppesen, W. O. Tatum, M. Brazdil, Y. Wang, *et al.*, "Automated seizure detection using wearable devices: A clinical practice guideline of the International League Against Epilepsy and the International Federation of Clinical Neurophysiology," *Clinical Neurophysiology,* vol. 132 (5), pp. 1173-1184(2021)
54. F. Mormann, R. G. Andrzejak, C. E. Elger, and K. Lehnertz, "Seizure prediction: the long and winding road," *Brain,* vol. 130 (2), pp. 314-333(2007)

Analysis of Intramuscular Coherence of Lower Limb Muscle Activities Using Magnitude Squared Coherence

K. Arunganesh, N. Sivakumaran, S. Kumaravel, and P. A. Karthick

Abstract Coherence analysis of electromyogram (EMG) signals from synergistic muscles is a typical technique for studying intramuscular synchronization and learning about the central nervous system's techniques for controlling motor task execution. The main aim of this study was to compare the intramuscular coherence of lower limb muscles during the various tasks. The study utilized a publicly available full-body mobile brain-body imaging database for the analysis. The database consists of simultaneously recorded electroencephalogram (EEG), electromyogram (EMG), electrooculogram (EOG), and inertial measurement unit (IMU) signals of ten healthy subjects (age range, 18–31 years old) during five steady locomotion, namely, level ground walking, stair descent, stair ascent, ramp descent, and ramp ascent. The EMG signals were recorded using active bipolar electrodes (fixed electrode distance of 20 mm) at 1000 Hz, and it was recorded from six sites bilaterally: tibialis anterior (TA), medial gastrocnemius (MG), vastus lateralis (VL), rectus femoris (RF), semitendinosus (ST), and biceps femoris longus (BFL). In this study, we have analyzed different lower limb muscle pairs such as TA-MG, TA-VL, and TA-BF with the help of magnitude squared coherence (MSC) method during the various locomotive tasks. In the present study, we found significant coherence in TA-MG and TA-VL muscle pairs in entire tasks. In addition, the TA-BF muscle pair coherence was found to have a very low value, particularly in stair ascending tasks. Our results may reflect the specific functional roles in

K. Arunganesh (✉)
Department of Instrumentation and Control Engineering, National Institute of Technology, Tiruchirappalli, Tamil Nadu, India

Periyar Maniammai Institute of Science and Technology, Thanjavur, Tamil Nadu, India

N. Sivakumaran · P. A. Karthick
Department of Instrumentation and Control Engineering, National Institute of Technology, Tiruchirappalli, Tamil Nadu, India

S. Kumaravel
Department of Orthopedics, Thanjavur Medical College and Hospital, Thanjavur, Tamil Nadu, India

M. Murugappan, R. Yuvaraj (eds.), *Biomedical Signals Based Computer-Aided Diagnosis for Neurological Disorders*, https://doi.org/10.1007/978-3-030-97845-7_12

corticospinal activation during walking. This provides an effective rehabilitation therapy concerning impaired gait in neurological conditions including Parkinson's disease, spinal cord injury, and stroke.

Keywords EMG · Intramuscular coherence · Magnitude squared coherence (MSC)

1 Introduction

The unique synchronous drive within two muscles or two segments of the same muscle is measured by EMG-EMG coherence (intramuscular coherence). EMG-EMG coherence is an effective method for analyzing the various muscle contraction since it is straightforward to measure, requiring only the recording of EMG signals and no system modification or excitation [1–3]. Intramuscular coherence studies can be used to measure corticomotor activity during a greater consistency like human gait, and variations in coherence can be used to monitor neuroplasticity [4]. The common corticospinal drive from the primary motor cortex to the muscles that have been employed to measure corticospinal drive as a result of a gait rehabilitation intervention is thought to be reflected by EMG-EMG coherence. The EMG-EMG coherence can be utilized to predict the impact of neuroplasticity-enhancing therapies like rigorous (robot-assisted) gait training or noninvasive brain stimulation [4].

In recent studies, the association between EMG coherence analysis and the level of correlation in synaptic input to motor neurons has been investigated [5, 6]. According to recent studies, integrating robot-assisted movements with traditional rehabilitation methods can improve a patient's performance [7, 8]. In addition to robotic assistance, the assessment of electrophysiological data provided by the neuromuscular system can help with rehabilitation by serving as a technique for adaptive recovery measurement. Electrical activity in muscle fibers reveals a lot of information about the affected neuromuscular networks that control movement [9, 10].

One of the most recently used approaches is the investigation of intramuscular coherence, which illustrates the relationship between two parts of the same muscle [11]. The behavior of EMG-EMG coherence during static stimulation of the tibialis anterior muscle has been studied recently, and it has been proposed as a quantitative diagnostic technique in the healing of chronic spinal injuries [12]. However, the repeatability of these investigations is restricted, limiting their applicability to diverse muscles and tasks. To overcome the limitations of the above studies, we propose the use of multitasking techniques that have been investigated to quantify intramuscular coherence.

The unique association among canonical correlation analysis (CCA) and the MSC range is taken advantage of in this research to develop a new MSC prediction model. The suggested technique has two advantages: first, it reduces noise in predictions at non-correlated frequencies; and second, it improves spectral resolution over Welch's method [13–15].

In this study, intramuscular coherence between lower limb muscle contraction of tibialis anterior (TA), medial gastrocnemius (MG), vastus lateralis (VL), and biceps femoris longus (BF) was analyzed using magnitude squared coherence (MSC). During walking activities, intramuscular coherence has been calculated between pairs of muscles such as TA-MG, TA-VL, and TA-BF. This method accurately represents intramuscular coherence and could be used to assess the functional recovery in robot-assisted gait rehabilitation.

2 Methods

The study utilized a publicly available full-body mobile brain-body imaging database for the analysis [16]. Ten healthy individuals (five male and five female; age range, 18–31 years old) participated in the experiment. A 64-channel EEG setup was used to record EEG signals from the scalp with a sampling frequency of 1000 Hz. The brain electrical activities were acquired using modified international 10–20 system with slight modifications. Surface EMG signals were recorded using active bipolar electrodes (fixed electrode distance of 20 mm) at 1000 Hz, and it was recorded from six sites bilaterally: tibialis anterior (TA), medial gastrocnemius (MG), vastus lateralis (VL), rectus femoris (RF), semitendinosus (ST), and biceps femoris longus (BF) (Fig. 1).

The participants were asked to perform five modes of steady locomotion, namely, level walking (LW), descent of stair (SD), ascent of stair (SA), descent of ramp (RD), and ascent of ramp (RA). During the experiments, the subjects were requested to walk at their own speed and to avoid unnecessary head, eye, and jaw movements. The subjects performed level walking, descended the ramp, ascended the staircase, and come to a halt at the conclusion of the first stage (forward direction). Then the

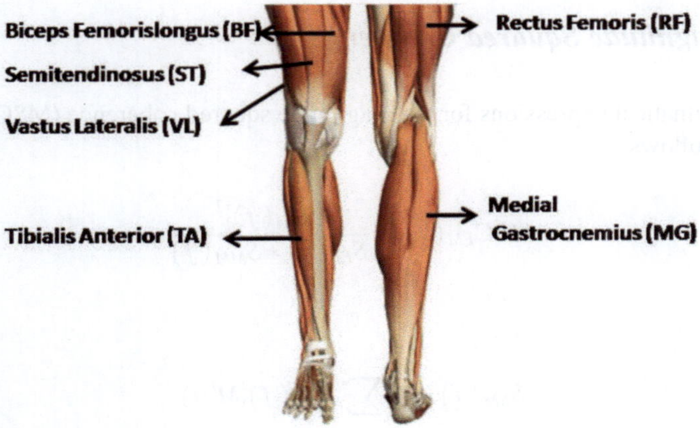

Fig. 1 Posterior view of the right leg

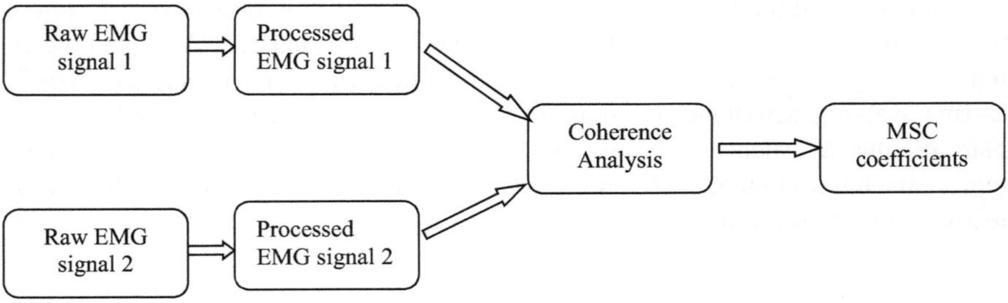

Fig. 2 Simplified work plan

subject turned in a 180° angle and performed stair descending, level ground walking, and ramp ascending in the reverse direction.

In the experiment both the forward and reverse direction of the course were considered as one complete trial. All the subjects were asked to complete 10 walking trials, and therefore there are 100 trials present in the database. Roughly, the time that has been taken for one task is around 3 min. During the walking trials, the EEG and EMG data were recorded simultaneously (Fig. 2).

Before starting the experiment, the EMG signals were evaluated for quality. To extract the signal envelope, the EMG data were re-referenced to bipolar derivations, then high-pass filtered above 30 Hz (using a FIR filter, order 226), rectified, and low-pass filtered below 5 Hz (using a FIR filter, order 846). The latencies of right and left heel strikes were aligned across trials using this approach.

In this study, we consider the EMG signals acquired from the tibialis anterior (TA), medial gastrocnemius (MG), vastus lateralis (VL), and biceps femoris longus (BF) muscles for the analysis. The custom-made programs are developed to preprocessing and coherence estimation in a MATLAB version 2017 environment for magnitude squared coherence (MSC).

2.1 Magnitude Squared Coherence

The mathematical expressions for the magnitude squared coherence (MSC) value is given as follows:

$$\mathrm{MSC}_{EM}(f) = \frac{[S_{EM}(f)]^2}{S_{EE}(f) * S_{MM}(f)} \tag{1}$$

where:

$$S_{EM}(f) = \frac{1}{n}\sum\nolimits_{i=1}^{n} E_i(f).M_i'(t) \tag{2}$$

$$S_{EE}(f) = \frac{1}{n}\sum\nolimits_{i=1}^{n} E_i(f).E_i'(f) \tag{3}$$

$$S_{MM}(f) = \frac{1}{n}\sum\nolimits_{i=1}^{n} M_i(f).M_i'(f) \tag{4}$$

The magnitude squared coherence between EMG_1 and EMG_2 can be calculated for the frequency f from their cross- and auto-power spectral density using Eqs. (1)–(4), where S_{EM} is the cross-power spectral density of EMG_1 and EMG_2 signal and S_{EE} and S_{MM} are the related power spectral density of EMG_1 and EMG_2, respectively. The Welch's averaged method is used to estimate power and cross-power spectral densities between S_{EE} and S_{MM}, where, $i = 1, \ldots, n$ is the number of data segments available for analysis and E' and M' denote complex conjugate. The use of 512-point segments with a sampling rate of 1000 Hz provided a 1.95 Hz frequency resolution in the coherence spectra.

The MSC consists of first dividing the two signals in "*x*" time windows of the same length (i.e., with the same number of samples) and then calculating each window's power spectral densities before averaging them to obtain final MSC values with respective sampling frequencies [17]. For each of the coherence value calculation, the significance can be checked corresponding to a desired level of confidence α with the following equation [18]:

$$CL(\alpha) = 1 - (1 - \alpha)^{\frac{1}{n-1}} \tag{5}$$

where n is the number of segments and CL is the level of coherence value to meet the α confidence level. We considered coherence to be significant above the 95% confidence limit ($\alpha = 0.95$), as there were 180 segments for coherence calculation. Coherence is expressed as a real number between 0 and 1, with 1 indicating a perfect linear association.

3 Results

Figure 3 shows the representation of muscle signals of EMG during the level walking (LW) task. It can be seen that the amplitude range of EMG signals is ±1 mV. Similarly, Fig. 4 shows EMG-recorded muscle signals during ramp descent (RD).

Figure 5 shows the magnitude squared coherence (MSC) between EMG-EMG signal pairs of tibialis anterior-biceps femoris longus (TA-BF), tibialis anterior-medial gastrocnemius (TA-MG), and tibialis anterior-vastus lateralis (TA-VL) during level walking (LW) task.

The coherence is found to be slightly higher in the muscle pair of TA-VL when compared with other muscle pair such as TA-MG and TA-BF. The maximum

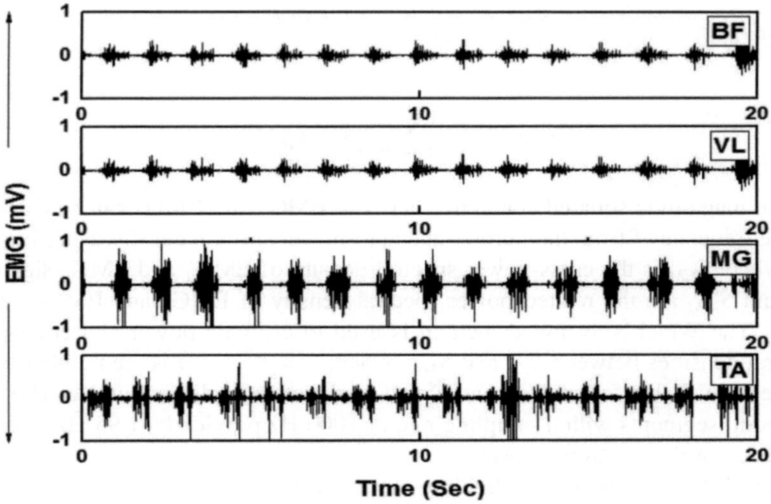

Fig. 3 Representative EMG signals from biceps femoris longus (BF), vastus lateralis (VL), medial gastrocnemius (MG), and tibialis anterior (TA), during level walking (LW) task

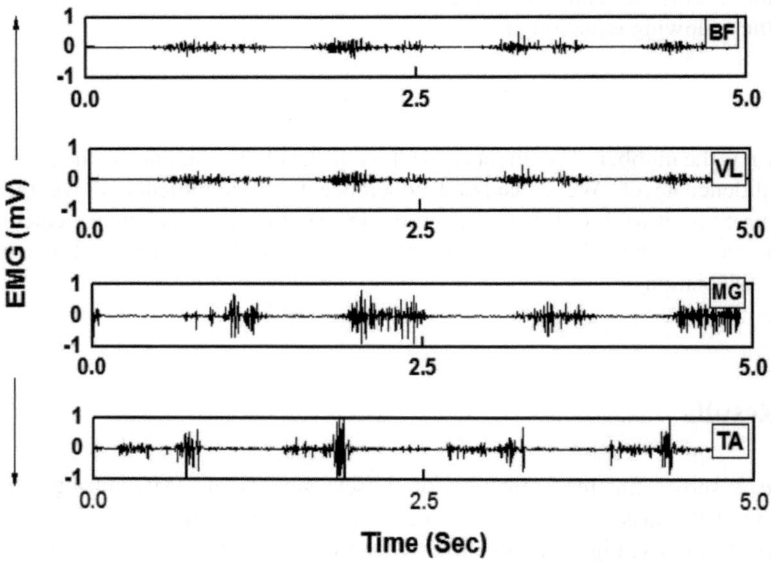

Fig. 4 Representative EMG signals from biceps femoris longus (BF), vastus lateralis (VL), medial gastrocnemius (MG), and tibialis anterior (TA), during ramp descent (RD) task

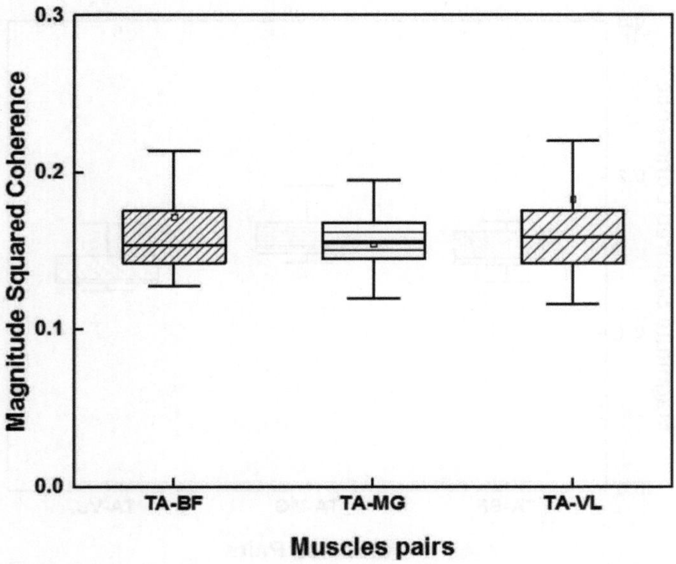

Fig. 5 Magnitude squared coherence between EMG signal pairs of tibialis anterior-biceps femoris longus (TA-BF), tibialis anterior-medial gastrocnemius (TA-MG), and tibialis anterior-vastus lateralis (TA-VL) during level walking (LW)

median value of MSC is 0.1612 for TA-VL, 0.1588 for TA-MG, and 0.1469 for TA-BF. The MSC of EMG-EMG signals pairs during the ramp descent (RD) is presented in Fig. 6. The median value of MSC associated with TA-MG is found to be higher than other two muscle pairs.

Figure 7 shows the MSC between EMG-EMG signal pairs of ramp ascending (RA) task. It can be noticed that the intramuscular coherence of TA-BF muscle pair is greater than other two muscle pairs and also there is a significant difference from each muscle pairs. The maximum median value of TA-BF pair is 0.1826 during the ramp ascending task. It is considerably higher than previous tasks such as LW and RD.

Figure 8 presents the intramuscular coherence between EMG-EMG muscle pairs during the stair descending (SD) task. It can be observed that the higher median value has found the TA-BF and TA-VL muscle pairs. Similarly, we have found the higher coherence value in the same muscle pair such as TA-BF and TA-VL in the stair ascending (SA) task shown in Fig. 9.

SPSS 26.0 was used for all statistical analyses. The threshold for statistical significance was fixed at $P < 0.05$. The data is presented as average values and standard deviance (SD). Table 1 represents maximum coherence value of all 10 subjects during the LW task. Similarly, Table 2 represents ramp ascend and descend tasks, and also Table 3 represents stair ascend and descend tasks.

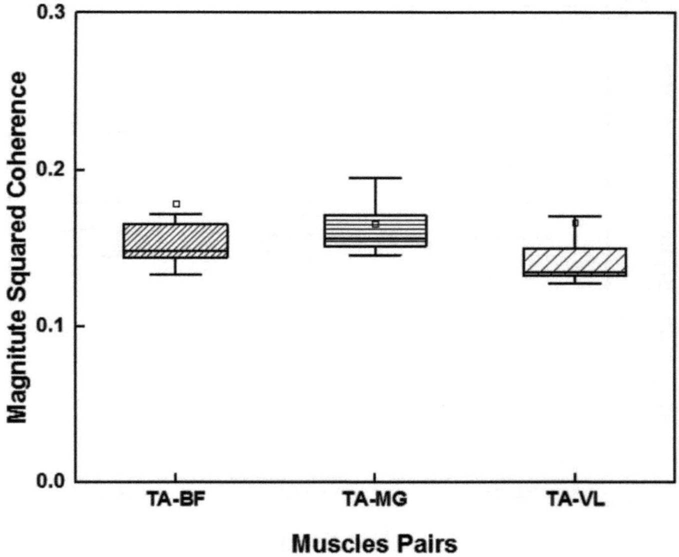

Fig. 6 Magnitude squared coherence between EMG signal pairs of tibialis anterior-biceps femoris longus (TA-BF), tibialis anterior-medial gastrocnemius (TA-MG), and tibialis anterior-vastus lateralis (TA-VL) during ramp descent (RD)

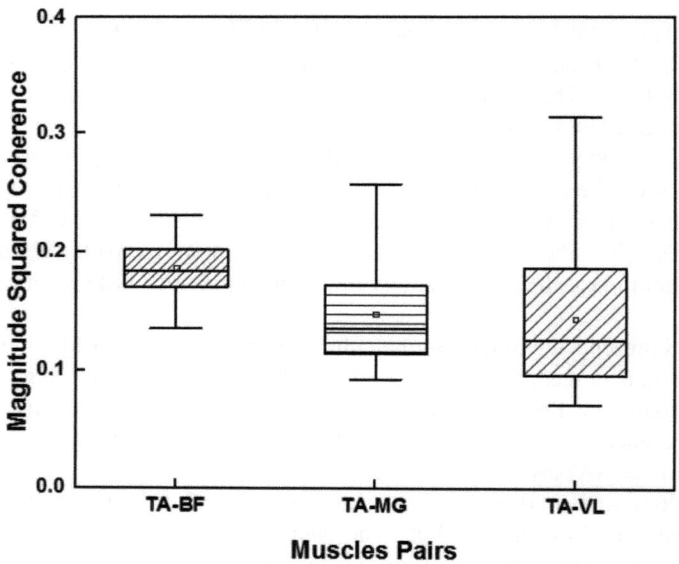

Fig. 7 Magnitude squared coherence between EMG signal pairs of tibialis anterior-biceps femoris longus (TA-BF), tibialis anterior-medial gastrocnemius (TA-MG), and tibialis anterior-vastus lateralis (TA-VL) during ramp ascent (RA)

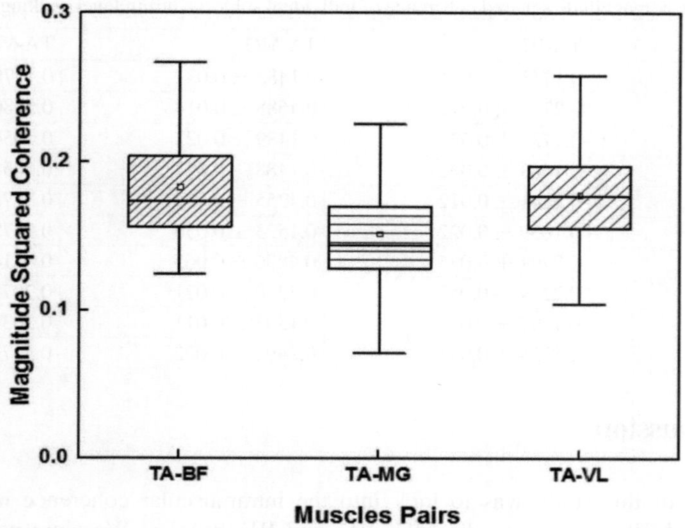

Fig. 8 Magnitude squared coherence between EMG signal pairs of tibialis anterior-biceps femoris longus (TA-BF), tibialis anterior-medial gastrocnemius (TA-MG), and tibialis anterior-vastus lateralis (TA-VL) during stair descent (SD)

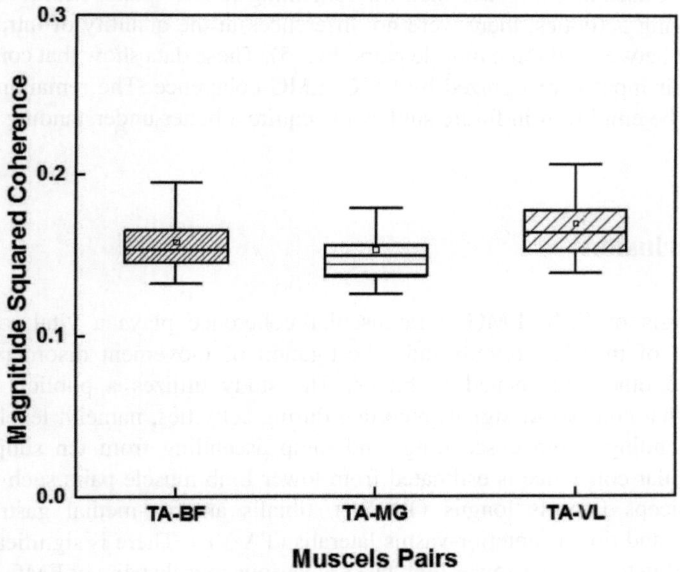

Fig. 9 Magnitude squared coherence between EMG signal pairs of tibialis anterior-biceps femoris longus (TA-BF), tibialis anterior-medial gastrocnemius (TA-MG), and tibialis anterior-vastus lateralis (TA-VL) during stair ascent (SA)

Table 1 Peak magnitude squared coherence of individual subjects during level walking (LW) task

Subjects	TA-BF	TA-MG	TA-VL
1	0.1713 ± 0.021	0.1483 ± 0.031	0.1579 ± 0.011
2	0.2734 ± 0.011	0.4688 ± 0.014	0.5280 ± 0.023
3	0.1737 ± 0.023	0.1489 ± 0.023	0.1654 ± 0.043
4	0.1663 ± 0.032	0.1488 ± 0.027	0.1654 ± 0.065
5	0.2341 ± 0.012	0.4255 ± 0.034	0.3772 ± 0.045
6	0.1639 ± 0.022	0.1523 ± 0.056	0.1678 ± 0.023
7	0.2791 ± 0.035	0.4896 ± 0.065	0.4814 ± 0.034
8	0.2234 ± 0.043	0.3210 ± 0.021	0.2971 ± 0.043
9	0.1790 ± 0.021	0.1593 ± 0.011	0.3031 ± 0.054
10	0.1754 ± 0.011	0.2490 ± 0.022	0.2173 ± 0.023

4 Discussion

The goal of this study was to look into the intramuscular coherence in the TA muscle, which is linked to the MG, VL, and BF muscles. We were particularly interested in how intramuscular coherence changed during tasks such as level walking, ramp ascending, ramp descending, stair ascending, and stair descending. The study's main findings are listed as follows. The TA-BF muscle pairs had much more intramuscular coherence than the remaining muscle pairs. Moreover, during level walking activities, there were no differences in the quantity of intramuscular coherence between all three muscle pairs (Fig. 5). These data show that considerable TA-BF pair input is recognized by EMG–EMG coherence. The remaining muscle pairs will be employed in future studies to acquire a better understanding.

5 Conclusion

The analysis of EMG-EMG intramuscular coherence plays a vital role in the estimation of muscle strength and investigation of movement disorders such as stroke and other gait-related problems. The study utilizes a publicly available database that consists of signals recorded during activities, namely, level walking, stair descending, ramp descending, and ramp ascending from ten subjects. The intramuscular coherence is estimated from lower limb muscle pairs such as tibialis anterior-biceps femoris longus (TA-BF), tibialis anterior-medial gastrocnemius (TA-MG), and tibialis anterior-vastus lateralis (TA-VL). There is significant coherence found in the various lower limb tasks at various muscle pairs of EMG signals. It can be very useful to identify the muscle pair contraction of the various dynamic task of the lower limb for muscle intension. The measurement of muscle contraction intramuscular coherence can reveal a lot about how neuromuscular control behaves during voluntary periodic contractions. The results of this study will help in the

Table 2 Peak magnitude squared coherence of individual subjects during ramp descent and ascent (RD and RA) task

Sub	Ramp descent (RD)			Ramp ascent (RA)		
	TA-BF	TA-MG	TA-VL	TA-BF	TA-MG	TA-VL
1	0.1741 ± 0.031	0.1483 ± 0.011	0.2494 ± 0.034	0.1533 ± 0.011	0.1390 ± 0.043	0.2194 ± 0.031
2	0.2735 ± 0.021	0.5281 ± 0.014	0.4681 ± 0.033	0.2350 ± 0.011	0.2015 ± 0.022	0.2599 ± 0.023
3	0.1737 ± 0.023	0.1489 ± 0.023	0.1537 ± 0.013	0.2432 ± 0.023	0.2356 ± 0.013	0.2836 ± 0.043
4	0.1663 ± 0.032	0.1488 ± 0.027	0.1655 ± 0.065	0.2572 ± 0.036	0.2913 ± 0.027	0.2310 ± 0.065
5	0.2342 ± 0.011	0.4010 ± 0.044	0.3771 ± 0.015	0.2286 ± 0.012	0.2602 ± 0.034	0.2287 ± 0.045
6	0.1734 ± 0.013	0.1523 ± 0.011	0.1678 ± 0.023	0.1308 ± 0.022	0.1452 ± 0.056	0.2180 ± 0.023
7	0.2792 ± 0.045	0.6296 ± 0.087	0.4814 ± 0.034	0.2326 ± 0.035	0.2292 ± 0.011	0.2124 ± 0.056
8	0.2234 ± 0.043	0.3210 ± 0.034	0.2972 ± 0.013	0.2249 ± 0.043	0.3142 ± 0.056	0.2056 ± 0.056
9	0.1701 ± 0.021	0.1594 ± 0.011	0.3032 ± 0.031	0.2665 ± 0.021	0.2563 ± 0.011	0.2506 ± 0.054
10	0.1755 ± 0.037	0.2491 ± 0.012	0.2085 ± 0.023	0.2004 ± 0.043	0.1709 ± 0.036	0.2180 ± 0.014

Table 3 Peak magnitude squared coherence of individual subjects during stair descent and ascent (SD and SA) task

	Stair descent (SD)			Stair ascent (SA)		
Sub	TA-BF	TA-MG	TA-VL	TA-BF	TA-MG	TA-VL
1	0.2381 ± 0.011	0.2792 ± 0.031	0.2102 ± 0.042	0.1669 ± 0.031	0.1853 ± 0.056	0.1612 ± 0.011
2	0.2726 ± 0.056	0.2853 ± 0.014	0.2357 ± 0.023	0.1652 ± 0.056	0.1927 ± 0.023	0.1699 ± 0.053
3	0.1672 ± 0.041	0.1984 ± 0.023	0.1597 ± 0.043	0.1737 ± 0.023	0.2208 ± 0.023	0.1880 ± 0.043
4	0.2372 ± 0.032	0.2580 ± 0.036	0.2106 ± 0.065	0.1563 ± 0.032	0.1761 ± 0.027	0.1618 ± 0.065
5	0.1711 ± 0.012	0.1928 ± 0.034	0.2892 ± 0.045	0.1625 ± 0.012	0.1646 ± 0.034	0.1691 ± 0.045
6	0.1466 ± 0.032	0.2083 ± 0.056	0.2776 ± 0.023	0.1532 ± 0.022	0.1746 ± 0.056	0.1569 ± 0.023
7	0.1273 ± 0.035	0.3104 ± 0.065	0.3018 ± 0.034	0.1742 ± 0.035	0.2241 ± 0.034	0.1818 ± 0.024
8	0.2254 ± 0.023	0.2533 ± 0.021	0.2256 ± 0.043	0.2736 ± 0.043	0.2696 ± 0.021	0.2343 ± 0.043
9	0.2386 ± 0.056	0.2561 ± 0.036	0.2911 ± 0.011	0.1664 ± 0.036	0.1789 ± 0.042	0.1837 ± 0.056
10	0.2387 ± 0.011	0.2201 ± 0.032	0.2224 ± 0.033	0.2830 ± 0.011	0.2635 ± 0.022	0.2095 ± 0.023

awareness of neural control mechanisms that underpin adaptability to varied walking situations, as well as the development of a logical walking rehabilitation program.

Acknowledgment The work is supported by grant ECR/2018/1370 of Early Career Research Award of Science and Engineering Research Board (SERB), India.

References

1. Grosse P, Guerrini R, Parmeggiani L, Bonanni P, Pogosyan A, et al. (2003) Abnormal corticomuscular and intermuscular coupling in high-frequency rhythmic myoclonus. Brain 126: 326–342.
2. Grosse P, Cassidy M, Brown P (2002) EEG-EMG, MEG-EMG and EMG-EMG frequency analysis: physiological principles and clinical applications. Clin Neurophysiol 113: 1523–153.
3. Hansen NL, Hansen S, Christensen LO, Petersen NT, Nielsen JB (2001) Synchronization of lower limb motor unit activity during walking in human subjects. J Neurophysiol 86: 1266–12761.
4. Rogers LM, Madhavan S, Roth H, Stinear JW (2011) Transforming neurorehabilitation of walking following stroke: The promise of non-invasive brain stimulation - a review. Restor Neurol Neurosci 29: 507–516. doi:https://doi.org/10.3233/RNN-2011-0625.
5. Ward, N. J., Farmer, S. F., Berthouze, L., and Halliday, D. M. (2013). Rectification of EMG in low force contractions improves detection of motor unit coherence in the beta-frequency band. J. Neurophysiol. 110, 1744–1750.
6. Yoshitake, Y., and Shinohara, M. (2013). Oscillations in motor unit discharge are reflected in the low-frequency component of rectified surface EMG and the rate of change in force. Exp. Brain Res. 231, 267–276.
7. Westlake, K.P., Patten, C.: Pilot study of lokomat versus manual-assisted treadmill training for locomotor recovery post-stroke. J. Neuroeng. Rehabil. 6(18) (2009).
8. Chang, W.H., Kim, Y.-H.: Robot-assisted therapy in stroke rehabilitation. J. Stroke 15(3), 174–181 (2013).
9. Negro, F., Holobar, A., Farina, D.: Fluctuations in isometric muscle force can be described by one linear projection of low-frequency components of motor unit discharge rates. J. Physiol. 587, 5925–5938 (2009).
10. Negro, F., Yavuz, U.S., Farina, D.: The human motor neuron pools receive a dominant slow-varying common synaptic input. J. Physiol. 594(19), 5491–5505 (2016).
11. Holobar, A., Minetto, M.A., Farina, D.: Accurate identification of motor unit discharge patterns from high-density surface EMG and validation with a novel signalbased performance metric. J. Neural Eng. 11, 016008 (2014).
12. Bravo-Esteban, E., Taylor, J., Aleixandre, M., Simón-Martínez, C., Torricelli, D., Pons, J.L., Avila-Martín, G., Galán-Arriero, I., Gómez-Soriano, J.: Longitudinal estimation of intramuscular tibialis anterior coherence during subacute spinal cord injury: relationship with neurophysiological, functional and clinical outcome measures, J. Neuroeng. Rehabil. 14(58) (2017).
13. L. L. Scharf and C. T. Mullis, "Canonical coordinates and the geometry of inference, rate and capacity," IEEE Trans. Signal Processing, vol. 48, no. 3, pp. 824–831, March 2000.
14. Y. Hua, M. Nikpour, and P. Stoica, "Optimal reduced-rank estimation and filtering," IEEE Trans. Signal Processing, vol. 49, no. 3, pp. 457–469, Mar. 200.

15. P. J. Schreier and L. L. Scharf, "Canonical coordinates for transform coding of random signals from noisy observations," IEEE Trans. Signal Processing, vol. 54, no. 1, pp. 235–243, Jan. 2006.
16. Brantley, J. A., Luu, T. P., Nakagome, S., Zhu, F. & Contreras-Vidal, J. L. Full body mobile brain-body imaging data during unconstrained locomotion on stairs, ramps, and level ground. Sci Data 5, 180133.
17. Amine Dehbaoui, Victor Lomné, Thomas Ordas, Lionel Torres, Michel Robert, and Philippe Maurine. Enhancing electromagnetic analysis using magnitude squared incoherence. IEEE Trans. Very Large Scale Integrated Circuits, PP: 1-5, 2011.
18. J. Rosenberg, A. Amjad, P. Breeze, D. Brillinger, D. Halliday, The fourier approach to the identification of functional coupling between neuronal spike trains, Progress in biophysics and molecular biology 53 (1) (1989).

Index

M. Murugappan, R. Yuvaraj (eds.), *Biomedical Signals Based Computer-Aided Diagnosis for Neurological Disorders*, https://doi.org/10.1007/978-3-030-97845-7

MIX
Papier aus verantwortungsvollen Quellen
Paper from responsible sources
FSC® C105338

If you have any concerns about our products,
you can contact us on
ProductSafety@springernature.com

In case Publisher is established outside the EU,
the EU authorized representative is:
Springer Nature Customer Service Center GmbH
Europaplatz 3, 69115 Heidelberg, Germany

Printed by Libri Plureos GmbH
in Hamburg, Germany